A Short
HISTORY OF SCIENCE
TO THE
NINETEENTH CENTURY

BY
CHARLES SINGER

DOVER PUBLICATIONS, INC.
Mineola, New York

31143007023790
509 Sin
Singer, Charles Joseph,
1876-1960.
A short history of
science to the nineteenth
century

Published in Canada by General Publishing Company, Ltd., 30 Lesmill Road, Don Mills, Toronto, Ontario.
Published in the United Kingdom by Constable and Company, Ltd., 3 The Lanchesters, 162–164 Fulham Palace Road, London W6 9ER.

Bibliographical Note

This Dover edition, first published in 1997, is an unabridged republication of the work first published by Oxford University Press in 1941 and then reissued, with corrections, by Oxford in 1943.

Library of Congress Cataloging-in-Publication Data

Singer, Charles Joseph, 1876–1960.
    A short history of science to the nineteenth century / by Charles Singer. — Dover ed.
        p.   cm.
    Originally published: Oxford : Clarendon press, 1941.
    Includes bibliographical references and index.
    ISBN 0-486-29887-6
    1. Science—History.   I. Title.
Q125.S58   1997
509—dc21   97-18159
                                                                CIP

Manufactured in the United States of America
Dover Publications, Inc., 31 East 2nd Street, Mineola, N.Y. 11501

'These times are the ancient times, when the world is ancient, and not those which we account ancient, *ordine retrogrado*, by a computation backwards from ourselves.'

Francis Bacon

'The whole succession of men through the ages should be considered as one man, ever living and always learning.'

Pascal

PREFACE

THIS book seeks to present, in simple form, the development of the conception of a rational and interconnected material world. It considers, therefore, both physical and biological, but not psychological, social, or abstract mathematical problems. A natural pause is reached with the acceptance, in the nineteenth century, of that classical body of scientific doctrine which is the normal foundation of modern scientific discipline.

So elementary a work can indicate only a very few out of many lines of thought, especially for the period since the Revival of Learning. In dealing with these later centuries I have had recourse to a type-system. Persons, movements, advances, and inventions are selected as illustrative examples. No two writers would make the same choice; mine has been determined largely with an eye to continuity in the narrative and, specifically, to the emergence of the doctrines of Energy, of Atomism, and of Evolution.

It is impossible to complete even the simplest account of any human activity extending over two and a half millennia without a sense of inadequacy. Many reasons make this peculiarly true for science. In constructing this book I have felt, in particular, the lack of accepted precedents as to method. There are few comprehensive histories of science; all are comparatively modern, and there is no consensus as to the lines on which such a work should be constructed. My own attempt is, I am aware, of an experimental nature.

I have been occupied upon this little book for far more years than the result may justify. Through all this time my wife and I have been engaged on complementary tasks and the work of each has made that of the other possible. Dr. Douglas McKie has been of assistance on many special points and has saved me from at least some errors. Moreover, for Chapter VIII, he has written most of Section 4 and some part of Section 5. Had he not done so the book would have been delayed yet longer. To him I express my grateful thanks.

I would like this volume to go as a greeting to two transatlantic colleagues, George Sarton and Henry Sigerist. With the former I have been in fraternal relations for half a lifetime; with the latter

Preface

for a time shorter only because he has had the advantage of having been born later. I owe much to the work and personality of both.

A word of advice to the reader. The argument is, at times, necessarily somewhat intricate and it tends to become more so as it proceeds. It can be more easily followed if the pattern of the narrative is held clearly in view. This can only be done by constant reference to the rather elaborate Table of Contents.

C. S.

April 1941

NOTE

I HAVE been able to make some corrections in this reprint. For indicating errors I am especially indebted to Prof. E. N. da C. Andrade, Dr. Julian Huxley, Dr. D. McKie, and Dr. F. J. North.

C. S.

March 1943

CONTENTS

vii

Contents

Contents

LIST OF FIGURES

List of Text-Figures

List of Text-Figures

List of Text-Figures

INTRODUCTION

Nature of the Scientific Process

1. What is Science?

'WHAT is meant by *science*?' is the question that will naturally be asked on opening this book. Yet this question, if answered at all, can hardly be answered at the outset. In a sense the book is itself an answer.

Science is often conceived as a *body of knowledge*. Reflection, however, will lead to the conclusion that this cannot be its true nature. History has repeatedly shown that a body of scientific knowledge that ceases to develop soon ceases to be science at all. The science of one age has often become the nonsense of the next. Consider, for example, astrology; or, again, the idea that certain numbers are lucky or unlucky. With their history unknown, who would see in these superstitions the remnants of far-reaching scientific doctrines that once attracted clear-thinking minds seeking rational explanations of the working of the world? Yet such, in fact, is their origin. So, too, we smile at the explanation of fossils as the early and clumsier attempts of an All-powerful Creator to produce the more perfect beings that we know ourselves to be. Yet such conceptions were legitimate stages in the development of modern geological theory, just as the scientific views of our own time are but stages in an agelong process that is leading to wider and more comprehensive conceptions of the nature of our world.

It therefore behoves the historian of science to be very charitable, very forbearing, very humble, in his judgements and presentations of those who have gone before him. He needs to remember that he is dealing with the work of erring and imperfect human beings, each of whom had, like himself, at best but a partial view of truth, but many of whom had a sweep of genius far beyond his own.

There is an unquenchable and irresistible thirst of the soul that demands an explanation of the world in which it finds itself. One expression of that eternal yearning is the formulation of religious systems. Akin to such aspiration is that of the historian, who also seeks law and order in the universe. History, like science, like religion, is a constant search for such law, which yet always just eludes the grasp. And if the historian hopes to be judged at all by posterity, he can but echo the epitaph:

> Reader, thou that passest by,
> As thou art so once was I;
> As I am so shalt thou be;
> Wherefore, reader, pray for me.

Time, still, like an ever-rolling stream, bears all its sons away. It is the stream itself and the spirit that dwells therein that we shall seek to study.

Science, then, is no static body of knowledge but rather *an active process* that can be followed through the ages. The sheer validity and success of the process in our own age has given rise to a good deal of misunderstanding of its nature and not a little misapplication of such terms as 'science' and 'scientific'. We hear of the *scientific methods* of some prize-fighter, and a book has been published on the *Science of the Sacraments*. There is nothing in the laws of this or any other country which forbids its citizens from giving to the words of their language such significance as they may choose, but *science* and *scientific* as employed in these connexions have no relation to the great progressive acquisition of knowledge with which we have here to deal. The very form of the adjective 'scientific' might give pause to those who would force the word to cover such topics as the skill of the boxer, or a knowledge of the theory and practice of the sacraments. By derivation *scientific* means *knowledge making*, and no body of doctrine which is not *growing*, which is not actually *being made*, can long retain the attributes of science.

2. Origins of the Scientific Tradition.

Science, then, is a process. But when did the process begin? It is as hard to answer this as to answer the question, When does a man begin to grow old? 'Before that I to be begun, I did begin to be undone.' Anthropologists perceive germs of the scientific process in the rudest races of mankind. When a child first begins to observe, he marks the differences of dress and manner in those

Nature of the Scientific Process

about him. The savage sees the action of living beings in the sway of the trees or the stir of the waters. Both generalize from imperfect experience. The baby calls every woman 'mummy' and every man 'daddy'. Both make imperfect attempts to deduce general rules or laws. The attempts of both, in their kind and in their degree, partake of the nature of science.

Man of the Old Stone Age lived on the flesh of the creatures he could slay. His dependence on the chase led him to observe the habits and the forms of the animals that he hunted. The magic in which he believed suggested to him that the mere representation

FIG. 1. Magdalenian drawings of bison with arrows embedded in the heart, from the cavern of Maux on the Ariège, S. France.

of these animals, in the act of being slain, might result in their falling within his power (Fig. 1). The accuracy and beauty of his paintings rouse the wonder and admiration of those who explore his caves. The exactness of the observations of the palaeolithic artist and the care exerted in the representation of the form, movements, and even the anatomy of animals certainly betray elements akin to the scientific process.

When man attained the agricultural stage, he felt the need of some means of fixing the time of onset of the seasons. In the tropics, where man first became human, the days do not lengthen and shorten with the change in relation of earth and sun. There the most natural and obvious means of calculating time is by changes of the moon. Her recurring appearances are still recalled in our calendars. Our *months* are but *mooneths* altered to fit our newer reckoning of time. Our *weeks* are but quarters of the 28-day cycle of the moon and recall her changes ('week', compare German *Wechsel*=change).

3

Introduction

As man spread beyond the seasonless tropical forest he came to inhabit regions where agriculture arose. There was now need for a calendar that should tell him when to sow and when to reap. The movements of the stars were found to bear a fixed relation to that of the sun and therefore of the seasons. Observations of a very early date that bear on their relationship have come down to us from the civilization that developed in the valley of the Euphrates and Tigris. Thus the demands of agriculture, the first occupation, after hunting, for which man became organized, led to the accumulation of knowledge and to processes of generalization. These, on their level, are certainly scientific.

A settled agricultural civilization demands tools. Technology developed. The age of stone passed into the age of metals. The treatment of ores and the working of metals called for a class with special knowledge. The development of rights in land demanded some sort of surveying. Greek tradition has it that the inundation of the Nile, by obliterating all landmarks, forced on the Egyptians an annual remeasurement of their fields. Thus *geo-metry* (literally *earth-measurement*) was born. The craft of the butcher, as well as the practice of sacrifice and the examination of the entrails of the victims for purposes of divination, led to some knowledge of the structure of the body. In these processes we may see the practical sources of sciences that we now call metallurgy, mathematics, anatomy.

As society became more complex, commerce developed. A system of numerical notation was now evolved. The ancient world presents us numerous such instances of invention fathered by necessity and mothered by experience. All have a like claim to be included in a history of science. Ultimately a work will be written which will include them all.

The older civilizations, which advanced thus far along scientific lines, all developed cultural and religious bonds which united their members into tribal and ultimately into imperial units. Looking back on the past and viewing it from the vantage point of our own civilization, we are struck with the failure of these ancient cultures to stress human individuality. In the earlier Biblical record the punishment or reward of a people for the shortcomings or virtues of a single member passes without remark.

4

Nature of the Scientific Process

Of none of the great primary discoveries which made social life possible has the name of the discoverer come down to us. The inventors and the successive improvers of the means by which fire can be made, of pottery, of the wheel, of the cutting-edge, of the bow, of the metals and their preparation, advanced mankind along the path which led to science. Yet their names, their dates, even their tribal affinities are utterly lost. So with the early thinkers. While we have ample record of the religious and ethical outlook of the peoples of the ancient world, we have none of that peculiarly individual product of the human intellect that in its later development we call *philosophy*, a product of which science is a part. We have no knowledge of those who first set out on the prime task of the philosopher, the individual endeavour to understand and to explain himself and his world. Even when prophet or priest seeks to deliver a message, he is always insistent that it is not his but another's; and not seldom that other is beyond our ken, for he is the Dweller above the Firmament.

Thus it happens that while we may discern science in these more ancient civilizations, no one has yet been able to give a continuous account of the development among them of scientific ideas; still less has it been possible to show how science influenced the modes of thinking of the ancient peoples. For a clearer view we must turn to another and later culture. In our survey of the history of science we therefore disregard the broken lights that are all that can be distinguished of the scientific elements in the once brilliant civilizations of the Empires of the ancient East. We open with the Greeks. It is not that the first men of science were Greeks— for they were not. But it is true that the first men of whom we have a record, who were conscious of science as a distinct process and who were conscious, too, that the process might be indefinitely extended, spoke a dialect of Greek and numbered themselves among the Hellenes.

FIG. 2. Western Asia Minor.

I. RISE OF MENTAL COHERENCE

The Foundations (about 600–400 B.C.): Ionia, Magna Graecia, Athens

1. Beginnings of Ionian Science and the Eastern School.

IN writing history it is commonly necessary to rely upon written documents. Without such records, the narrative is always imperfect and often incoherent. The earliest scientific documents that we possess that are in any degree complete are in the Greek language. They were composed about 500 B.C. Our story starts about a century before that date.

It is certain that Greek science in its origin was dependent on traditions that came from more ancient civilizations, notably from Egypt and Mesopotamia. On this the Greeks themselves insisted. They have been confirmed by modern discoveries. Documents of Egyptian and Mesopotamian origin have been brought to light which take back the scientific disciplines of medicine and mathematics at least a thousand years behind the earliest Greek records of these studies.

The Greeks were themselves immigrants. They first invaded the eastern shores of the Mediterranean as a mixed host about 1400 B.C. The main impact of invasion fell on continental Greece. Tribal streams passed also eastward to the sea coasts and islands of Asia Minor and westward to Sicily and Southern Italy. Chief among the Asiatic Greeks were the Ionians, who colonized the shores of the Aegean from Ephesus in the north to Halicarnassus in the south. Yet farther south settled the Dorians (Fig. 2). South Italy and Sicily were colonized secondarily both from Greece and Asia Minor (Fig. 8). It was among the Ionians that the first great scientific movement arose. Dorian elements, however, crept into it at an early date.

The Ionians were very favourably placed for the reception of foreign ideas. Eastward they were in relations with the ancient Mesopotamian culture. This was invaded in the sixth century by a people from yet farther East, the Persians, who left a permanent mark on all contemporary civilizations. Their influence is to be discerned in the New Testament where we read of the *Magi*

7

Rise of Mental Coherence

(*Authorized Version* 'wise men', *Matthew* ii. 1), a Persian word that has given us our term *magic*. Persia was the most vigorous power of the age and brought new contacts to the Ionians. Further, the Ionians were a maritime and trading people. Through their regular sea traffic suggestions came to them from Egypt, the most ancient and settled of all civilizations. Ionians traded, too, with Phoenicia and reached as far as India whence some of their ideas were derived.

It was, in general, a time of travel, of movement, of the breakdown of old and of the rise of new civilizations. Such was the stage, such the atmosphere of change in which science became first clearly distinguished. We see science emerging into the light of historic day in the person of the Ionian Greek Thales.

Though the son of a Phoenician mother, THALES (*c.* 624–565 B.C.) was a citizen of the Ionian city of Miletus. Tradition tells that he was a man of great sagacity, exhibited no less in politics and commerce than in science. He suggested a federal system for the cities of Ionia and made a fortune as a merchant.

In the course of his business Thales visited Mesopotamia and Egypt. In the former country he learned of the 'Saronic cycle', that is to say the interval of eighteen years and eleven days, a multiple of which the observations of ages by temple star-gazers had shown to be usual between eclipses of the sun.[1] Knowledge of this enabled the shrewd traveller to make a lucky forecast of the eclipse visible at Miletus in 585 B.C. His prediction drew much attention. It may well be that the impression thus created directed the attention of the Greeks to the advantages that might accrue from systematic observation of nature. At any rate, they always reputed Thales to be the father of that study.

Further achievements of Thales were chiefly of a geometrical nature. Now it is important here to recall that the Greeks did not invent geometry. They could and did gather some knowledge of the subject from their neighbours in the Nile Valley. The Egyptians, however, had hardly reached beyond an empirical usage of certain

---

[1] *Saros* from a Babylonian word *saru* (Greek *saros*) for the number 3,600, i.e. (60)² and hence for a period of 3,600 years. The application of the word to the cycle of 223 lunations (18 years 11 days) is a modern misunderstanding. The word is, however, now firmly fixed in scientific nomenclature.

special relations of figures, and especially of triangles and rectangles, of pyramids and spheres. Thus, for instance, the Egyptians knew that the square on the longest side of a right-angled triangle is equal to the sum of the squares on the other two sides; but they knew it only for such special cases as that in which the sides are in the ratio 3, 4, and 5; thus $5 \times 5 = 3 \times 3 + 4 \times 4$ (Fig. 3). Again,

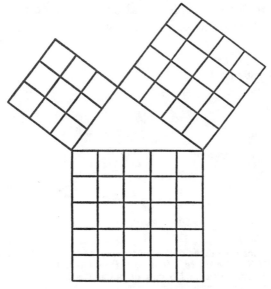

FIG. 3. Special case of squares on sides of right-angled triangle.

they could estimate the cubic contents of a pyramid, but only of a pyramid of a certain definite type with a certain definite number of sides sloped at a certain definite angle.[1] Thales succeeded in generalizing such special cases. He thus discovered that the angles at the base of an isosceles triangle are equal; that when two straight lines cut one another the opposite angles are equal; that the angle on the circumference of a circle subtended by the diameter is always a right angle; that the sum of the angles of a triangle is equal to two right angles; that the sides of triangles with equal angles are proportional.

[1] The question as to how far the Egyptians generalized mathematical conceptions is still under discussion.

9

Thales, moreover, succeeded in applying such knowledge. He was able, for example, by a simple application of the principle of similar triangles, to determine the distance from the shore to a ship at sea (Fig. 4), and to measure the height of a pyramid by comparing the length of its shadow with that cast by an object of known height. Such problems had been tackled before his time. But Thales not only sought to enunciate them clearly and to solve them demonstrably but also to widen and generalize them so as to lay bare their essential nature.

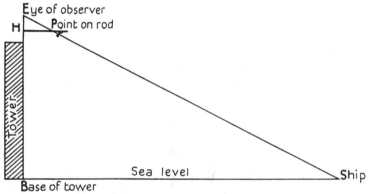

FIG. 4. Thales measures distance to ship at sea. Triangle *EHP* similar to triangle *EBS*. Therefore *EH* is to *HP* as *EB* is to *BS*. Since *EH*, *HP*, and *EB* are all measurable *BS* can be calculated.

As with every Ionian thinker, the ultimate object of the thought of Thales was to find a formula for all things. He thus set himself the task of discerning constancy amidst the diversity and variety of nature. This is but to say that his science was a part of his philosophy. To the general question 'Of what is the world made?' he would answer 'Water', meaning thereby some mobile essence, changing, flowing, without distinctive shape or colour, yet presenting a cycle of existence passing from sky and air to earth, thence to the bodies of plants and animals, and back to air and sky again. But his real place in the history of science is better brought out by the more concrete statement that in his mathematical work we have the first enunciation, as distinct from implicit acceptance, of natural laws.

The Foundations: Ionia, Magna Graecia, Athens

Following on Thales, a long line of Asiatic Greeks, mostly of Miletus, contributed to the extension of the conception of natural law. Thus ANAXIMANDER (611–547 B.C.), a Miletan pupil of Thales, took much interest in geography. He was the first among the Greeks to represent the details of the surface of the earth by maps. The idea of map-making was known in Egypt, where plans of

FIG. 5. Egyptian map of gold mines. New Kingdom.

particular districts or objects as mines, houses, and temples were being drawn up as early as 1400 B.C. (Fig. 5). Anaximander, however, sought to convey a concrete picture of the surface of the earth as a whole. The suggestion doubtless came from Mesopotamia, where simple diagrams of this sort were being made in his time. From Babylon also he introduced the sun-dial. It consisted in essence of a *gnomon*, a fixed upright rod, the direction and length of the shadow of which can be measured hour by hour. The records of these make it possible to determine the movements of the sun as well as the dates of the two *solstices* (the shortest and longest days) and of the *equinoxes* (the two annual occasions when day and night are equal). Anaximander was thus led to develop his own astronomical conceptions. He was the first to speculate on the size and distance of the heavenly bodies. The earth was for him a flat disk in the centre of all things. Sun, moon, and stars are enclosed in opaque rings, rotating with the earth as centre. We see them only through vents in these rings.

ANAXIMENES (born *c.* 570 B.C.), another Miletan, extended Anaximander's ideas, especially in astronomy.

The ultimate essence of all things he regarded as 'air' rather than the 'water' of Thales. This air was linked up with that essence which is essential to life. He called it *pneuma*—literally *breath*—and held that in a sense the universe itself was alive: 'As our soul, being air, sustains us, so *pneuma* and air pervade the whole World'.

At about the same date CLEOSTRATUS of Tenedos, who lived rather outside the Ionian zone, made two important contributions to astronomy. One was an improvement in the calendar, involving a better measure of the solar year. The other was the knowledge of the signs of the zodiac which he introduced from Mesopotamia. Zodiacal signs are frequently encountered upon Mesopotamian boundary stones and indicate the time of year at which the stones were erected (Fig. 50).

Among the Greeks of Asia Minor towards the end of the sixth century B.C. there was not only considerable speculative activity, but also the sum of positive knowledge was being systematically increased. The process was encouraged by the roving character of the Asiatic Greeks. Active and daring seamen, they brought back to their homes accounts of many of their adventures by land and sea.

Of these early explorers, the most distinguished was HECATAEUS, also of Miletus (born *c.* 540 B.C.). He visited Egypt, the provinces of the Persian Empire, Thrace and Lydia. He penetrated the Dardanelles and explored the coasts of the Black Sea. About 500 B.C. he adventured westward to the Gulf of Genoa and as far as Spain, reaching Gibraltar. There he had been preceded by the Phoenicians, who set up to their god Melkarth a great column on either side of the Strait. Later writers identified Melkarth with Hercules, and the gateway of the Mediterranean came to be called the 'Pillars of Hercules'. Hecataeus collected his experiences into a geographical handbook (Fig. 6). He is memorable for that

FIG. 6. The World as conceived by Hecataeus, *c.* 500 B.C.

scepticism of the marvellous which is a hall-mark of the man of science and a condition for scientific progress. He detested mythology. 'The stories of the Greeks', says Hecataeus, 'are in my opinion no less absurd than numerous.'

About the turn of the sixth into the fifth century, the character of Ionian thought was modified by closer contact with Persia. That power, under its great Emperor Darius I (522–486 B.C.), was advancing steadily westward. The weak and quarrelsome little

Asiatic Greek States were coming under its shadow. The Persian service attracted many of their citizens, who brought back to their native homes further knowledge of the world. Among the more typical of these venturers was the physician DEMOCEDES of Cnidus (born *c.* 540 B.C.). The peninsula of Cnidus was the seat of the most ancient medical school of which we have any record.

After travelling widely in Greek lands, Democedes became the medical attendant of the Persian monarch. Later he was employed as a spy to explore the coasts of Greece. He escaped from this service, however, and settled in the Greek colony of Croton, in the instep of Italy. Here he devoted himself to writing a treatise on medicine, the first Greek work on that subject of which we have tidings. Croton became an important scientific centre.

Thus, as time wore on, Ionian thinkers came more closely into contact with other civilizations. Their work becomes increasingly sophisticated. Philosophy is no longer the product of the leisure hours of business men, of sailors, or of physicians. Thinking has become a profession.

Amongst the great Ionians who concerned themselves exclusively with philosophy was HERACLEITUS of Ephesus (*c.* 540–475 B.C.). He is specially remembered for his view that 'everything is in a state of flux'. Change is the only reality. 'There's nothing is and nothing was, but everything's becoming.' Fire, the most changeful of elements, is the origin and image of all things. Living creatures are formed of a mixture of the changeful essences of which fire and air are types. Nothing is born and nothing dies. The illusions that we call birth and death are but a rearrangement of these unresting elements.[1]

Very different from the point of view of Heracleitus was that of his younger contemporary, the Miletan LEUCIPPUS (flourished *c.* 475 B.C.), founder of the atomic doctrine of matter. That theory has had a wide influence in both ancient and modern times. It has been associated with the attitude towards the world known sometimes as 'philosophic materialism'.

[1] The thought of Heracleitus bears a certain resemblance to that ascribed to the founder of Buddhism who was his contemporary. Whether one derived from the other or both from a common source is a matter which future research may decide.

The Foundations: Ionia, Magna Graecia, Athens

Leucippus—of whom we know little—is overshadowed by his pupil, DEMOCRITUS (c. 470–c. 400 B.C.) who was perhaps also of Miletus. This Democritus was a contemporary of Socrates (470–399 B.C.; p. 31), though the outlook of the two men is in the strongest possible contrast. For Democritus, very different to Heracleitus, all things were made up of solid concrete *atoms*, together with the space or *void* between them. We should note that this void has as much claim to be regarded as a primary reality as the atoms themselves. The atoms are eternal, invisibly small, and cannot be divided. (The word *atom* means 'indivisible'.) They are incompressible and homogeneous. They differ from one another only in form, arrangement, and size, that is to say only quantitatively, not qualitatively. The qualities that we distinguish in things are produced by movement or rearrangement of these atoms. Just as atoms are eternal and uncaused, so also is motion, which must, of its nature, originate in preceding motion. As everything is made up of these unchangeable and eternal atoms, it follows that coming into being and passing away are but a seeming, a mere rearrangement of the atoms. The beings that you and I think we are, are but temporary aggregations of atoms that will soon separate to enter into the substance of other beings. And yet, in ages of time, perhaps, we shall be re-formed, when it may so fall out that our atoms come together again. Thus history may repeat herself endlessly.

At first sight the positive teaching of Democritus and the concrete character of his atoms suggest a 'common-sense' philosophy that might be set against the Heracleitan vagueness. It must be remembered, however, that the atoms of Democritus were in no sense the product of experimental investigation. His atoms, like their motion and like the void in which they moved, were alike hypotheses and based on no sort of exact knowledge or experience. His teaching has obvious parallels with more modern scientific doctrines concerning the 'indestructibility of matter' and the 'conservation of energy', but the parallels are more apparent than real. Despite the positive trend of the thought of Democritus, his followers—known as 'Epicureans' after his most distinguished adherent, EPICURUS of Samos (342–270 B.C.)—showed little tendency to extend the range of scientific ideas.

Much of the spirit of Ionia is summed up in the life and writings of HERODOTUS of Halicarnassus (c. 484–425 B.C.). The native town of this remarkable man was within the limits of the Persian Empire at the time of his birth, and he remained a Persian subject till he was well into his thirties. From an early date his inquiring spirit led him to travel. He explored Greece and Asia Minor thoroughly, visiting many of the islands of the Greek Archipelago.

FIG. 7. The World as known to Herodotus.

He made the long and difficult journey from Sardis in Lydia, near the modern Smyrna, to Susa, the Persian capital (Fig. 7). He travelled next to Babylon ; then he explored the coast of the Black Sea and penetrated into Scythia and Thrace. His journeys were extended westward, and he visited Italy and Sicily. Southward from his home he passed into Syria, sojourned at Tyre, saw something of Palestine, and made a long stay in Egypt. Wherever he heard of anything curious or interesting, he stayed for a while and noted what he saw. Finally he joined a Greek colonizing party that settled in Italy. He spent the rest of his life preparing his delightful *History*.

Herodotus does not concern himself with the world as a whole, but he gives an excellent idea of the geographical knowledge of his day. His careful observations on the nature and habits of different peoples entitle his work to be regarded as the first treatise on the science of man. He is thus the father of anthropology, as he is also the father of history. Many of his allusions to the beliefs and practices of the time help us to check the early records of the history of science.[1]

2. *The Pythagoreans and the Western School.*

From a very early date Greeks had penetrated westward and

FIG. 8. Western Greek Colonies.

had established colonies in Southern Italy and Sicily, *Magna Graecia* as the area came to be called (Fig. 8). The intellectual activity of these western colonies played an important part in the development of Greek science. The most influential of the western scientific movements was that of the 'Pythagoreans'.

The founder of this school or sect, PYTHAGORAS (born *c.* 582 B.C.), was by birth an Ionian of Samos. He travelled widely. About 530 he settled at Croton, where a Dorian colony had been estab-

[1] Herodotus is especially responsible for the view that Greek institutions were derived from Egypt.

lished. There he founded his brotherhood or sect, which persisted long after him. He left nothing in writing, and the veil of mystery which his followers drew over themselves often prevents us from ascribing the scientific advances which they made to their actual originators.

From the hazy philosophical outlook of the Pythagoreans there emerge certain ideas which have exerted a profound influence. Foremost is their peculiar teaching on the subject of numbers. These were held to have a real and separate existence outside our minds. The use by the Greeks, as by the Hebrews, of letters to express numbers gave an especial currency to this conception, which was capable of, and received, all sorts of mystical and magical application. An example will readily come to the mind in connexion with 666 'the number of the beast' in the book of *Revelation* (xiii. 18). There was a similar Pythagorean tendency to ascribe an objective independence to the divisions of time. Again a Biblical illustration is to hand:

> 'Job cursed the day.
> Let that day perish wherein I was born,
> Let it not be joined unto the days of the year.'
> *(Job* iii. 1–6.)

The word *mathematics* itself—which means simply 'learning'— was given its special relationship to numbers by the Pythagoreans.[1] Aristotle tells us in his *Metaphysics* that

'the Pythagoreans devoted themselves to mathematics. They thought that its principles were the bases of all things. In numbers they saw many resemblances to the things that exist and are coming into being—one modification of number being *Justice*, another *Reason*, another *Opportunity*—almost all things being numerically expressible. Again they regarded the attributes and ratios of the musical scale as expressible in number. They therefore regarded numbers as the elements of all things, and the whole heaven as a musical and numerical scale. The very arrangement of the heavens they collected and fitted into their scheme. Thus, as 10 was thought to be perfect and to comprise in itself the whole nature of numbers,

[1] Greek *mathēsis* 'learning', *mathētēs* 'disciple', so used in New Testament, *mathēmatikos* 'fond of learning', so used by Plato and Aristotle. The word *mathematics* did not enter the English language till the late sixteenth century. The curious plural form is an elliptical expression for 'mathematic sciences' and has no foundation in Greek.

they said that the bodies which move through the heavens were ten in number; but since the visible heavenly bodies are but nine, they invented a counter-earth.' (See Philolaus, p. 21.)

The conception seems very fanciful to us now. Nevertheless fancies of this type have been repeatedly of value in the history of science. The human mind, it must be supposed, is somehow attuned to the processes of nature. We live in a world that is susceptible of mathematical expression. Thus the theoretical investigations of mathematicians correspond in some degree to the findings of the physicists and astronomers. Such is the nature of things, though why this should be so is a mystery. Perhaps it is not even the business of science to discuss this mystery. But consciousness of a correspondence between the workings of our minds and the workings of nature is illustrated by this doctrine of the Pythagoreans. Their conception of the 'harmony of the spheres'—on which Aristotle touches in the above passage—was related to an interest in music. It proceeded from the observation that the pitch of musical notes depends on a simple numerical ratio in the length of the chords struck. This numerical ratio, it was held, corresponded to the distances of the heavenly bodies from the centre of the world.

The beautiful conception of a world bound together in a harmony has captivated the imagination of poets in every age. There was a time

> When the morning stars sang together
> And all the sons of God shouted for joy.
> (*Job* xxxviii. 7.)

It is the dullness of the ear of flesh, so the Middle Ages would have had us believe, that prevents us from hearing still these glorious notes. The Christianity, which set off body against spirit, at times would claim to catch the heavenly tones;

> soft stillness and the night
> Become the touches of sweet harmony.
>
> .   .   .   .
>
> There's not the smallest orb which thou behold'st
> But in his motion like an angel sings,
> Still quiring to the young-eyed cherubins;
> Such harmony is in immortal souls;

But, whilst this muddy vesture of decay
Doth grossly close it in, we cannot hear it.
(*Merchant of Venice*, Act V, Sc. i, ll. 56–65.)

The Pythagorean habit of giving character and qualities to numbers becomes more intelligible to us if we remember that for the Greeks mathematics was, in effect, geometry. Thus, to take

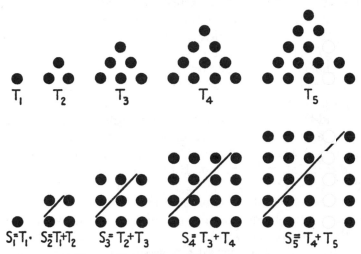

FIG. 9. Triangular and square numbers.

a prominent example, the Pythagoreans distinguished the series

1, 1+2, 1+2+3, 1+2+3+4, 1+2+3+4+5 · · ·

as *triangular* numbers, and they exhibited geometrically the interesting fact that the sum of any two consecutive *triangular numbers* is a *square number* (Fig. 9).

The so-called 'Pythagorean theorem', that is that the square on the hypotenuse of a right-angled triangle is equal to the sum of the squares on the other two sides (Fig. 3), was referred by the ancients to Pythagoras himself. The Pythagoreans erected a system of plane geometry in which were formulated the principal theorems which concern parallels, triangles, quadrilateral and regular polygonal figures and angles. They discerned many important properties of prime numbers and progressions. In particular they worked out a theory of proportion which involved both commensurables and incommensurables. This was of great importance as providing

The Foundations: Ionia, Magna Graecia, Athens

the link between arithmetic and geometry. They recognized at least four types of proportion. Thus:—

arithmetical proportion $a-b = b-c$

geometrical proportion $a : b :: b : c$

harmonic proportion $a-b : b-c :: a : c$

musical proportion $a : \dfrac{2ab}{a+b} :: \dfrac{a+b}{2} : b$

The most striking mathematical achievement of the Pythagorean thinkers is perhaps their attainment of a conception of the nature of *irrational quantities*, such, that is, as are not expressible by ordinary numbers. With the imperfect mathematical notation of the time, however, great algebraical advance was impossible, and irrational numbers could not be algebraically represented (compare p. 189). Greek mathematics was thus forced to preserve its geometrical bias. The Greeks, in fact, constantly resorted to geometric methods when we should prefer algebraic. A very simple example will suffice. The equation $(x+y)^2 = x^2+2xy+y^2$ was geometrically proved by reference to such a figure as the adjoining (Fig. 10).

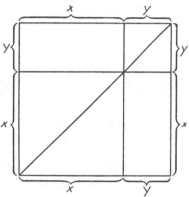

FIG. 10. The Pythagorean presentation of the equation $(x+y)^2 = x^2+2xy+y^2$.

Led by their mystical view that the sphere is the perfect figure, just as 10 is the perfect number, the Pythagoreans introduced the conception that the earth and the heavenly bodies are spheres. This important advance is among the many in the history of science in which the formation of general ideas on theoretical grounds has preceded and not followed practical observation.

An interesting astronomical hypothesis was put forward by the Pythagorean PHILOLAUS of Tarentum (c. 480–400 B.C.). He abandoned the theory that the Earth is the mid-point of the universe, and supposed that it is similar to the other planets in its movements, and that all revolve round a central fire. This fire, he held, is invisible to us, since the part of the earth which we inhabit is ever turned away from it. To balance his system he invented a *counter-earth*, bringing his spheres of the movable heavenly bodies up to the sacred number 10, that is to say, Sun, Moon, Earth, five planets, Counter-earth, and sphere of the stars. Philolaus was the first to publish a book on Pythagorean doctrine. It was used by Plato in the composition of the *Timaeus* (p. 34). The conception by Philolaus of a moving earth and central fire influenced Copernicus (p. 180).

Another Pythagorean development was destined to influence thinkers in after ages in a very curious way. Manipulating equilateral triangles and squares in three dimensions, the Pythagoreans dis-

| 4 | 6 | 8 | 12 | 20 sides |

FIG. 11. The five Platonic bodies.

cerned four 'regular solids', that is figures with all their sides and angles equal. These four were the regular 4-sided pyramid (*tetrahedron*), the 6-sided *cube*, the 8-sided *octahedron*, and the 20-sided *icosahedron*. They were taken to represent the four elements of the physical world, earth, air, fire, and water. Later was discovered the geometrical mode of constructing regular *pentagons* or 5-sided plane figures. One of the Pythagoreans found that these could be built into a fifth regular solid, the 12-sided *dodecahedron*. In the absence of a fifth element this was taken to represent the universe. The five possible regular solids became later known as the 'Platonic bodies'. They played a large part in subsequent philosophical and mathematical development, much of it very fanciful. Kepler's thought about the Platonic bodies in the sixteenth century suggested the first modern unitary theory of the universe (p. 200–6) (Figs. 11 and 61).

From the regular pentagon it was easy to pass to the 5-pointed star or *pentagram*, formed by an endless line joining alternate angles of a pentagon. The Pythagoreans used the pentagram as a secret sign of recognition. It thus started on its career of mystery, passing into magic and humbug. For Pythagoreans and Platonists it expressed completeness, health, well-being. Among lesser souls it degenerated into the commonest and most banal of charms. No evil could pass it! Faust has a pentagram on the threshold of his study which prevents Mephistopheles from leaving it. The history of the pentagram provides a type of the degradation that science has repeatedly suffered (Fig. 12).

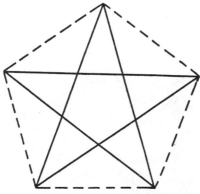

FIG. 12. The 'magic pentagram', a continuous line or 'endless knot' formed by producing the sides of a regular pentagon both ways or by joining its alternate angles.

It was not only in cosmical and mathematical speculation that the western colonies exhibited their intellectual activity. During the fifth century B.C. there developed among the Greeks in Italy and Sicily a remarkable naturalistic art. Painters closely observed and represented the parts and structures of animals (Fig. 13). This naturalistic tendency is reflected by the Italo-Greek scientific thinkers. Among them, ALCMAEON of Croton (c. 500 B.C.), a pupil of Pythagoras, extended the scientific field to living things. He began the practice of scientific dissection. He discovered the nerves that proceed from the brain to the eyes. He described those passages connecting mouth and ear, through which, if the nose be pinched and the cheeks blown out, air is driven into the ear-drums. These tubes were next investigated by the anatomist Eustachi, after whom they are now called *Eustachian tubes.* Eustachi lived in Italy more than twenty-two centuries after

Alcmaeon! Alcmaeon believed that these tubes carried the *pneuma* (see Anaximenes, p. 12).

An important Western thinker, upon whom Pythagoras had influence, was EMPEDOCLES of Agrigentum in Sicily (*c.* 500–*c.* 430 B.C.). He held that the blood is the seat of the mysterious *innate heat*, an idea taken from folk belief that 'the blood is the life' (*Deuteronomy* xii. 23). This innate heat he closely identified with the soul. He held the heart to be the centre of the system of blood-vessels through which the innate heat, or essential factor of life, is distributed to the bodily parts. Thus for the followers of Empedocles the heart was the special seat of life. This idea passed to Aristotle (p. 44).

Sargus vulgaris

Crenilabrus mediterraneus

FIG. 13. Paintings of fish on plates from Magna Graecia of fourth century B.C. They are very exactly drawn and the species can be identified.

The teaching of Empedocles led to curiosity as to the distribution of the blood-vessels. Our first coherent account of these is the work of DIOGENES of Apollonia in Crete (*c.* 430 B.C.), who was greatly influenced by the thought of Empedocles and his school (Fig. 14).

Empedocles supposed that Love and Strife alternately held sway over all things. Everywhere there was opposition and affinity. In matter itself the so-called *four elements* could be distinguished as exhibiting these relationships. All matter was held by him to be made up of the four essential elements—*earth, air, fire,* and *water.* These were in opposition or alliance to one another.

24

Thus water was opposed to fire, but allied to earth. Each of the elements was, moreover, in its turn compounded of a pair of

FIG. 14. The vascular system as described by Diogenes of Apollonia about 400 B.C. He described a system of vessels penetrating the whole body, proceeding from great medial trunks, and he distinguished arteries from veins as regards form, function, and distribution.

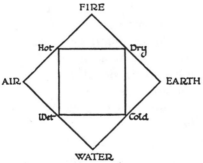

FIG. 15. The Four Elements and Four Qualities of Empedocles.

the four 'primary qualities', heat and cold, moisture and dryness (Fig. 15). These qualities exhibit affinity and opposition as do the elements.

It must not be imagined that such philosophers as Empedocles

25

thought that the 'elements' were the substances that we know by the names of earth, water, air, and fire on our earthly sphere. Here we find the elements only in combination. Thus the substance we know as water contains, according to the theory, a preponderance of elemental water, but contains also small amounts of the other three elements. The element water forms only the essence of water, an essence that we human beings can never apprehend.

This doctrine has left its mark on our language. We still speak of a storm as 'the raging of the elements'; we wear coats 'to protect ourselves from the elements'; and we think of 'elemental forces'. We still read the passage in *Galatians* in which St. Paul adjures us not to 'turn again to the weak and beggarly elements' (*Galatians* iv. 9); nor have we difficulty in understanding references to a 'fiery nature' or to an 'aerial spirit'. These things come to us from Empedocles, and they come through Aristotle (p. 48) and the Athenian School.

3. *Fathers of Athenian Science.*

By the middle of the fifth century B.C. both the Eastern and the Western schools of Greek thought were overshadowed by Athens, now the intellectual centre of the Greek world. An important factor in this concentration was the Ionian ANAXAGORAS (488–428 B.C.) of Clazomenae. He came to Athens (464 B.C.) burning with scientific zeal, and attracted the attention and friendship of the statesman Pericles (490–429 B.C.) and of the poet Euripides (480–406 B.C.), both of 'whom he inspired with his own love of science. From Socrates (p. 31) he differed profoundly. Much of the course of thought in later ages may be traced to this divergence, for Plato was the philosophic heir of Socrates while Aristotle took much from Anaxagoras.

Anaxagoras developed an obscure and difficult philosophic system which involved rational theories concerning many celestial phenomena. He gave scientific accounts of eclipses, meteors, and rainbows. The sun was a vast mass of incandescent metal, the light of the moon was reflected from it, and other heavenly bodies were stones rendered white hot by rotation. Such interpretation outraged the religious opinion of the day, and he was prosecuted for impiety. Defended by Pericles and acquitted, he yet found it prudent to withdraw to his native Asia Minor. Thus early began the persecution of scientific doctrine opposed to current religion. The intellectual conditions in the Athenian metropolis were very different from those in the colonies of Ionia and Magna Graecia. In Athens the greater complexity of life was making itself felt. The systematic accumulation of knowledge was beginning to render a little old-fashioned those who 'took all knowledge to be their province'. The eloquence of the popular educators known as 'sophists' entertained and attracted the volatile Greeks beyond anything else. But many of the sophists were little but professional talkers, and few or none had any direct acquaintance with scientific matters, which were left to another class. Thus something in the nature of scientific specialization began to appear. The movement affected especially two departments, medicine and mathematics. By a curious chance, the two typical exponents of these disciplines bore the same name and came from neighbouring and similarly named islands. They were the physician, Hippocrates of Cos, and the mathematician, Hippocrates of Chios.

HIPPOCRATES THE PHYSICIAN was born about 460 B.C. on the island of Cos just inside the Dorian Zone. He came of a family of physicians. Both on his own island and on the opposite peninsula of Cnidus (p. 7) medical schools had long been established. It was their destiny to transform the tradition that had developed there into a scientific procedure. The change afterwards became traditionally associated with the name of Hippocrates.

Hippocrates led a wandering life, following his profession in Thrace, in the neighbourhood of the sea of Marmora, on the island of Thasos, at Athens, and elsewhere. He had many pupils, among whom were his sons and sons-in-law. He is said to have died in his hundredth year, an appropriate age for a great physician! This is almost all we know of his personal history. Yet it is impossible to exaggerate the influence on medicine of the picture that was early formed of him. Learned, observant, humane, with a profound reverence for the claims of his patients, but possessed of an overmastering desire that his experience should benefit others; orderly and calm; anxious to record his knowledge for the use of his brother physicians and for the relief of suffering; grave, thoughtful,

27

and reticent; pure of mind and master of his passions; such is the image of the father of medicine as it appeared to his successors.

While the philosophers developed the conception of a rational world, it was the physicians, typified by Hippocrates, who first put the rational conception to the test of experience. It was they who first consciously adopted the scientific procedure which, in its relation to medicine, is sometimes called the 'Hippocratic Method'.

The method of the Hippocratic writers is that now known as 'inductive'. Without the vast scientific heritage that is ours to-day; with but a small number of recorded observations and those from scattered and little organized experiences; surrounded by all manner of bizarre religious cults which recognized no adequate relation of cause and effect; above all, constantly urged by the exuberant genius for speculation of their own people whose intellectual temptations they shared, the Hippocratic physicians remained, nevertheless, patient observers of fact, sceptical of the marvellous and the unverifiable, hesitating to theorize beyond the facts, yet eager to generalize from actual experience. There are few types of mental activity known to us that cannot be paralleled among the Greek writings. Careful and repeated return to verification from experience, expressed in a record of actual observations, has been rare at all times in history. It is wonderful that so many Greek works have come down to us expressing this attitude. A large proportion of these are by Hippocratic authors.

It is true that the Greeks had scientific forebears (p. 7). It is probable that they borrowed, more frequently than we know, from other civilizations. But the 'Religion of Science' of these early physicians, the belief in the constant and universal sequence of cause and effect in the material world, was theirs before all other men. The first prophet of that religion was Thales. The first writings on that religion bear the name of Hippocrates. The first great exponent of that religion whose works are still substantially intact is Aristotle (p. 39).

The Hippocratic writings, important for the history of medicine, are even more significant for the conception that they contain of the nature of science itself. This conception is beautifully expounded in a treatise on the *falling sickness*, or epilepsy. In those days the affliction was regarded as a divine visitation, a 'sacred disease'.

28

The Foundations: Ionia, Magna Graecia, Athens

A Hippocratic writer composed a book on it, in which he sets forth the proper attitude of the scientific man towards such claims. It is a monument of the rational spirit, and is perhaps the first book in which there is clear opposition between the claims of science and of religious tradition.

In our own time natural events are not always treated, even by educated men, in the spirit of the Hippocratic writers. Both leases and insurance certificates have still sometimes a clause as to the type of accident to which the lawyers refer as an 'act of God'. The type of these acts of God has altered in the course of ages. They used to include, for instance, infectious disease. Our word 'plague' is from a Latin word meaning a *blow* or *stroke* which comes to us from the days when the 'plague-stricken' were held to be stricken by God himself. The legal term 'act of God' still includes the action of tempest and of lightning. Yet the attitude of the Hippocratic work called the *Sacred Disease*, written more than 400 years before the birth of Christ, is very different:

> 'As for this disease called divine, surely it has its nature and causes, as have other diseases. It arises—like them—from things which enter and quit the body, such as cold, the sun and the winds, things ever changing and never at rest. Such things are divine or not—as you will, for the distinction matters not—and there is no need to make such division anywhere in nature, for all are alike divine or all are alike human. All have their antecedent causes which can be found by those who seek them.' [Slightly paraphrased.]

We have spoken of the belief in the constant sequence of cause and effect as a 'religion' (p. 28), since it was—and perhaps still is—essentially a matter of faith. In Hippocratic times there was as yet no large body of exact observations by which the operations of nature could be exactly forecasted, save only the astronomical record. Thus the regularity of the astronomical sequences was, by an act of faith, set forth as the type to which all nature should accord. The heavenly bodies herald those regularly recurring changes of season which determine the lives of men. It is but a step to regard them as the causes of those changes and thus to treat them as gods. The step was often taken and the planets still bear the names of deities.

29

Rise of Mental Coherence

HIPPOCRATES OF CHIOS, the mathematician (*c.* 430 B.C.), was the first to compile a work on the *Elements of Geometry*. This title has made a household word of his successor, Euclid (p. 57). Hippocrates of Chios is the first known mathematical 'specialist'. He began life as a business man. Chance brought him to mathematics. He came on a law-suit to Athens. That city was rapidly becoming the centre of learning, and the provincial Hippocrates had now an opportunity to consort with philosophers. His

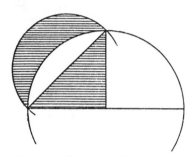

FIG. 16. Lune of Hippocrates of Chios.

real abilities rapidly asserted themselves, and he began to devote himself with ardour to mathematical pursuits.

The work of Hippocrates of Chios may be illustrated by one of his most acute investigations. It gives an idea of the standard to which mathematics had attained in Greece about 400 B.C. Hippocrates discovered that the *lune* bounded by an arc of 90°, and by a semi-circle upon its chord, is equal in area to the triangle formed by the corresponding chord with the centre as its apex (Fig. 16). The lune—a figure bounded by curves—being thus equated with a figure bounded by straight lines, its area can be ascertained. He discovered two other lunes of which the areas could be similarly expressed. Finally, he discovered a particular lune which, when added to a circle, enables the whole to be represented geometrically as a square. This lune by itself cannot, however, be squared, and so the method cannot be used for squaring the circle. These remarkable researches became misrepresented and tradition told tha Hippocrates had succeeded in the impossible geometrical task of squaring the circle! His proofs, in fact, imply great familiarity with advanced geometric methods. They are based on the theorem, which he himself proved, that circles are to one another as the squares of their diameters.

Thus by the end of the fifth century not only had philosophical thought taken a scientific turn, but science itself had emerged as a preoccupation of men set aside from their fellows. Two depart-

30

ments, medicine and mathematics, had become well differentiated. Astronomy had been the special interest of such philosophers as Pythagoras (p. 17), Philolaus (p. 21), Empedocles (p. 24), and Anaxagoras (p. 26). This earlier phase of Greek thought terminated in the fifth century with a thinker of a very individual type.

The name of SOCRATES (470–399 B.C.) is associated with a great intellectual revolution, perhaps the greatest that the world has seen. His overwhelming preoccupation was with conduct. For him 'Knowledge is Virtue'. The attitude of Socrates towards the sciences of his day has been set forth by his pupil Xenophon (430–350 B.C.), who tells that

> 'with regard to astronomy Socrates considered a knowledge of it desirable to the extent of determining the day of the year or of the month and the hour of the night; but as for learning the courses of the stars, occupying oneself with the planets or inquiring about their distance from the earth or about their orbits or the causes of their movements, to all these he strongly objected as a waste of time. He dwelt on the contradictions and conflicting opinions of the physical philosophers . . . and, in fine, he held that speculators on the Universe and on the laws of the heavenly bodies were no better than madmen.'

The triumph of the Socratic revolution depressed for a while both science and physical philosophy. But out of the conflict between the Socratics and the physical philosophers arose the main streams of later Greek thought. These two streams derive their titles and their tendencies from the two gigantic figures that occupy the stage during the fourth century, the age of Plato and Aristotle.

II. THE GREAT ADVENTURE

Unitary Systems of Thought: Athens, 400–300 B.C.

1. *Plato and the Academy.*

THE thought of PLATO (427–347 B.C.), like that of his master Socrates, was dominated by the ethical motive. Convinced, like Socrates, that Truth and Good exist and that they are inseparable, he embarked on an inquiry which had as its object to expose, account for, and resolve into one comprehensive theory the discrepancies of ordinary thinking. During this process he developed a doctrine destined to be of great moment for the subsequent relation of scientific thought with that which comes under the heading of religion and philosophy. It is the so-called *Doctrine of Ideas*. The nature of this doctrine and the manner in which Plato reached it have been briefly set forth by his pupil, Aristotle.

'In his youth', says Aristotle, 'Plato became familiar with the doctrine of certain philosophers that all things perceived by the senses are ever in a state of flux and there is no knowledge concerning them [see Heracleitus, p. 14]. To these views he held even in his later years. Socrates, however, busied himself about ethical matters, neglecting the world of nature, but seeking the universal in conduct. He it was who fixed thought for the first time on definitions. Plato accepted his teaching but held that the problem of what was to be defined applied not to anything perceived by the senses but to something of another sort. His reason was that there could be no real definition of things perceived by the senses because they were always changing. Those things which could alone be defined he called *Ideas*, and things perceived by the senses, he said, were different from these *Ideas* and were all called after them.' (Aristotle's *Metaphysics*.)

Thus concepts, things of the mind, became for Plato something more concrete, while our impressions of the material universe, percepts, became something more vague. It is as though the word 'horse' were to suggest to the mind not Ned or Dobbin or even a cart-horse or a carriage-horse but a generalized being that is approximately expressed by the biologist's definition of the *species* horse. Further this 'Idea' of the species was more truly an entity than any individual horse. The Platonic 'Idea' contained in it the conception of form, for only in the Idea was the form separated from matter. The conception is put epigrammatically by Plato in the phrase 'the Soul is the place of forms',[1] that is, of those forms which can be defined.

Plato expresses a great admiration for mathematical principles, and he regards mathematics as exhibiting that type of certitude and exactness to which other studies should conform. Mathematics indeed relies for its material upon something of the nature of Plato's Ideas. It might be expected, therefore, that mathematics would appeal to him. Many of Plato's thoughts assume a mathematical guise. He exhibits at times a view which seems to approach that of Pythagoras, who had attached a moral and spiritual value to numbers (p. 18). Plato thus tended to respect a science in the degree to which it had progressed in the mathematical stage of its development. The heavenly bodies evinced, in the opinion of those Pythagorean days, the exemplars of perfect geometric forms (p. 22). For astronomy—especially on its theoretic as distinct from its observational side—Plato had therefore a high regard. Indeed, for many of his Greek followers mathematics became identified with astronomy. We think of astronomy as a field for the *application*, the Platonists rather for the *exemplification* of mathematics.

The attitude of Plato was less favourable to those sciences, other than astronomy, to which we nowadays habitually apply our mathematics. On the non-mathematical sciences he smiled even less. He repudiated the theories of such thinkers as Democritus, who not only denied the existence of mind as a separate entity but also assumed the universe to be the result of accident (p. 15). Such a universe was hardly susceptible of exact presentation. In ultimate analysis the position of Democritus was a denial of the validity of philosophy. On the other hand, Plato speaks with respect of Hippocrates the physician, the very type of the scientific man in antiquity—Hippocrates of whom a follower said 'he was the first who separated science from philosophy'.[2] Plato's respect for Hippocrates, however, did not tempt him to follow in

---

[1] The phrase is not found in the extant works of Plato but is quoted by Aristotle in the *De anima*.

[2] Celsus, *De re medica*, Introduction.

his footsteps. Nor is this surprising, for, firstly, Plato assigned a relatively unimportant place to phenomena and, secondly, his mind was too full of a greater vision to enable him to lend himself to the tedium of the pursuit of the inductive method.

Nevertheless, the greatest of thinkers could not refrain from producing some general theory of the universe of phenomena. The work in which this appeared, the dark and difficult *Timaeus*, is under strong Pythagorean influence (p. 22). Its spokesman is a member of that sect. Its very darkness and difficulty provide an unintentional appeal for that patient, impartial objective process of observation and record that is the very foundation of science. The *Timaeus* demonstrates how knowledge can be degraded, even by Plato, in the relentless endeavour to ascribe a meaning to all parts of the universe. The work displays the Platonic mood at its weakest.

The trend of Platonism in general and of ancient Platonism in particular has normally been away from observational activity, even when friendly to mathematics. There are, however, many and evident exceptions and, moreover, Platonism has often been helpful to science in the presence of an entrenched and static Aristotelianism.

It has been said that 'everyone is by nature a disciple either of Plato or of Aristotle'. There is much truth in this. Aristotle himself set forth the difference between the two attitudes, reduced to its simplest expression. In his great work, the *Physics*, Aristotle discusses the use of mathematical formulae. The objects studied in the physical sciences, he says, do present, of course, planes, lines, and points. Such planes, lines, and points are the subjects also of mathematical study. How, then, are we to distinguish the procedure of mathematics from that of the true physical sciences which often invoke mathematics?

To this, Aristotle answers that the mathematician does indeed study planes, lines, and points, but he studies them as mental abstractions and not as the 'limits of a physical body'. The objects of mathematics, though in fact inseparable from a physical, movable, and therefore changeable body, are studied in abstraction from that change to which all material things are subject. This process of abstraction necessarily involves error. The mistake made by Plato's theory of Ideas, says Aristotle, is that of attempting to exclude from his consideration of matter those conceptions in which are involved the very nature of matter, though not that of mathematical objects. Thus odd and even, straight and curved, number, line, figure—all these can be studied wholly out of connexion with the change or movement inseparably connected with material things. They are subjects for the mathematician. Such things as flesh, bone, man, nay, even inorganic nature, minerals and earths, sounds and colours, heat and cold, cannot be so studied. They are subjects for the man of science. Change is indeed an essential part of nature, fundamental to real existence, as Thales, the father of science, had seen (p. 8) and Heracleitus with his 'being as becoming' had emphasized (p. 14). Yet change has to be ignored in pure mathematical investigation. This principle of change or movement prevents nature from ever really repeating herself, while in mathematical conceptions one unit is exactly like another.

We may see the contrasted effects of the Platonic and the Aristotelian attitudes in the scientific works of the two great philosophers. So far as science is concerned, it is by their fruits that we must know them. Plato has shrouded his views in the *Timaeus*. From the deceptive shadows seen in the twilight of that work he has elevated into picture form, from an 'Idea', a mechanism that never was on land or sea. On the other hand, in the great biological works of Aristotle we have a magnificent series of first-hand observations and positive studies to which, in each succeeding generation, naturalists still return with delight, with refreshment, and with respect.

The importance of Plato, so far as the subsequent development of science is concerned, is thus to be sought chiefly in the depart·ment of mathematics. Plato was, in fact, an accomplished mathematician and had had Pythagorean teachers. The 'Platonic bodies', the five regular solids which have equal sides and equal angles, were known to the Pythagoreans (p. 22). Plato describes them in the *Timaeus*, exhibiting full understanding of them. There are many other passages in his writings which show mathematical penetration, nor is it easy to overrate his influence upon later mathematical developments. We may consider it under four headings:

The Great Adventure

(a) It is through Plato that mathematics obtained, and retains, a place in education. In the abstractions of mathematics he saw an instrument for the training of logical thought. The study of mathematics was thus for him the portal to philosophy. 'Let none who has not learnt mathematics enter here' was inscribed over the entrance to his school, the Academy.

(b) The hand of Plato may be traced in the actual course of mathematical development. To his logical teaching the body of mathematical knowledge owes the systematic structure and logical finish that have since distinguished it. This factor exhibited itself in his pupils and his spiritual descendants. Such a work as Euclid's *Elements* is in essence a product of Plato's thought and of Plato's school (p. 37). It is certainly no overstatement that, through Euclid (p. 57), every schoolboy is nowadays a student of Plato.

(c) The inspiration of Plato can be traced very clearly also in the history of astronomy. He early came to regard the irregularities of planetary motion as inconsistent with his view of the essential perfection of the universe. These movements had, in his opinion, to be explained as somehow compounded of simple circular movements, a conception that he derived from his Pythagorean teachers (p. 21). Plato accordingly set his pupils to seek out rules by which the movements of the heavenly bodies could be reduced to a system of circles and spheres. This was the main task of astronomers from his time to Kepler (p. 200)—a stretch of two thousand years! During all those centuries the hand of Plato ruled astronomy. Here Aristotle (p. 39) is but a pupil of Plato as Plato is of Pythagoras.

(d) Plato may be said to have made a positive contribution to science of first-class importance. It cannot be said that this is wholly his creation, since the germs of it are to be found among the Pythagoreans, but its formal introduction is Plato's work. It is the method of assuming that a problem is solved and working back from it until a statement is reached, the truth or falsehood of which is already known. Thus may be discerned whether the problem is, in fact, soluble or not, and indications may be forthcoming as to the general direction of the solution and whether there are any limitations to it. The method is set forth in the *Meno*.

36

Euclid often used this method and it is current in modern elementary geometry.

There is a curious Platonic conception that is perhaps a mere by-product of his thought but was yet fraught with consequences for after ages. The Pythagorean Timaeus, in Plato's dialogue of that name, pictures the universe as a living thing with a soul penetrating its body. The passage is well summarized by Aristotle:

> 'Timaeus tries to give a physical account of how the soul moves its body. The soul is in movement and the body moves because it is interwoven with it. The Creator compounded the soul-substance out of the elements and divided it according to the harmonic numbers (p. 18) that it might have an innate perception of harmony and that its motion might be with movements well attuned. He bent its straight line into a circle. This he divided into two circles united at two common points. One of these he divided into seven circles [that is the orbs of the seven planets] in such wise that the motions of the heavens are the motions of the soul.' (*De anima.*)

This view of the universe gave a framework for the Neoplatonic conception that the structure of the universe foreshadowed that of man. Thus arose the doctrine of the intimate relation of *macrocosm* ('great world') and *microcosm* ('little world', that is, Man). This doctrine permeated medieval Christian thought (p. 123).

Plato's school, under the name of the Academy, persisted for many centuries, but was chiefly occupied with philosophical discussion. One of his first disciples to distinguish himself in science was EUDOXUS (409–356 B.C.) of Cnidus, the founder of observational cosmology. Eudoxus had also studied with the Pythagoreans. Under the stimulus of Plato he made advances in mathematical theory, but occupied himself chiefly with examining the heavens. Among his achievements is his remarkably accurate estimate of the solar year as 365 days and 6 hours. His most influential contribution was his view that the heavenly bodies move on a series of concentric spheres, of which the centre is Earth, itself a sphere. Eudoxus had observed the irregularities in the movements of the planets. To explain these he supposed each planet to occupy its own sphere. The poles of each planetary sphere were supposed to be attached to a larger sphere rotating round other poles. The secondary spheres could be succeeded by tertiary or quaternary spheres according to mathematical and observational needs. For Sun and Moon Eudoxus found three spheres each sufficient. In the explanation of the movements of the other planets, four spheres each were demanded. For the fixed stars one sphere sufficed. Thus twenty-seven spheres in all were demanded. These spheres—save that of the fixed stars—were treated by Eudoxus not as material but in the manner of mathematical constructions.

CALLIPUS of Cyzicus, a pupil of Eudoxus and friend of Aristotle,

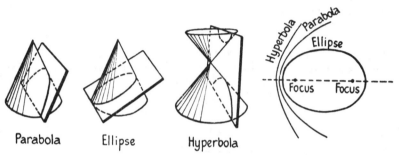

Parabola   Ellipse   Hyperbola

FIG. 17. Types of curve obtained by section of cones by planes.
(Compare Figs. 26 and 58.)

observed movements of the heavenly bodies and irregularities unknown to his master. To explain these he added yet further spheres, making thirty-four in all. The Eudoxan theory thus modified was adopted by Aristotle (p. 47).

HERACLEIDES of Pontus (*c.* 388–315 B.C.), a pupil of Plato, contributed to astronomy a suggestion that the Earth rotates on its own axis once in twenty-four hours, and that Mercury and Venus circle round the Sun like satellites. His teaching led on to that of Aristarchus (p. 59).

Important for subsequent mathematical developments was MENAECHMUS, another pupil of Eudoxus. Menaechmus initiated the study of conic sections. He cut three kinds of cone, the 'right angled', the 'acute angled', and the 'obtuse angled', by planes at right angles to a side of each cone. Thus he obtained the three types of conic section which we now call by the names allotted to them by his Alexandrian successor Apollonius (p. 70)(Fig. 17).

Many others of Plato's followers made contributions to pure mathematics, and, in the sense which we have discussed (p. 36), all subsequent mathematicians are Plato's spiritual heirs. There is also evidence of a certain amount of botanical activity in the Academy, and some physiological theories which became popular in later centuries may be traced to Plato. Platonism passed into Christianity early, mainly through St. Augustine, so that the Christian Middle Ages, until the twelfth century, were mainly Platonic. The later school of philosophy known as 'Neoplatonism' also profoundly influenced Christianity (pp. 121–5).

2. *Aristotle.*

ARISTOTLE (384–322 B.C.) was born at Stagira, a Greek colony a few miles from the northern limit of the present monastic settlement of Mount Athos. His father was physician to the monarch of Macedon. At seventeen Aristotle became a pupil of Plato at Athens. On his master's death in 347 he crossed the Aegean Sea to reside in Lesbos, an island off the coast of Asia Minor. In 342 he became tutor to the young prince Alexander of Macedon. He remained in Macedon till 336 when Alexander started his career of conquest that was to alter the face of the world. Aristotle then returned as a public teacher to Athens. There he owned a garden known as the *Lyceum*, whence the word has derived its special significance. In it he established his famous school afterwards called the *Peripatetic* (Greek, 'walking around'), for he had there his favourite *Peripatos* or cloister where he lectured.

Aristotle's writings cover the whole area of knowledge. The earliest are biological. These were written, or at least drafted, during his residence in Asia Minor (347–342). Most of his other works were produced during his second period at Athens (335–323), in the twelve years that preceded his death. We must always remember that the whole of Aristotle's science, and indeed the whole cast of his mind, was deeply influenced by his biological experience.

Regarded from the modern scientific standpoint, Aristotle appears at his best as a naturalist. His first-hand observations are on living things, and his researches on them establish his claim to be regarded as a man of science in the modern sense. In his

39

great work, *On the Parts of Animals,* he sets forth what he regards as the relation between 'physics'—which is for him a general description of the universe—and the study of living things.

'Of things constituted by nature', he says, 'some are ungenerated, imperishable, eternal; others subject to generation and decay. The former are excellent beyond compare and divine, but less accessible to knowledge. The evidence that might throw light on them, and on the problems which we long to solve respecting them, is furnished but scantily by our senses. On the other hand, we know much of the perishable plants and animals among which we dwell. We may collect information concerning all their various kinds, if we but take the pains.

'Yet each department has its own peculiar charm. The excellence of celestial things causes our scanty conceptions of them to yield more pleasure than all our knowledge of the world in which we live; just as a mere glimpse of those we love is more to us than the grandest vista. On the other side we may set the certitude and completeness of our knowledge of earthly things. Their nearness and their affinity to us may well balance the loftier interest of the things of heaven, that are the objects of high philosophy.

'But of a truth every realm of nature is marvellous. It is told that strangers, visiting Heracleitus (p. 14) and finding him by the kitchen fire, hesitated to enter. "Come in, come in", he cried, "the gods are here too." So should we venture on the study of every kind of creature without horror, for each and all will reveal something that is natural and therefore beautiful. Absence of haphazard and conduciveness of all things to an end are ever to be found in nature's works, and her manner of generating and combining in ever-changing variety is of the highest form of the Beautiful.' [Somewhat paraphrased.]

Though it cannot be claimed that Aristotle was an evolutionist in the sense that he regarded the different kinds of living things as actually related by descent, yet there can be no doubt that he fully realized that the different kinds can be arranged in a series in which the gradations are easy. His scheme was a 'Ladder of Nature' (Fig. 18) as it came to be called by later naturalists. Thus he writes in his *History of Animals*:

'Nature proceeds by little and little from things lifeless to animal life, so that it is impossible to determine the exact line of demarcation, nor on which side thereof an intermediate form should lie. Thus, next after lifeless things in the upward scale, comes the plant.

Of plants one will differ from another as to its amount of apparent vitality. In a word, the whole plant kind, whilst devoid of life as compared with the animal, is yet endowed with life as compared with other corporeal entities.' Indeed, there is observed in plants a continuous scale of ascent toward the animal.'

The peculiar principle that Aristotle invoked to explain living phenomena we may call 'soul', translating thereby his word

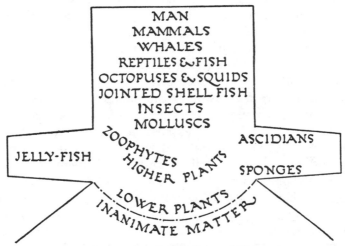

MAN
MAMMALS
WHALES
REPTILES &FISH
OCTOPUSES & SQUIDS
JOINTED SHELL FISH
INSECTS
MOLLUSCS

JELLY-FISH

ZOOPHYTES
HIGHER PLANTS

ASCIDIANS

SPONGES

LOWER PLANTS
INANIMATE MATTER

FIG. 18. Aristotle's Ladder of Nature.

psyche. His teaching on that topic is to be found in his great work *On the Soul* usually cited by its Latinized title *De anima.* He thinks of things as either 'with soul' or 'without soul' (*empsychic* or *apsychic*). His belief as to the relationship of this soul to the matter in which it is embodied is difficult and complicated, but he tells us that 'Matter is identical with potentiality, form with actuality, the soul being that which gives the form or actuality in living things'. Thus for Aristotle 'soul' is not a separate existence. In this he differs from his master Plato and no less from early Christianity which, through St. Augustine (p. 123), borrowed much from Plato. Aristotle believes, too, that the soul works ever to an end, and that

'As every instrument and every bodily member subserves some partial end, some special action, so the whole body must be destined

The Great Adventure

to minister to some fuller, some completer, some greater sphere of action. Thus an instrument such as the saw is made for sawing, since sawing is a function, and not sawing for the saw. So, too, the body must somehow be made for the soul and each part thereof for some separate function to which it is adapted.' [*Parts of Animals*, somewhat paraphrased.]

Aristotle is thus a *vitalist* (Latin *vita*, 'life') and a *teleologist* (Greek *telos*, 'end', 'object'), that is to say, he believes that the presence of a certain peculiar principle is on the one hand essential for the exhibition of any of the phenomena of life, while on the other hand it serves to integrate all such phenomena towards the emergence of the perfect living individual. The Democritans, to whom Aristotle was opposed, believed that all the actions of living things were the result of the interaction of the atoms of which they were composed (p. 15). Thus life, for the Democritans, was capable of mechanical expression. They were *mechanists*. The division between *vitalist* and *mechanist* extends throughout the history of science and still separates students of living things.

Living things are for Aristotle the type of existence, and existence as a whole presents, according to him, evidence of design.

'Everything that nature makes is a means to an end. For just as human creations are the products of art, so living objects are manifestly the products of an analogous cause or principle. . . . That the heaven is maintained by such a cause, there is, therefore, even more reason to believe than that mortal animals so originated. For order and definiteness are even more manifest in the celestial bodies than in our own frame. . . . Thus Nature is marvellous in each and all her ways.' [*Parts of Animals*, greatly abbreviated.]

Aristotle attempted to analyse the nature of generation, of heredity, of sex. His are the first presentations of many such topics which are to-day discussed by naturalists. There is an amazing variety and depth in his biological speculations. These have a permanent value and are constantly cited by biologists of our own time.

Aristotle's psychological studies are only partly within our purview. The psychological questions with which we are concerned come mostly into his discussion of the nature of life. 'Of natural bodies,' he says, 'some possess life and some do not; where by life we mean the power of self-nourishment and of independent growth and decay.' It should be noted that in the Aristotelian sense the egg or germ is not at first a living thing, for in its earliest stages and before fertilization it does not possess 'soul' even in its most elementary form.

In a famous passage from his work *On the Soul* Aristotle says:

'The term life is used in various senses. If life be present in but a single one of these senses, we speak of a thing as alive. Thus, there is intellect, sensation, motion from place to place and rest, the activity concerned with nutrition, and the processes of decay and growth. Plants have life, for they have within themselves a faculty whereby they grow and decay. They grow and live so long as they are capable of absorbing nutriment. In virtue of this principle [the vegetative soul] all living things live, whether animals or plants, but it is sensation which primarily constitutes the animal and justifies us in speaking of an *animal soul*. For, provided they have sensation, creatures even if incapable of movement are called animals. As the nutritive faculty may exist, as in plants, without touch or any form of sensation, so also touch may exist apart from other senses.'

Apart from these two lower forms of soul (*a*) the *vegetative*, or nutritive and reproductive, and (*b*) the *animal*, or motile and sensitive soul, stands (*c*) the *rational* or conscious and intellectual soul that is peculiar to man.

The possession of one or more of the three types of soul, vegetative, animal, and rational, provides in itself a basis for an elementary form of arrangement of living things in an ascending scale. In fact the basis of Aristotle's 'Ladder of Nature' (p. 40) is really psychological, depending on the character of soul or mind. It is characteristic of Aristotle's method that the various departments of investigation should thus interlock.

In the closest possible association with Aristotle's biological views stand his innumerable and admirable observations. Among the more striking are the following:

(*a*) A series of records of the life and especially the breeding habits of a large variety of animals. About 540 species are discussed.

(*b*) Embryological investigations of the developing chick, which has ever since been the classic object for such investigations.

(*c*) Accounts of the habits and development of the octopuses and squids which have, in some cases, been surpassed only in modern times.

(*d*) Anatomical descriptions of the four-chambered stomach of the ruminants, of the complex relationships of the ducts and vessels in the mammalian generative system and of the mammalian character of the porpoises and dolphins, all unsurpassed until the sixteenth century.

(*e*) Accounts of exceptional modes of development of 'fish. Among them is one of a species of dogfish of which the young is linked to the womb by a navel cord and placenta, much in the manner of a mammal. Nothing has contributed more to Aristotle's scientific reputation in modern times than the rediscovery of this phenomenon.

(*f*) As a result of his embryological investigations Aristotle attached very great importance to the heart and vascular system. He came to regard the heart as 'the first to live and the last to die',[1] a conception which passed to the Middle Ages and was current until the eighteenth century.

(*g*) A lasting addition to the technique of scientific instruction was made by Aristotle in introducing diagrams to illustrate complex anatomical relations. Some of his diagrams can be restored from his descriptions (Fig. 19).

Most of Aristotle's biological work reads like that of a modern naturalist, for his methods are closely similar to those of our own time. But when we turn to examine Aristotle's view of the universe we encounter not only a different method of work but a mode of thought so diverse from ours that we can neither understand nor sympathize with him without some special study. The intellectual revolution of the insurgent century (Ch. VII) resulted in complete destruction of the Aristotelian physical philosophy. Modern science is the product of that revolution, and it is difficult for us to go behind it in our thinking.

We are all of us brought up from early years with the idea of the 'uniformity of nature', that is that the same causes always and everywhere produce the same results. Thus, for instance, we think of astronomers exploring the heavens and discovering

---

[1] This sentence is often given as a quotation from Aristotle. It occurs, however, nowhere in his writings, though the idea is to be found there.

new facts about worlds other than our own. We assume, and we are justified in assuming, that in the starry spaces there rule the general physical laws which we have learned on our

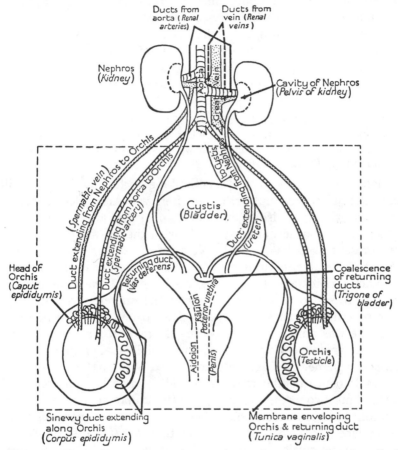

FIG. 19. Generative and excretory systems of a mammal as described by Aristotle. The part framed in a dotted rectangle restores a lost diagram prepared by Aristotle and described in his *Historia animalium*. The legends in brackets are the modern scientific terms, the others transliterations or translations of Aristotle's terms.

earth. On this principle astronomers deduce, for instance, the exact chemical constitution of many of the stars. Did we question ourselves on this matter, we might, perhaps, ask how, if the physical laws that we know on earth did not prevail in the stars, could astronomers make discoveries at all? But this law of uniformity that we take for granted was by no means obvious to Aristotle. To him heaven was not only different from earth, but its ways were incommensurate with the ways of earth.

Aristotle knew nothing of the book of Isaiah. But his philosophical distinction between the rules of heaven and of earth made a special appeal to the Church fathers and to his medieval followers who had read that book. It was brought nearer to them by a superb and oft-quoted passage,

> 'My thoughts are not your thoughts, neither are your ways my ways, saith the Lord. For as the heavens are higher than the earth, so are my ways higher than your ways, and my thoughts than your thoughts.' (*Isaiah* lv. 8, 9.)

Isaiah, like Socrates (p. 31), was thinking of the moral order in his contrast of heaven and earth. So, often, was Aristotle. But Aristotle was thinking also of other kinds of order, and it is with the other kinds of order, and especially with the physical order, that our present work has to deal. We must remember, however, that for Aristotle all the kinds of order were related to each other.

When Aristotle had completed his biological works he applied himself to set forth a general view of the universe which should link together its various aspects. The structure of the material universe was among these aspects. He revised his account over and over again, seeking to fit his earlier biological findings into his general scheme. We are only concerned with that scheme in so far as it concerns the material world. Aristotle's physical and astronomical conceptions, however, were unlike his biological conceptions in being untouched by profound personal knowledge and experience. Regarded scientifically they are far inferior to his biological conclusions. Nevertheless it was Aristotle's physical and astronomical conceptions that influenced the centuries which followed, while his biological works were neglected and ultimately forgotten, to be rediscovered in relatively modern times.

Aristotle, like Plato, exhibits in his physical scheme some Pythagorean tendencies. Especially he emphasized the circle and the sphere as the most 'perfect' figures and therefore those on which the world is modelled. Thus he was led to regard the heavens as a series of concentric spheres arranged round our earth as a central body (Fig. 20). These spheres he described, however, as crystalline, mechanizing them from the mathematical scheme of Eudoxus (p. 37). Around our earth was the sphere of the atmosphere and around that spheres of pure elemental nature, being, from within outward and in order of density, earth (or rather

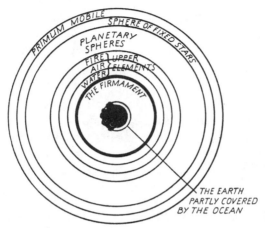

FIG. 20. The Universe of Aristotle as conceived by a medieval writer.

earthy exhalation) water, air, and fire. These spheres of pure elements are as inaccessible to us as the heavens themselves. Next, outward beyond the sphere of elemental fire, lies the region of a yet more mysterious substance, the *ether* (Greek 'shining') which enters into the composition of the heavenly bodies. Yet farther out are in succession the seven spheres, each of which carries a planet, while beyond is the eighth sphere which bears the fixed stars. Finally, beyond all others, is the sphere whose divine harmony causes the circular revolution of the whole celestial system.

Such was the basis of the system that was to control for two thousand years the view that men took of Nature. We may thus summarize the system, its history, and its fate:

(a) *Matter is continuous.*

In taking this view Aristotle opposed Democritus and sided with Socrates and Plato. The followers of Democritus and of his disciple Epicurus, who took an atomic view of matter (p. 14), were associated with doctrines which were peculiarly abhorrent to the early and medieval Church. The atomic theory was the only alternative to Aristotle's conception of matter. Thus criticism of Aristotle on this point drew theological odium on itself. The atomic theory, we shall therefore see, passed into the background for many centuries.

(b) *All mundane things are made up of four 'elements', earth, air, fire, and water, which, in their turn, contain the four 'qualities', heat, cold, dryness, and moisture, in binary combination* (Fig. 15).

This view of matter was taken from Empedocles (p. 24) and is probably of yet more ancient origin. It is the Aristotelian expression of the Pythagorean conception of all things being in a state of love or hate—fire, for instance, being opposed to water but allied to air. The doctrine of the four elements was almost unquestioned until the seventeenth and lasted until the end of the eighteenth century. It fitted well with Christian and Moslem thought and became a part of orthodox medieval theology.

(c) *Stars and planets move with uniform circular velocity in crystalline spheres, centred round the earth. Each sphere is subject to the influence of those outside it.*

This general conception is of Pythagorean origin (p. 21). Aristotle did little but borrow it from Eudoxus, mechanize it, and fit it into a general system of philosophy. His scheme, or some modification of it, held its ground till the time of Kepler in the seventeenth century (p. 200).

(d) *Circular movement is perfect since the circle is the perfect figure. Circular movement represents the changeless, eternal order of the heavens. It is contrasted with rectilinear movement which prevails on this our changing and imperfect earth.*

'Where imperfection ceaseth, heaven begins.'

Here again are Pythagorean influences. The basis of the conception is that while heavenly bodies appear to circle round us, bodies on earth tend to fall or rise. Newton at the end of the seventeenth century succeeded in expressing the movements of the heavenly bodies in known and experimentally demonstrated

48

terms. Until his time the differences between the behaviour of earthly and heavenly bodies remained a puzzle or paradox or both.

(*e*) *The Universe is limited in space in the sense that it is contained within an outer sphere. It is unlimited in time in the sense that it is subject neither to creation nor destruction as a whole.*

The finiteness of the Universe both in space and time became necessary to all the theological systems of the Middle Ages and notably to that of the Western Church. It was effectively unquestioned till the time of Bruno (died 1600). Thus Aristotle himself could not be completely accepted. The philosophical return to the conception of a Universe infinite both in space and time is a landmark in the history of science (p. 186).

It has been urged against Aristotle that he obstructed the progress of astronomy by divorcing terrestrial from celestial mechanics, for he adopted the principle that celestial motions were regulated by their own peculiar laws. He thus discouraged astronomical observation, placed the heavens beyond the possibility of experimental research, and at the same time impeded advance in the knowledge of mechanics by his assumption of a distinction between 'natural' and 'unnatural' motion. For two thousand years the general outline of the world as set forth by Aristotle remained the orthodox view. It was dangerous even to question it. How far was Aristotle responsible for this intellectual tyranny? To this question there are many answers, of which we shall adduce but four.

(*a*) It was not Aristotle who introduced the distinction between celestial and terrestrial physics. Such distinction had been taken for granted by his predecessors. The Pythagoreans, for example, had made much of them. In fact by his exposition of a positive and tangible scheme he gave a new interest to the study of nature.

(*b*) It is unfair to bring his own greatness as a charge against Aristotle. All our conceptions of the material world—'scientific theories' as we call them—should be but temporary devices to be abandoned when occasion demands. This is a proposition which Aristotle himself puts forth. In expounding the motions of the planets he advises his readers to compare his views with those that they themselves reach. That his scheme lasted for two thousand years without effective criticism is no fault of his. It is rather evidence that the men who followed him were dwarfs compared with 'the master of those who know'.

(c) Some of Aristotle's reasons against what we now regard as the form of our world are, in fact, valid. Thus he argues against the motion of the earth. Such movement, if it existed, should, he considered, produce *apparent motion* among the fixed stars. This is a just objection. It was only met in the nineteenth century by the demonstration of interstellar motion. The reason that this was not previously detected is that the vast distance of the heavenly bodies from us makes this apparent motion so small that excessively delicate instruments are needed.

(d) We need to remember that the rigidity of the Aristotelian scheme lay not in itself but in the interpretation given to it; especially in the Middle Ages. By linking the theories of Aristotle with their own religious views, men of those times introduced a bitterness into the debate concerning the validity of the Aristotelian scheme that had nothing to do with its philosophical or scientific value.

3. *Peripatetics, Stoics, and Epicureans.*

It is improbable that his connexion with Alexander was of any service to Aristotle himself.[1] There can be no doubt, however, that the great conqueror was a friend of learning and that important investigations were initiated by him. Thus he made an attempt to survey his empire by employing a special force whose duty it was to maintain the condition of the main roads. The services of these men were available for scientific purposes, such as the collection of data bearing on the natural history of the districts where they were at work. Investigations were also made by certain of Alexander's commanders, notable by his admirals, NEARCHUS and ANDROSTHENES. Portions of their botanical and geographical works are preserved.

Aristotle's own work was continued by his school, the Peripatetics, of whom the best-known was the long-lived THEO-

---

[1] A number of statements to the contrary can be found in writings of later classical antiquity. None, however, bears critical scrutiny.

Unitary Systems of Thought: Athens, 400–300 B.C.

PHRASTUS (372–287 B.C.) of Eresus in the island of Lesbos. Though a pupil of Aristotle he lived to be contemporary with the first generation of Alexandrian science (Ch. II). He made important botanical researches and continued Aristotle's work in Aristotle's spirit. It is interesting to observe that he exhibits the same 'evolutionary' bias that characterizes the biological work of his master. In one of his great botanical treatises Theophrastus observes that 'where there is growth there is life. Wherefore we should observe these things not for what they are but for what they are becoming. And, moreover, though some be peculiar, yet the general plan can everywhere be traced and is never lost.'

Ancient science suffered from lack of a scientific terminology. This defect Theophrastus attempted to remedy in his own chosen department of botany. For his technical terms he did not rely, as do we, on an ancient and classical language, but sought rather to give special meanings to words in current use. Among such words were *carpos*, 'fruit', and *pericarpion*, 'seed vessel'. From Theophrastus are derived the modern botanical definitions of *fruit* and of *pericarp*. Many Theophrastan plant-names also survive in modern botany.

The botanical works of Theophrastus are the best arranged biological treatises that have survived from antiquity. They contain many acute and accurate observations. Among these are his clear and exact distinction between monocotyledons and dicotyledons. Interesting, too, is his attempted distinction of sex in plants, an attempt which is only successful in the case of the palms. Of those plants, as Herodotus tells us, the ancient Babylonians had the same idea.

Another younger contemporary of Aristotle was AUTOLYCUS of Pitane (*c.* 360–*c.* 300). He worked at his native town and at Sardis, and expounded the geometry of the sphere for astronomical and geographical purposes. A pupil of Aristotle who worked on somewhat the same lines was DICAEARCHUS (*c.* 355–*c.* 285). He employed himself on physical geography and wrote a description of the world accompanied by a map. He, too, worked on information derived from Alexander's officers and was the first to draw a parallel of latitude across a map. This was used merely as a convenient dividing line. It extended from the Pillars of Hercules

(p. 13) due east along the Taurus and 'Imaus' (Himalaya) ranges to the Eastern Ocean.

It is appropriate to mention here the explorer PYTHEAS of Marseilles (*c.* 360–*c.* 290 B.C.) though he was not of the Peripatetic school. The itinerary of his remarkable voyage can be traced with some exactness. He left Marseilles about March 320 B.C. and made for Spain, followed the coast through the pillars of Hercules to Cadiz and then along the Atlantic seaboard as far as Cape Ortegal. From there he struck across the ocean to Ushant and on to Cornwall. He next sailed round Great Britain and, returning to Kent, crossed to the continental side of the English Channel and followed the North Sea coast to the mouth of the Elbe. From there he turned north following the Scandinavian coast as far as Trondhjem at about latitude 63. After having put forth thence into the open sea, he turned back along the way he had come and reached Marseilles towards the end of October of the same year.

Pytheas was a good astronomer, and made a number of observations of latitude, among others of his native place Marseilles, which he fixed with remarkable accuracy. He was the first of the Greeks who arrived at any correct notion of the tides, indicating their connexion with the moon and its phases.

One of the best-known of the earlier Peripatetics was the Thracian, STRATO of Lampsacus (*c.* 300 B.C.). He reduced the formation of the world to the operation of natural forces. He recognized nothing beyond natural necessity and, while retaining opposition to atomism, he sought to explain all the functions of the soul as modes of motion.

After the first generation the Peripatetic school devoted itself to preserving or to commenting upon the work of its founder. It exhibited no scientific originality, and from about 300 B.C. onward Athens ceased to be a great scientific centre. Two of the later Peripatetics are, however, of some importance for the history of science. One, ANDRONICUS of Rhodes, was about contemporary with Christ. He prepared a critical text of the works of Aristotle which was probably closely similar to that which we now possess. The other was the Cilician ALEXANDER of Aphrodisias (*c.* A.D. 200). He was an industrious commentator whose writings, much used by the Neoplatonists (p. 122), were the foundation of the Arabian commentaries (p. 129 et seq.) and through them of many of the Latin Aristotelian commentaries. Soon after Alexander's death the Peripatetic was merged into the Neoplatonic school (p. 122).

Contemporary in origin with the Peripatetics was the philosophical school called *Stoic*, from a *stoa* or corridor of the marketplace at Athens, where its members used first to meet. The Stoics stressed the operation of natural forces in the manner of Strato the Peripatetic (p. 52). They differed from the Peripatetics, however, in emphasizing the interaction of all different parts of the material world. Thus, while there are reasons for everything in nature, it is also true that everything in nature is among the reasons for the rest of nature. All existence is capable of acting or being acted upon so that 'force' the active and 'matter' the passive principle pervade each other. With this doctrine of 'universal permeation' there is no real difference between matter and its cause. The conception of Deity becomes indistinct and blended with that of 'reason' or 'law' which is but an aspect of a pantheistic system.

Important for the history of science was the Stoic cosmology. From 'primitive being' or *pneuma* there separated the four elements in succession, fire first, earth last. The remaining *pneuma* is the 'ether' (p. 47). From these five factors arose a universe on the Aristotelian model. In the world which has thus been formed we, who are parts of it, must obey the inevitable laws. But this world will again decay and dissolve into elements and finally into primitive being or *pneuma*. Our individual souls are part of the universal *pneuma*, temporarily separated therefrom. In the embryo the soul is still in the 'vegetative' stage. It becomes successively 'animal' and 'rational' (p. 43) but joins, in the end, the universal *pneuma*.

So far as human relations and human conduct go, the key to Stoicism is *fate*. The Stoic schooled himself to disregard the inescapable, the nature of which came to be tested by astrology (p. 63). He devoted himself to the development of his own soul through duty, awaiting inevitable absorption into the world-soul.

The Stoic school maintained itself in Athens, Rhodes, and Alexandria. It attained no great importance till Roman Imperial times, but then became the prevalent faith of the upper class

The Great Adventure

(p. 94). Among its exponents were the poet Cleanthes of Assus (*c.* 250 B.C., p. 116), the meteorologist Aratus of Soli (*c.* 260 B.C., p. 116) and the Bithynian scholar POSIDONIUS of Apamea (135–50 B.C.). The latter, as an exponent of Stoicism, was anxious to demonstrate the interrelations of different parts of the universe. He was thus attracted to the discussion of the influence of the Moon on the tides. He also made estimates of the size of the Sun in excess of those of any other ancient writer. Posidonius was a friend and admirer of Cicero (p. 118) and thus links Greek with Roman Stoicism.

A rival sect to Peripatetics and Stoics was that of the Epicureans refounded in 307 B.C. by EPICURUS of Samos (342–270). The thought of Epicurus was based on the atomism of Democritus (p. 15) and to a less extent on Anaxagoras (p. 26). Epicurean philosophy was traditionally divided into the three branches of logic, physics, and ethics. Beyond a discussion of atomic doctrine, however, the school exhibited little interest in phenomena, and Epicurus himself deprecated scientific pursuits.

Epicurean philosophy spread rapidly and widely in Asia and Egypt. About 150 B.C. it established itself at Rome where its ablest exponent was Lucretius (*c.* 95–55 B.C., p. 95).

The warring of these sects—Peripatetic, Stoic, Epicurean—seems a trivial incident as against the great constructive thought of Plato and Aristotle. With Aristotle we have parted with the first and most active stage of ancient scientific thought. In estimating his place in the history of science we may say that

(*a*) He represents the final stage of the 'Great Adventure', the attempt to represent the world as a whole and as a unitary system.

(*b*) He provided a philosophic synthesis which, in more or less modified form, satisfied intellectual aspirations from his own time until the seventeenth century.

In that philosophical system there remained two great breaks in continuity. One hiatus was between celestial and terrestrial physics. This first began to be filled by the workers of the 'Insurgent Century' from Bruno (p. 185) to Newton (p. 248). The other gap was between the world of the living and of the not-living. The Epicurean philosophy attempted to fill the breach in ancient times by the introduction of a 'mechanist' system (p. 42). The Christian Church in medieval times, repudiating with vigour the Epicurean solution, accepted the breach as part of the divine order of the world. The physiologists in modern times, beginning with van Helmont (p. 231), Descartes (p. 221), Borelli (p. 239), and Sylvius (p. 240) in the seventeenth century, have been seeking to resolve it ever since.

In leaving the heroic age of Greek science we would again emphasize the 'universal' character of the philosophical attempt that we call the 'Great Adventure'. The scientific activity of the age partook of the nature of what we should now term 'philosophy'. The object of each investigator was to fit his observations and the laws that he deduced into some general scheme of the universe. From their day to ours philosophy has continued her attempt thus to storm the bastions of heaven. But with the new age that we have to discuss, there was a failure of nerve in that great frontal attack. Science, becoming gradually alienated from philosophy, begins to proceed by her own peculiar method of limited objectives. The first series of these attempts resulted in the 'Great Failure', the story of which we shall trace through two thousand years (Chs. III, IV, V). Nerve fails first, as with the Alexandrian school (Ch. III), next Inspiration falters under the Roman Empire (Ch. IV), lastly Knowledge itself fades in the Middle Ages (Ch. V). At length there is a rebirth. The science of the Renaissance—in which we still live—began again to proceed by the method of limited objectives (Ch. V). How that method differed from that with which the Great Failure is associated is a matter which we shall have to discuss.

III. THE FAILURE OF NERVE

Divorce of Science and Philosophy (300 B.C.–A.D. 200):
Alexandria

1. *Early Alexandrian Period (300–200 B.C.).*

WHEN Alexander died (323 B.C.), his Empire broke into fragments
(Fig. 21). Egypt was seized by one of his generals, named Ptolemy,
and the Ptolemaic dynasty endured for three hundred years. Its
members were mostly able and intelligent men and women. The

FIG. 21. Break-up of Alexander's Empire.

first of the line established the tradition of learning. The second
founded a library and museum at Alexandria. That city became
the centre of the scientific world. Learned men flocked to it and
were supported by funds provided by the Ptolemaic rulers. The
school continued very active for a couple of centuries. By 100 B.C.,
however, it was beginning to languish, and by A.D. 200 in rapid
decay, though there was spasmodic scientific activity until
about 400.

The Alexandrian library in its earlier stages had many distinguished curators. Most were literary men, but some, such as
Eratosthenes (p. 70) and Apollonius (p. 69), were also men of
science. From 300 B.C. to A.D. 200 most eminent men of science
were teachers at Alexandria. A few, notably Archimedes and
Galen, were less intimately linked with the Egyptian metropolis.
Yet even they were pupils of the school and corresponded with

Divorce of Science and Philosophy: Alexandria

Alexandrian teachers. Greek science from about 300 B.C. onward is thus not inadequately described as 'Alexandrian science'. Alexandria was not, however, entirely without rivals as a seat of learning. The most prominent were the island of Rhodes and the city of Pergamum in western Asia Minor. Of the enmity between Alexandria and Pergamum there is an interesting reminder in our language. The Alexandrian books were written on rolls prepared from *papyrus* reeds, whence our word *paper*. To prevent Pergamum from acquiring copies of their literary treasures, the jealous Ptolemies put an embargo on the export of papyrus. The Pergamene kings, cut off from a valued import, sought to improve the preparation of skins, the Asiatic medium for writing. Thus was developed the *membranum pergamentum* which has reached our language as *parchment*.

It is characteristic of Alexandrian science that it developed along the lines of 'specialities'. These came to lose their relation to general philosophic thought with which they had hitherto been linked. It is convenient to consider Alexandrian science in three chronological divisions; an *early period* containing the first and second generations of the school to nearly 200 B.C., a *middle period* to about the birth of Christ, and a *late period* to the complete decline of the school. Archimedes (p. 63) demands individual discussion.

The early Alexandrian period is noteworthy for the fact that mathematics at once assumed a prominent and independent position. Among the first to be called to the Alexandrian Academy was the illustrious mathematician EUCLID (*c.* 330–*c.* 260). He was trained at Athens, probably by a pupil of Plato. His most famous work, the *Elements of Geometry*, has determined all subsequent teaching. Perhaps no book save the Bible has been so much studied. For the next twenty-two centuries parts of the *Elements*, and especially the first six of its thirteen books, were the customary introduction to geometry. Even though the work has recently been superseded in the schools, the newer forms of geometrical teaching are based on their Alexandrian predecessor.

To what extent was Euclid's work original? Elementary works on geometry had already been written by other authors, notably by Hippocrates of Chios (p. 30). Before Euclid, it had been generally agreed to base geometry on the straight line and circle. The properties of the right-angled triangle and the doctrine of proportion for both commensurables and incommensurables (p. 20) had been investigated. Some properties of conic sections were known (p. 38). Philosophers were familiar with the five 'Platonic bodies' (pp. 22, 35), The solution of such problems in solid geometry as the relation between the volume of a cone or pyramid and that of the cylinder or prism circumscribed around it had been attained. To all this mathematical activity Euclid certainly added advances in arrangement, in logical sequence, in form of presentation, and in completeness. His treatise displaced all that had gone before it, and rapidly assumed the position which it has since held.

Although Euclid's great work is called the *Elements of Geometry*, its subject-matter extends far beyond what is now regarded as geometry. Thus three of its thirteen books are devoted to the theory of numbers. In particular they contain the proof that no limit can be set to the number of prime numbers. This is a matter of importance in view of the great attention focused on the prime numbers by previous mathematicians such as the Pythagoreans and Plato and by subsequent mathematicians, notably by Eratosthenes (p. 70), Euler (p. 265), Lagrange (p. 266), and Gauss (p. 277).

Euclid's tenth book expounds the dominating concept of *irrational quantities*, thus opening up a thought-world of which the facts cannot be given tangible expression. The Pythagoreans (p. 21) had already broken into that world, and of it both Plato and Aristotle had had a Pisgah sight, but Euclid was the first to attempt any systematic exploration of it. It should be noted, however, that Euclid and his Greek successors distinguished sharply between *irrational quantities* and *irrational numbers*. In the theory of proportion as developed in Euclid's fifth book, the basis of the theory of irrational numbers is laid but is not developed. For its exposition the world had to wait until Descartes (p. 221) showed the deep unity of the long separated fields of number and form.

Euclid was a voluminous writer. Many of his works are lost, others survive in Arabic translation or in interpolated or corrupted texts. Of those lost we should particularly like to have his work,

On Fallacies, which dealt with the causes of error in geometrical research. Other of his works dealt with astronomy, optics (p. 80), and music.

ARISTARCHUS of Samos (*c.* 310–230 B.C.) taught at Alexandria soon after Euclid. He was himself the pupil of a disciple of Strato (p. 52). The peculiar views of Aristarchus on the position of the Earth among the heavenly bodies have earned him the title of the 'Copernicus of Antiquity'. He extended the view of an earlier philosopher that the Earth rotates about its own axis (p. 26) by maintaining that the Sun itself is at rest, and that not only Mercury and Venus but also all the other planets, of which the

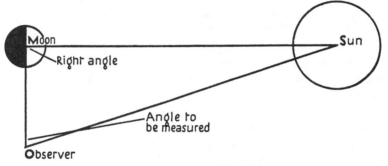

FIG. 22. Aristarchus measures relative distances of Sun and Moon from Earth.

Earth is one, revolve in circles about the Sun. It is interesting to observe that this view of Aristarchus brought on him the same charge of impiety as had descended on the head of Anaxagoras (p. 27) two centuries earlier.

We owe to Aristarchus the first scientific attempt to measure the distances of the Sun and Moon from the Earth, and their sizes relative to each other (Fig. 22). He knew that the light of the Moon is reflected from the Sun. When the Moon is exactly at the half, the line of vision from the observer on the Earth to the centre of the Moon's disk M must be at right angles to the line of light passing from the centre of the Sun's disk S to the centre of the Moon's disk M. Now the observer can measure the angle that the Sun and Moon form at his own eye O. With a knowledge of the two angles at M and O the relative lengths of the sides OS and

59

OM can be determined. This gives the relative distances of Sun and Moon from the observer.

The difficulty lay in determining exactly the angle at *O*. A very small error here makes a very great difference in the result. Aristarchus estimated this angle as 87 degrees when the reality is 89 degrees 52 minutes. In the resulting calculation he estimated the Sun as 18 times more distant than the Moon, instead of over 346 times more distant!

If we have the relative distances of Sun and Moon from the observer, the relative sizes of these bodies can be estimated, provided that we know the relative sizes of their disks, as they appear to an observer on the Earth. On this basis Aristarchus calculated that the Sun was seven thousand times larger than the Moon. Here further observational errors were introduced, and the ratio is very far from the truth. Nevertheless Aristarchus perceived that while the Moon is smaller than the Earth, the Sun is enormously greater. This fundamental relationship may well have affected his thought, for it seems inherently improbable that an enormously large body would revolve round a relatively minute one.

Contemporary with Aristarchus at Alexandria were other astronomers who recorded the positions of stars by measurements of their distances from fixed positions in the sky. Thus they defined the position of the more important stars in the signs of the zodiac, near to which all the planets in their orbits pass. They thereby facilitated accurate observations and record of the movements of the planets. Their observations were used by later astronomers, notably by Hipparchus (p. 76).

The philosophy which was the parent of science among the Greeks interested itself in three main aspects of the material world: (*a*) number and form and their relation to each other and to material objects, (*b*) the form and workings of the universe, and (*c*) the nature of man. In Alexandria, where science had freed itself from philosophy (p. 57), it was thus to be expected that the systematization of mathematics and astronomy would be accompanied by a similar development in the basic studies by which alone medicine can continue its progressive scientific tradition.

It was during the first generation at Alexandria that anatomy and physiology became recognized disciplines. The earliest important medical teacher of the school was HEROPHILUS of Chalcedon (flourished *c.* 300 B.C.), contemporary with Euclid. He began the practice of dissecting the human body publicly. In describing the anatomy of man he compared it with that of animals. He recognized the brain as the centre of the nervous system, and he regarded it as the seat of the intelligence. The name of Herophilus is still attached to certain parts of the brain. One is called by modern anatomists the 'winepress of Herophilus'. It is the meeting-place of four great veins at the back of the head. Their arrangement reminded him of the handles of a press. Herophilus was the first to distinguish clearly between veins and arteries. He observed that arteries pulsate, in which respect, among others, they differ from the veins. Their movement, however, he did not ascribe to the heart's action, but wrongly considered that it was natural to the arteries themselves.

A little younger than the anatomist Herophilus was the physiologist ERASISTRATUS of Chios (*c.* 280 B.C.), who also taught at Alexandria. He was an atomist and a follower of Democritus (p. 15), but his physiology was based on the idea that every organ is a complex of a threefold system of vessels—veins, arteries, and nerves—extending by ever more minute branching beyond the reach of vision. In those days, and for long afterwards, the nerves were regarded as hollow. Their imaginary cavities were thought to convey the hypothetical 'nervous fluid', much as the arteries and veins carry blood.

Erasistratus, like Herophilus, paid particular attention to the brain. He distinguished between the main brain, or *cerebrum*, and the lesser brain, or *cerebellum*. He observed the convolutions in the brain of both man and animals, and associated their greater complexity in man with his higher intelligence. He made experiments on animals which led him to distinguish between the posterior nerve-roots of the spinal cord, which convey sensations from the surface of the body, and the anterior nerve-roots which convey the motor impulses. This discovery was forgotten or neglected till the time of Sir Charles Bell (1774–1842) in the nineteenth century (p. 365).

Erasistratus also observed the lacteals, those lymphatic vessels that convey the white, milk-like fluid—the so-called 'chyle'—

derived from the food in the intestine, to the liver. The lacteals were seldom mentioned again until the Italian Gasparo Aselli (1581–1626) recorded them in the seventeenth century. They play a very important part in the animal economy.

A word must be said as to the views of Erasistratus on the general working of the animal body. He supposed that air is taken in by the lungs and passes to the heart. Here, as he held, it enters the blood and is changed into a peculiar kind of *pneuma* or spirit— the 'vital spirit'—which is sent to the various parts of the body by the arteries. It is carried to the brain, among other parts, and is there further altered into a second kind of pneuma, the 'animal spirit'. This animal spirit reaches different parts of the body through the nerves, which he wrongly regarded as hollow. The physiological system of Erasistratus was further developed by Galen, who, however, advanced great objections to the views of his forerunner (p. 90).

After the first generation anatomical enthusiasm at Alexandria waned. We may refer to three special points concerning it and concerning Alexandrian science in general:

(*a*) The names of Herophilus and Erasistratus are linked with the terrible charge of having dissected living men. Historians who have investigated the charge are satisfied that it is false.

(*b*) Erasistratus considered the *pneuma* that circulates in the body to be ultimately drawn from the air, or *pneuma* of the great world. This gave a physiological basis to the philosophical conception of the spirit of man as part of the world-spirit. Such a conception is frequently encountered in later writings, as, for example, in the works of the Stoic school (2nd cent. A.D.) such as those of the Emperor Marcus Aurelius or in the so-called 'Hermetic' writings (3rd cent. A.D.). Physiology and philosophy thus reacted on each other.

(*c*) In the third century B.C. Alexandria was an important Jewish centre. Parts of the Old Testament had been rendered from Hebrew into Greek by about 250 B.C. Greek contacts went far toward rationalizing the Hebrew view of nature. Thus, while earlier Biblical literature contains many references to divine intervention in the course of nature, the *Wisdom Literature* of Alexandrian date equates natural law with divine ordinance. In some passages the various types of Greek philosophy are set over against this Hebrew view. Among the Greeks various 'first principles' had been adopted. Thales had proposed 'water' (p. 10), Heracleitus 'fire' (p. 14), Pythagoras the 'circling stars' (p. 18), Anaximenes 'air' (p. 12), yet other philosophers vague essences that may be rendered 'wind' or '*pneuma*' (p. 12). Finally the new astrological science coming in from Babylon suggested the complex mathematical order of the heavenly bodies, which *signalled* the seasons, as *controlling* the seasons and through them the lives of men. A Jewish work written in Alexandria about 100 B.C. inveighs against all these views:

> 'Surely vain were all men in their natures, and without perception of God Who could not, from the good things that are seen, know Him that is; Neither by giving heed to the works did they recognise the Workmaster, But either fire [Heracleitus] or wind or the swift air [Anaximenes] Or circling stars [Pythagoras] or raging water [Thales] or the lights of heaven [astrology]
> They deemed the gods that govern the world.'
>
> (*Wisdom of Solomon* xiii. 1–2.)

The influence of Greek science can similarly be traced into the domain of Hebrew physiological conceptions. Thus, for instance, the seat of the understanding in the *Wisdom Literature* is usually placed in the heart. This is Aristotelian and contrary to Herophilus and Erasistratus, who placed the seat of intelligence in the brain. It is also opposed to the older Hebrew view (e.g. *Psalm* xvi. 7) which placed it in the liver. In several places, too, the Alexandrian "Wisdom Literature" as well as the New Testament writings (e.g. 2 *Peter* iii. 10, *Galatians* iv. 8–9) set forth the Greek doctrine of the four elements.

2. *Archimedes. Rise of Mechanics.*

ARCHIMEDES (287–212 B.C.) of Syracuse in Sicily was the greatest mathematician of antiquity. His life was entirely devoted to scientific pursuits, and his work is so fundamental that it affects every department of science. He was himself the son of an astronomer and on intimate terms with King Hiero of Syracuse. He visited Alexandria, where he met successors of Euclid. His whole work is instinct with a human element. Moreover, despite his absorption in science, he was not above applying his knowledge

63

The Failure of Nerve

to practical matters. Thus his name is remembered in connexion with the Archimedean screw for raising water (Fig. 23). It is said that he invented it during a visit to Egypt, and it is still in wide use there. The use of the screw as a means of applying mechanical force was unknown before Archimedes and was probably suggested by his device. He also contrived war engines for the defence of his native city against the Romans. Accounts of these and of his other mechanical devices are extant, but he himself wrote no works on them.

FIG. 23. Screw of Archimedes.

The writings of Archimedes show a generous appreciation of the mathematical achievements of others. He had friendly personal relationships with his younger contemporaries, notably Eratosthenes (p. 70). His lofty intellect, his compelling lucidity, and his terseness of exposition, made a profound impression on his fellow mathematicians. His mechanical skill must have been of a high order, for we hear also of his 'planetarium', a sphere of the heavens with models of the Sun, Moon, Earth, and planets, whose movements were displayed with an elaboration of detail that showed even eclipses.

A well-worn story tells of one application of the knowledge of Archimedes to practical affairs. The tyrant Hiero, on gaining power in Syracuse, vowed a golden crown to the gods. He contracted for its manufacture and weighed out the gold. The contractor duly delivered a crown of correct weight. But a charge was made that some gold had been abstracted and an equivalent weight of silver substituted. Hiero invoked Archimedes to put the matter to the test. While it was on his mind, Archimedes happened to go to the bath. On getting in, he observed that the more of his body was immersed, the more water ran over the top. This suggested the solution. Transported with joy he rushed home shouting 'Eureka! Eureka!' ('I have found it, I have found it!') What he had found was, in effect, the conception of specific gravity.

He made two masses of the same weight as the crown, one of gold, the other of silver. Next he filled a vessel to the brim and dropped in the mass of silver. Water ran out equal in bulk to the silver. The measure of this overflow gave the bulk of silver. The same was done with the gold. The smaller overflow corresponding to the gold was, of course, as much less as the gold was less in bulk than the silver, for gold is heavier than silver. The same operation was now done with the crown. More water ran over for the crown than for the bulk of gold of like weight, less than for the bulk of silver. Thus was revealed the admixture of silver with the gold. Archimedes had, in effect, obtained the relative *specific weights* of gold, silver, and of the mixture of the two, by comparing the relative amounts of water displaced by the same weight of the three. The scientific aspect of the subject is set forth in his work *On Floating Bodies*. This is the first record of the scientific employment of what we should call in modern parlance specific weights', though, of course, long before Archimedes, men must have been well aware that some substances were relatively heavier than others.

This question of the scientific use or development of a piece of common knowledge is important for the history of science. Discussion of it throws some light on the nature of the scientific process. Thus to Archimedes the ancient world owed a general exposition of the doctrine of levers (Fig. 24). This must not be taken to mean that Archimedes invented the lever any more than that he had discovered some bodies to be heavier than others. Levers in various forms were used from remotest antiquity, and an intelligent ape will use a stick as a lever. But it is one thing to use or even to contrive a device, and another to lay bare its exact mathematical principles and to follow them to their theoretical applications and conclusions. Important in this connexion is the statement of Archimedes of the possibility of moving a weight, however large, by a force, however small—a valuable theoretical application of levers. His saying is often recalled, 'Give me but a place to stand, and I can move the world.' He demonstrated this with a compound lever by which, with only the slightest effort, he was able to move a laden ship. Archimedes no more invented levers than the Greeks invented science. But science owes to the Greeks its formal and conscious development as a

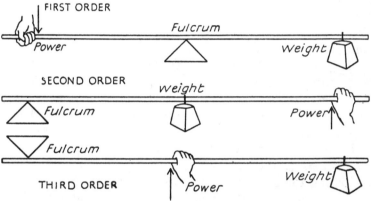

FIG. 24. The three orders of lever.

discipline and a method (p. 5), and the doctrine of levers owes to Archimedes its first formal and systematic exposition as susceptible of exact analysis. Formal and systematic exposition is a main task of science and without it knowledge cannot rise into the realm of science.

Perhaps the earliest work of Archimedes that we have is that *On Plane Equilibrium*. In this some fundamental principles of mechanics are set forth as rigorous geometric propositions. The work opens with his famous 'postulate': 'Equal weights at equal distances are in equilibrium; equal weights at unequal distances are not in equilibrium but incline toward the weight at the greater distance.' This is, in effect, the principle of the steelyard. It led him in the end to the discovery of the centre of gravity in a variety of geometric figures.

Among the mathematical achievements of Archimedes a very

high place must be given to his methods of measuring the areas of curved figures and surfaces. The simplest expression of this effort, 'squaring of the circle', had been broached by Hippocrates of Chios (p. 30). Eudoxus (p. 37), in estimating the volume of certain solid bodies, had propounded a method that involved in its essence the idea of 'limits'. This idea had been used by Euclid for a particular proposition of his twelfth book. Archimedes, however, employed limits systematically. This doctrine is of the utmost practical and historical importance, since it has formed a main foundation of modern mathematical development. It is essential to the 'calculus' as developed by Newton (p. 252) and Leibniz (p. 265). The calculus in its turn has been the starting-point for the development of many types of mathematical research.

The principle of the doctrine of limits can be expressed very simply. A square can be inscribed within a circle. Of such a figure two propositions are obvious:

(*a*) The sum of the sides of the square is less than the circumference of the circle:

(*b*) The area of the square is less than the area of the circle.

It is quite easy to double the number of sides and make an eight-sided figure, still inscribed within the same circle. Proposition (*a*) and (*b*) remain true but the difference is smaller in each case. We can go on doubling the number of sides to 16, 32, 64, 128, 256 or to any higher number. The more we increase the number of sides the more nearly will the sum of the sides and the area of the inscribed figure approach the circle. 'In the limit', when its sides are so small as to be no more than points, the polygon may be conceived as becoming the circle. Archimedes realized that this limit can never be reached but that it can be approached as nearly as we wish (Fig. 25).

Archimedes proves that the area of a circle is equal to that of a triangle of base equal to the *circumference* of the circle and of height equal to the *radius* of the circle. To calculate this area it is necessary to find the ratio between circumference and diameter, In estimating this ratio Archimedes sought the limit approached by the sides of regular polygons both inscribed and circumscribed on the circle. The limits for their ratio to that of the diameter he found to lie between $3\frac{10}{71}$ and $3\frac{10}{70}$. The latter has, since his day, been generally accepted as a good approximate value of the quantity known as π.

In his *Quadrature of the Parabola* Archimedes relates that he had been led by the study of mechanics to the solution of the problem of finding the area of a segment of a parabola, and that he had then obtained geometric proof of the correctness of his solution. His method resembles that which he adopted for the circle, namely to take both an inscribed and a circumscribed figure in

FIG. 25. Doctrine of limits.

relation to the curve under investigation. The two rectilinear figures are, as it were, compressed one from within and the other from without until they coincide with the curvilinear figure.

This mode of procedure, as well as that of using mechanics for the solution of problems afterwards demonstrated by geometry, leads us to the consideration of an extremely interesting treatise by Archimedes, the nature of which is suggested by its title *On Method*.

For the most part, Archimedes, like other Greek men of science, gives us only his final results. He gives us his proofs, but does not tell us how he reached them. In the *Method*, however, Archimedes, addressing Eratosthenes (p. 70), recalls the mathematical discoveries which he had sent on a former occasion and proceeds to inform him that he is now sending a description of the way in which he elicited them.

In essence the 'Method' consists in the application of two principles. The first is that a plane figure may be regarded as an aggregate of an infinite number of parallel lines with certain common properties. The second is the consideration of the respective weights of the two plane figures as drawn on paper whose area has to be compared. The process is also applied to demonstrate relationship between the areas of solid figures considered as aggregates of an infinite number of parallel planes. It amounts to a practical solution of problems of the relation between areas or volumes of two figures by analysis, mechanical or other, after which the philosopher returns to a synthetical mathematical process. He thus gains by experiment some insight into the solution before he seeks its mathematical demonstration.

Finally we may mention the remarkable system used by Archimedes for expressing very large numbers. It is so efficient that it enables any number to be expressed, up to that which, in our notation, would require eighty thousand million million ciphers. Archimedes expressed the opinion that his system was adequate to express the number of grains of sand that it would take to fill the universe! He therefore called his work the *Sand Reckoner*. From his calculation of the size of the universe, we get our idea of the cosmic conceptions of Archimedes. He knew the view of Aristarchus (p. 59) that the universe was heliocentric, the Earth revolving round the Sun in a comparatively unimportant orbit.

The sum of the contributions to knowledge by Archimedes is enormous. With his character, his humanity, his width of interest, his simplicity of exposition, and his unity of purpose, no mathematician of any age has commanded such general sympathy and respect.

3. *Middle Alexandrian Period (200–0 B.C.).*

A worthy Alexandrian successor of Archimedes was APOLLONIUS (fl. 220 B.C.) of Perga in Asia Minor (not to be confused with Pergamum). He studied under successors of Euclid at Alexandria and also at Pergamum. Apollonius is specially remembered for his

Conic Sections, a subject which he developed greatly and placed on a new footing.

Apollonius built on the work of Menaechmus (p. 38). That writer had derived the three types of conic section from three types of right cone. Apollonius showed, however, that all the three types of conic section can be derived from the same cone, whether right or scalene (Fig. 17). He established the terms *ellipse, parabola,* and *hyperbola* to denote the three types of section previously indicated by the angle of the cone of origin. The general geometric laws which give the properties of conic sections come to us, like the nomenclature of these figures, from Apollonius.

Archimedes and Apollonius between them originated two of the great problems which have ever since occupied geometers. The first is the quadrature of figures outlined by curves. This gave rise in due course to the infinitesimal calculus. The second is the theory of conic sections. This gave rise in due course to the theory of geometrical curves of all degrees.

The Ptolemies, in their zeal for learning, did not forget geography. Ptolemy III Euergetes (247–222 B.C.) rendered the greatest service to the science by his encouragement of ERATO-STHENES (c. 276–c. 194 B.C.), the librarian at Alexandria, and the most learned man of antiquity. His most important investigation, the measurement of the globe of the Earth, was performed by an operation of beautiful simplicity. Eratosthenes started from the three propositions (Fig. 27):

(*a*) That at Syene on the Nile (the modern Aswan) at noon on midsummer day an upright rod casts no shadow;

(*b*) That Syene is 5,000 stadia from Alexandria;

(*c*) That Syene is directly south of Alexandria.

Now, it is clear that, if we consider the Earth as a sphere, then the ratio

$$\frac{\text{Angle at centre subtended by 5000 stadia}}{\text{Four right angles}} = \frac{5000 \text{ stadia}}{\text{Circumference}}$$

The problem is, therefore, to determine the angle at the centre subtended by 5,000 stadia. But if on midsummer day the shadow cast by an upright rod at Alexandria is measured, then we shall be able to estimate the angle which the Sun's ray makes with the rod. Since, however, the Sun is so vastly distant from the Earth,

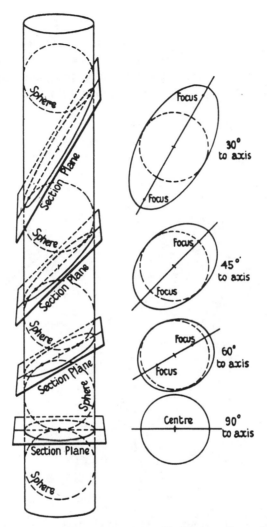

FIG. 26. The circle as special case of the ellipse, shown by series of sections through a cylinder. The cylinder of the diagram exactly contains a series of spheres; the points of contact of these with the section planes are the foci. The left figure is pictorial while the curves on the right give the true shape of the sections.

With a slightly more complex diagram the same relations may be shown in a series of sections through a cone, the cylinder being itself a special case of the cone (compare Fig. 58).

the Sun's ray at Alexandria is in effect parallel to the Sun's ray at Syene. Therefore the angle that the Sun's ray makes with the rod is equal to the angle subtended by 5,000 stadia at the Earth's centre. There is thus but one unknown—the Earth's

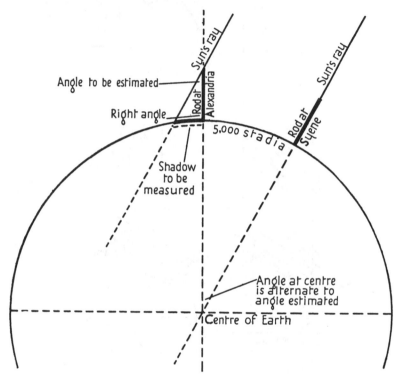

FIG. 27. Eratosthenes measures the earth.

circumference—in our equation. The circumference of the Earth thus obtained is a very fair estimate.

Having measured the Earth, Eratosthenes proceeded to consider the known parts of it. Here, in common with almost all ancient geographers, he fell into an error, or rather a self-imposed limitation. Eratosthenes regarded the habitable world as placed wholly within the northern hemisphere and forming only about a third of that. Again following his predecessors, Eratosthenes considered that the habitable world was longer than it was broad.

Divorce of Science and Philosophy: Alexandria

He estimated that the distance from the Atlantic to the Eastern Ocean was 78,000 stadia (that is, about 7,800 geographical miles), and from the parallel of the Cinnamon Land (Taprobane or Ceylon) to the parallel of Thule was 38,000 stadia. As Eratosthenes estimated the circumference or equator of the Earth at 250,000 stadia, he was able to estimate the circumference at the parallel of the Pillars of Hercules (p. 13), which he knew was also that of Rhodes (latitude 36°) (Fig. 28).

This fundamental parallel passed, as he erroneously thought, through other important points—the westernmost point of Spain, for example, and the southern points of Italy and Greece and along the Taurus mountains. At this parallel the total circumference of the world he estimated at 200,000 stadia. The rest was sea, so that, as he observed, 'if it were not for the vast extent of the Atlantic one might sail from Spain to India along the same parallel'. This is the first suggestion for the circumnavigation of the globe.

At right angles to the important parallel of Rhodes, Eratosthenes determined a north–south line between Alexandria and Syene. This line, produced northward, he regarded as passing through Byzantium and, beyond, to the mouth of the river Borysthenes (now called the Dnieper). Southward, he considered that it passed to Meroë, and then along the Nile to the Sembritae.

Both these fundamental lines contain several errors of allocation. Their determinations, together with those on other parallels of latitude and lines of longitude, are, however, sufficiently accurate for the construction of a map of the Mediterranean area recognizably similar to one based on modern knowledge (Fig. 28).

Eratosthenes exhibited great ability as a mathematician. He advanced the knowledge of prime numbers, a subject to which Archimedes had paid much attention. The famous *sieve of Eratosthenes* is a device for eliciting these numbers. Write down all integers in their natural succession. Then strike out all the multiples of 2, then the remaining multiples of 3, then those of 5, &c., through the other prime numbers (Fig. 29). The properties of prime numbers have attracted mathematicians in all ages, and it is astonishing how some simple rules concerning them have not been rationally explained to this day. Thus it is now well over a

FIG. 28. The World according to Eratosthenes.

century since it was remarked that every even number is the sum of two primes. This has been verified up to 200,000,000, but no proof is yet forthcoming.

Mathematical advance in Alexandrian times made possible a great development of astronomical theory. The discussion of the

FIG. 29. The Sieve of Eratosthenes.

supposed rotation of the celestial spheres and of the movements of the heavenly bodies gave rise to a nomenclature, parts of which have survived to our day, but parts of which have been modified by the Arabian and other authors through whose hands the Greek mathematical works have passed (p. 147).

The astronomical observer regarded himself as being in the centre of the vast heavenly sphere bearing the fixed stars. He considered the Earth so small that his distance from its centre was as nothing to his distance from the celestial boundary. Of this celestial sphere he could only see half, for the other hemisphere was hidden from him by the opaque Earth. The limiting circle thus imposed on his vision was the *horizon* (from a Greek word meaning 'to bound' or 'to limit'). This horizon formed a *great circle* on the heavenly sphere. He recognized, too, the celestial *poles* or points on the sphere pierced by the axis about which the

75

heavens seem to turn. On the sphere he marked out the *meridian,* which passes through the *zenith* (a word of Arabic origin) and the poles. The great circle at right angles to the line joining the poles was the *equator.* Starting from these elementary conceptions the Alexandrian observers worked out their whole astronomical system (Fig. 30).

Besides measuring the size of the Earth Eratosthenes also made

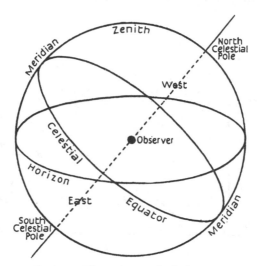

FIG. 30. The astronomical elements.

a remarkably accurate measurement of the angle which the circle of zodiacal constellations makes with the celestial equator, in other words a measurement of the *obliquity of the ecliptic.* His estimate works out at 23 degrees 51 minutes. This is only seven minutes from the truth.

The greatest astronomer of antiquity was HIPPARCHUS of Nicaea (*c.* 190–120 B.C.). He worked at Rhodes, where he erected an observatory and made most important researches. He developed trigonometry by which numerical calculations can be applied to figures drawn on either plane or spherical surfaces. The study is of great value to astronomy.

Hipparchus made numerous accurate astronomical observations. He also collected and collated the records of previous observers to see if astronomical changes had taken place in the course of the ages. There were available to him records of his Alexandrian and earlier Greek predecessors, and also those of the yet more ancient Babylonian astronomers. As a result of these comparisons he gave to the world two brilliant astronomical conceptions. (*a*) One of these, the precession of the equinoxes, was of permanent value. (*b*) The other, his theory of the movements of the planets and notably of the Sun and Moon, was of value to subsequent generations for the calculation of eclipses.

(a) *Precession of the equinoxes.* In 134 B.C. Hipparchus observed a new star in the constellation Scorpio. This suggested to him that he should prepare a catalogue of star positions. He therefore drew up a list of upwards of a thousand stars, each of which was given its celestial latitude and longitude. The constellations to which Hipparchus referred these stars are those which are to-day generally accepted. He showed great foresight in recording a number of cases in which three or more stars were in a line, so that astronomers of subsequent ages might detect changes in their relative positions.

Hipparchus proceeded to compare his observations with others of about 150 years earlier. He found that in this lapse of time there had been changes in the distance of the stars from certain fixed points in the heavens. The changes were of a kind that could only be explained by a rotation of the axis of the earth in the direction of the apparent daily motion of the stars. This causes the equinoxes to fall a little earlier each year. The knowledge of this *precession of the equinoxes* and of the rate at which it takes place was necessary for the progress of accurate astronomical observation. The complete cycle of precession takes 26,000 years.

(b) *Theory of motion of the planets.* When Hipparchus came to examine the apparent movement of the planets he had before him two theories, namely, that of 'epicyclic motion' and that of 'excentric motion'. Certain of his predecessors—notably Apollonius of Perga (p. 69)—had suggested the epicyclic view (Fig. 31). According to this each planet moves on a circle the centre of which moves on another circle, the centre of which is the centre of the Earth. Others of his predecessors had set forth the view of excentric motion. According to this the planet moves around the Earth but in a circle whose centre is not at the centre of the Earth. This secondary centre may also be represented as moving on a circle. Hipparchus explained the behaviour of the sun by a fixed and the moon by a moving excentric. (The geometric results of moving excentric and epicycle are identical.)

The epicyclic view finally prevailed through the mediation of the astronomer Ptolemy (p. 83). The theory of the excentric

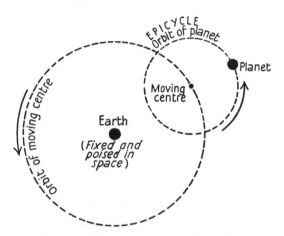

FIG. 31. To illustrate epicyclic motion.

motion of the Moon and to a less extent of the Sun, as enunciated by Hipparchus, was, however, of great service in that calculations based on it accorded much more closely with actual observations than did calculations based on any older doctrine of their movements. From the time of Hipparchus onward eclipses of the Moon could be predicted within an hour or two. Eclipses of the Sun could be predicted less accurately.

The Middle Alexandrian period, so brilliant in its development of the mathematical sciences, is disappointing when we come to consider its biological achievement. Of true scientific biology, apart from medicine, there was very little. The tradition almost died with Theophrastus (p. 51). With one exception the writings with biological bearing that have come down to us from the middle period are trivial. The exception is the herbalist CRATEUAS

Divorce of Science and Philosophy: Alexandria

(*c* 80 B.C.), who had the merit of introducing the systematic representation of plants by figures rather than by description. This method, important still, was doubly valuable in the absence of a system of botanical nomenclature. The plants figured by Crateuas were all of medical application. Copies of his figures

FIG. 32. 'Pheasant's eye', *Adonis aestivalis*, as represented by Crateuas about 80 B.C. and preserved in a MS. derivative of about 500 A.D.

have survived. They are of interest as the earliest specimens of scientific draughtsmanship (Fig. 32), and the tradition that they created can be traced through the ages to our own time.

In more purely medical matters illustration is perhaps also the main contribution of the middle Alexandrian period. The medical writings of the time were mainly commentaries on the works of the *Hippocratic collection*. Copies of the sketches of operations

79

and bandaging by APOLLONIUS of Citium (*c.* 100 B.C.) have survived, and give a good index of the conditions under which ancient
medical practice was conducted.

4. *Late Alexandrian Period to 200 A.D.*

Egypt became a province of the Roman Empire in 50 B.C.
Alexandria's achievement had now become an episode in her
history. There remained little native power of initiative, but some
scientific curiosity and considerable compilatory capacity. Creative
efforts—as those of Strabo (p. 100), of Ptolemy (p. 83), and of

FIG. 33. Hero's magic jug. As the thumb is pressed on or released from
the hole in the handle, the jug will pour or not.

FIG. 34. Hero's steam-engine. The globe is pivoted on tubes rising from
the boiler. It revolves by the reaction of the issuing steam.

Galen (p. 90)—were forthcoming only in response to definite
imperial needs.

An ingenious writer of the age was one HERO of Alexandria
(*c.* A.D. 100). He applied himself to entertaining contrivances
and sometimes to practical devices rather than to high scientific
themes. His *Pneumatica* describes many conjuring tricks. Thus
the principle of the siphon is applied to a jug from which water
pours or not at will (Fig. 33). Most famous of his toys was a globe
which whirls by force of steam—the first suggestion of a steam-
engine (Fig. 34). In his *Mechanica* he shows understanding of the
cogwheel, of rack and pinion, of multiple pulleys, of transmission
of force from a rotating screw to an axis at right angles to it, and

to the combination of all these devices with levers (Fig. 35). Hero records advances in optics. The oldest treatise on the mathematical aspect of that subject is by Euclid (p. 57), who considered that light moves in straight lines and believed vision to be something that goes forth from the eye. Hero showed that when light is reflected from a surface, it is at an angle equal to the angle of incidence. One of his surveying instruments depended for its working on the equality of these angles. His Dioptra (Fig. 36)

FIG. 35. Hero's mechanical repertoire.

served many purposes for which the theodolite is now used. Hero was also particularly ingenious in his use of water-levels in surveying.

Attempts were made to study refraction, that is, the behaviour of light in passing from one medium into another of different density, as from air into glass or water. The bent appearance of oars or rods dipped in water must have been observed very early, CLEOMEDES (first century A.D.) referred to the same principle the fact that an object, lying in an opaque basin and just obscured by the brim, could be rendered visible by pouring in water. He applied this principle to the atmosphere and suggested that the Sun, even when below the horizon, might be visible under certain circumstances (see Fig. 37). It is remarkable that he failed to give a practical application to this view of atmospheric refraction, for he disbelieved statements of his predecessors that in certain lunar eclipses the Sun seems to be still above the horizon while the eclipsed Moon rises in the east.

SIGHT

SIGHT

VERTICALLY
ROTATING
TOOTHED DISK

SCREW TURNING
TABLE
VERTICALLY

HORIZONTALLY
ROTATING
TOOTHED DISK

SCREW
TURNING
TABLE
HORIZONTALLY

FIG. 36. Hero's 'Dioptra' for taking angles as in levelling, estimating heights or distances between far-off points, etc. The circular graduated table has two sights, movable about its centre on a rigid arm. The table is supported by a column which can be rotated on its axis by a fixed screw working on a toothed disk. The table rests directly on a second toothed disk which can be rotated in a vertical plane by a second screw fixed to the column.

That some beginning had already been made of the science which deals with the eye as an optical instrument we learn from a work by a medical writer, RUFUS of Ephesus (c. A.D. 100). He had a fairly accurate conception of the structure of the eye. Some of the names which he applied to parts of this organ have survived

Divorce of Science and Philosophy: Alexandria

in modern scientific nomenclature. Rufus is the first to describe the eye as possessing a *lens*; he speaks of it as 'lentil-shaped'. A late Alexandrian writer, DIOPHANTUS (perhaps of about A.D. 180), is important as the best ancient exponent of algebra. His work on that subject was commented on by HYPATIA of Alexandria, the only woman mathematician of antiquity. She was murdered by Christian fanatics in 415. The work of Diophantus is the first that employs signs systematically. He gives symbols for the unknown, for powers, for minus, for equality, and so forth. He solves equations of the first, second, and, in one instance, of the third degree. He sets forth a method for finding two or more square numbers the sum of which is a given number, while each of the two approximates to the same number. The device he adopts is a method of approximation to limits. Thus in dividing 13 into two square numbers each of which is to be greater than 6 he reaches the result that the sides of the required squares are $\frac{258}{101}$ and $\frac{257}{101}$. Diophantus solved other comparable problems.

Not only was Greek algebra far behind Greek geometry but it was also far less influential on later mathematical development. Thus the work of Diophantus did not appear in print until 1575 and then only in a Latin translation. It was, therefore, without effect on the revival of mathematics in the sixteenth century. With Diophantus creative Greek mathematics comes to an end.

PTOLEMY of Alexandria (flourished A.D. 170),[1] who provided the final astronomical and geographical syntheses of antiquity, contributed also to the knowledge of optics. He not only knew that luminous rays in passing from one medium to another are deflected, but he actually measured the angle of deflection. Applying the known principle of the refraction of light, Ptolemy points out that the light of a star on entering the earthly atmosphere and on penetrating to the lower and denser parts must at each stage be gradually bent or refracted (Fig. 37). Thus it will appear to be nearer the zenith than is actually the case.

The great work of Ptolemy known as the *Almagest* has proved

---

[1] Not to be confused with the Ptolemies, kings of Egypt.

The Failure of Nerve

one of the most influential of all scientific writings. The very name has a history. The Greeks called the work the *megale syntaxis*, i.e. 'great composition'. The later translators from the Greek into Arabic, either from admiration or carelessness, converted the positive *megale* into the superlative *megiste*. Thus it became in Arabic *Almagisti*, whence Latin *Almagestum* and colloquial *Almagest*.

The *Almagest*, a work of utmost skill, was of the highest signifi-

FIG. 37. Refraction of ray by atmosphere makes the apparent position of a star nearer the zenith than the real position. The atmosphere, following Ptolemy, is represented as ending abruptly.

cance for mathematical development. It has provided the foundations of trigonometry, both plane and spherical. Its basic cosmic conceptions, however, Ptolemy certainly derived from his predecessors. Thus he invoked epicycles (p. 77) to explain the movements and behaviour of the planets, employing them to resolve some errors and inconsistencies of Hipparchus. He retained, however, *excentrics* (p. 78) to explain the movements of the Sun and Moon.

Among the contents of the *Almagest* is an account of the construction of the *astrolabe* (Fig. 38), the chief astronomical instrument of ancient and medieval times. It was, in essence, a device for determining the angle of elevation of a heavenly body. Ptolemy used the instrument to obtain the distance of the Moon by *parallax*. The method is substantially that still in use and is, in principle, very simple (Fig. 39). If in one place Z, the Moon is at zenith, then a line passing from the Moon at that place passes also through the centre of the Earth C. If an observer O takes at the same time the elevation of the centre of the Moon M, then we know the angle

84

at *O* of the triangle *MOC*. If we know the distance from *O* to *Z* we can calculate the angle at *C*. We thus know the three angles and therefore the relative lengths of the sides of the triangle *MOC*. Thus we can determine the ratio of *CM* to *CO*. Ptolemy thus estimates the Moon's distance to be 59 times the radius of the

FIG. 38. A simple form of Astrolabe. It consists essentially of a sus-pended disk graduated in degrees around the centre of which turns a limb with a sight at each end. The adjustment of the sights on to a heavenly body gives its elevation.

Earth, which is not very far from the truth. Working on an eclipse method of Hipparchus he estimates the Sun, however, to be only 1,210 Earth radii distant. This number is about one-twentieth of the true reckoning. He tells us that he has no means of estimating the distances of the lesser planets, but he follows tradition in accepting rapidity of motion as the main test of nearness. Thus from within outward his universe consists of Earth, Moon, Mercury, Venus, Sun, Mars, Jupiter, Saturn. This scheme was passed on to the Middle Ages (Fig. 40).

Ptolemy's other great work was his *Geographical Outline*. This was essentially a product of the knowledge brought by the expan-

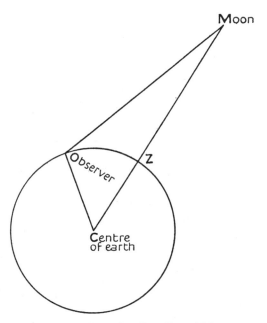

FIG. 39. Measuring Parallax of Moon.

sion of the Roman Empire. He studied itineraries of Roman officials and merchants. Thus he may be said to have preserved for us a summary of Roman knowledge of the Earth's surface, presented, however, in a form quite beyond the capacity of any Latin geographical writer. Ptolemy may well have had access to the great map prepared by Vipsanius Agrippa at Rome (p. 102).

Ptolemy developed his own manner of representing the curved surface of the Earth on a plane surface. In his scheme of 'projection' the parallels of latitude are arcs of concentric circles, the centres of which are at the North Pole. Chief among the parallels are the Equator and the circles passing through Thule, through Rhodes, and through Meroe. The meridians of longitude are represented by straight lines which converge to the Pole[1] (Fig. 41). He delineates in this manner the whole of the then known world. Its boundaries are: on the north, the ocean which surrounds the British Isles, the northern parts of Europe, and the unknown land in the northern region of Asia; on the south, the unknown land

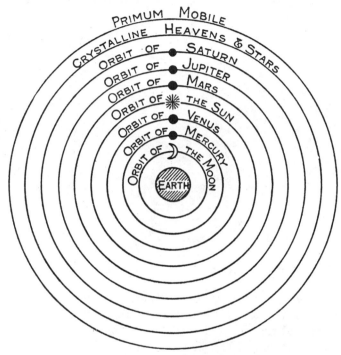

FIG. 40. The Ptolemaic World-System.

which encloses the Indian Sea, and the unknown land to the south of Libya and Ethiopia; on the east, the unknown land which adjoins these eastern nations of Asia, the Sinae (Chinese) and the people of Serica, the silk-producing land; on the west, the great Western Ocean and unknown parts of Libya. The portion of the Earth thus surveyed covers in length a hemisphere and in breadth between 63° north latitude and 16° south latitude.

[1] He has another scheme of projection in which the meridians are also curved.

As originally written Ptolemy's geography was furnished with maps. These have long since disappeared, but as Ptolemy gives the latitude and longitude of the places that he mentions his charts can be reconstructed. A peculiar interest attaches to the map of Britain, which can thus be put together (Fig. 42). Scotland is bent eastward with its axis at a right angle to that of England. This

FIG. 41. Ptolemy's Map of the World showing his scheme of projection.

is an unusual degree of error for Ptolemy. It is probable that he was here working not on the records of travellers, but on maps of the island, and that he had made the error of fitting the map of Scotland on to that of England on the wrong side!

Ptolemy exhibits the final extension of scientific geography in the Empire. How far the average educated citizen of the Empire was able or willing to appreciate science in general and geography in particular is another matter. It was the attitude of the Romans and especially of the Roman ruling class to things of the mind that determined the fate of science and with it, perhaps, the fate of the Empire. To estimate the attitude of the Roman to science we must turn to geographical works in Latin (pp. 102–4).

The *Almagest* of Ptolemy was translated into Latin in the later twelfth century and his *Geography* in the fifteenth. Thus they could not directly influence the earlier Middle Ages during which a simpler cosmic scheme based on Aristotle prevailed. In the later Middle Ages conflict between the views of Aristotle and those of Ptolemy became of considerable importance for the history of science.

The picture presented by the exact sciences of the late Alexan-

FIG. 42. The British Isles according to Ptolemy.

drian period is that of a number of minor works followed by one great synthesis and then a steady decline. We have seen this for astronomy and geography. It is repeated for the biological and medical sciences. In those departments we need only note the figures of Dioscorides and Galen.

PEDANIUS DIOSCORIDES of Anazarba in Asia Minor was an army surgeon who served in his own country under Nero. He wrote a work on drugs. It consists of short accounts of plants arranged, however, on a system that has hardly any reference to the nature of the plants themselves. The descriptions given are often terse and striking, and sometimes include a few words on the habits and habitats of plants. This elaborate pharmacopoeia was early illustrated in the style of Crateuas (p. 78), and some fine copies of these figures have come down to us.

The history of the work of Dioscorides reveals it as one of the most influential botanical treatises ever penned, despite the absence from it of anything like general scientific ideas. It provided most of the little botanical knowledge that reached the Middle Ages. It furnished the chief stimulus to botanical research at the time of the Renaissance. It has decided the general form of every modern pharmacopoeia. It has determined a large part of modern plant nomenclature, both popular and scientific.

The great biological and medical synthesis of antiquity was made by GALEN (A.D. 131–201) of Pergamum (p. 57). In his youth he visited Alexandria and other centres of learning, collecting all the knowledge of the day. Later he proceeded to Rome where almost all the rest of his very active life was passed.

In Galen's time the dissection of the human body had fallen into desuetude. The knowledge of anatomy had therefore declined. He made, however, accurate anatomical and physiological studies on a number of animals. Among these was the Barbary ape, the structure of which is not very far removed from that of man. Galen also made numerous dissections and experiments on living animals. He was thus able to evolve a complete and very ingenious physiological system. This was generally accepted by later antiquity and did not begin to be undermined until the work of Vesalius (p. 177) in the sixteenth century.

The basic principle of life in the Galenic philosophy was a spirit, or *pneuma*, drawn from the world-spirit by the act of breathing (compare Erasistratus, p. 61). It entered the body through the wind-pipe and so passed to the lung and thence (through the 'vein-like artery', which we now call the *pulmonary vein*) to the left ventricle, where it encountered the blood (Fig. 43). But what was the origin of the blood? To this question his answer was most ingenious, and the errors that it involved remained till the time of Harvey. Galen believed that chyle, brought from the alimentary tract by the portal vessel, arrived at the liver. That organ, he considered, had the power of elaborating the chyle into venous blood, and of imbuing it with a second spirit, or pneuma, innate in all living substance so long as it remains alive. This pneuma was called the *natural spirit*. Charged with natural spirit derived

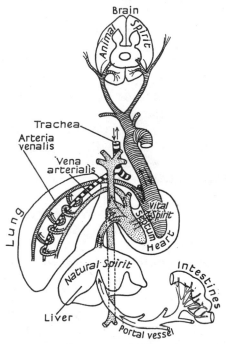

FIG. 43. Galen's Physiology.

from the liver, and with nutritive material derived from the intestines, the venous blood, Galen believed, was distributed by the liver throughout the venous system which arises from it, ebbing and flowing in the veins.

One great main branch of the venous system was the cavity that we now call the right ventricle of the heart. For the venous blood that entered this important branch, the right side of the heart, the Galenic scheme reserved two possible fates. The greater part remained awhile in the ventricle, parting with its impurities, which were carried off (by the 'artery-like vein'—now called the *pulmonary artery*) to the lung, and there exhaled. These impurities being discharged, the venous blood in the right ventricle ebbed back again into the general venous system. A small portion of this venous blood from the right side of the heart followed a different course. This small portion trickled through minute channels in the interventricular septum and entered the left ventricle drop by drop. There it encountered the pneuma brought thither from the outside world by the wind-pipe (through the 'vein-like artery'). These drops of venous blood in contact with the air in the left ventricle became elaborated into a higher type of pneuma, the *vital spirit*. Charged with this, the dark venous blood became fully developed bright arterial blood which was distributed through the arteries to all parts of the body.

Of the arteries, some went to the head, and thereby vital spirit was brought to the base of the brain. Here the arterial blood was minutely divided and became charged with yet a third pneuma, the *animal spirit*. This was distributed by the nerves, which were supposed to be hollow (Fig. 43).

The whole knowledge possessed by the world in the department of physiology, nearly all the biological conceptions, most of the anatomy, much of the botany, and all the ideas of the physical structure of living things from the third to the sixteenth century were contained in a small number of works of Galen. The biological works of Aristotle and Theophrastus lingered precariously in a few rare manuscripts in the monasteries of the East; the output of hundreds of years of Alexandrian and Pergamene activities was utterly destroyed; forgotten were the Ionian biological works, of which fragments have marvellously survived; but the vast, windy, ill-arranged treatises of Galen lingered on. Translated into Latin, Syriac, Arabic, and Hebrew, they saturated the intellectual world of the Middle Ages. Commented on by later Greek writers, who were in turn translated into the same list of languages, they were yet again served up under the names of other Greek writers in the Middle Ages and later.

What is the secret of the vitality of these Galenic biological conceptions? The answer can be given in four words: *Galen was a teleologist.* He believed that everything is made by God to a particular and determinate end (*telos* = 'end', aim'). Moreover, Galen's teleology is of a kind which happened to fit in with the

Divorce of Science and Philosophy: Alexandria

prevailing theological attitude of the Middle Ages, whether Christian, Moslem, or Jewish. According to Galen, everything which exists and displays activity in the human body is formed by an Intelligent Being on an intelligible plan, so that the organ in structure and function is the result of that plan. 'It was the Creator's infinite wisdom which selected the best means to attain his beneficent ends, and it is a proof of his omnipotence that he created every good thing according to his design, and thereby fulfilled his will.' To know man you must therefore know God's will. This attitude removes the foundation of scientific curiosity. After Galen there is a thousand years of darkness, and both medicine and biology almost cease to have a history. Men were interested rather in the will and purpose of God than in natural phenomena.

In leaving the Alexandrian period we may touch on one activity the influence of which has béen peculiarly persistent. Antiquity had a very highly developed and exact technology, but the attempts to rationalize it are lost. That such there were can be inferred from the earliest traces of 'alchemy' that reach us from Alexandrian sources from about 100 A.D. onward. Alexandrian alchemy is very sophisticated and clearly the result of a long evolution. The surviving texts are all in Greek and carry marks of Christian, Jewish, Neoplatonic, Gnostic, Greek, Egyptian, and perhaps Persian elements. Many names are associated with this strange literature, of which two are worth recording. One is MARY THE JEWESS, who is still remembered in the steam bath of our laboratories, the *bain Marie* of French chemists. The other is ZOSIMUS, the first alchemist who can be treated as an historic figure. He flourished about 300 A.D. These Alexandrian alchemical texts contain ideas that persisted to the very dawn of modern chemistry. It is fairly certain that the early Arabic alchemists (pp. 132–4) had access to far móre Alexandrian material than now exists.

IV. THE FAILURE OF INSPIRATION

Science the Handmaid of Practice (50 B.C.–A.D. 400):
Imperial Rome

1. Development of the Roman Attitude to Nature.

THE scientific idea, the conception of a reasonable universe, came
to the peoples of central Italy much later than to the Greeks of the
eastern Mediterranean and of southern Italy. Moreover, science
with the Romans always remained somewhat of an exotic. Rome
established her protectorate throughout the eastern Mediterranean
soon after 200 B.C. The influence of Greek ideas on Roman
civilization thenceforth grew rapidly. All educated men came to
learn Greek and were inevitably affected by Hellenic philosophy.
Yet despite the stimulus of Alexandrian thought, the Latins
produced no great creative men of science.

The prevalent attitude towards nature among the Latin-speak-
ing governing classes, whether Italian or provincial, was best
expressed by the *Stoic* creed. The Epicurean philosophy gained
fewer adherents among them. The Stoic system laid great stress
on correct conduct and duty. It was based on a rigid conception of
the interrelation of the different parts of the world. It provided
little stimulus for the acquisition of new knowledge or for anything in the way of research (p. 53). Thus, in place of knowledge
accumulating progressively on a basis of a wide and far-reaching
theory, we get, under Stoicism, either a type of exact but intellectually motiveless observation, or a rejection of all knowledge
not of practical importance. The dogmatism of Epicurean teaching was even less favourable to scientific research than was the
Stoic outlook.

There have been many attempts to explain why the Romans did
not continue the scientific works of the Greeks. It has been said
that the Roman mind could find no time from conquest and
administration to attend to scientific matters. This will not
explain the situation, for there were Romans who were able to
answer the no less exacting claims of philosophy and literature.
The matter, in fact, lies deep in the Roman character and tradition.
It was related to the ethics of the favourite Roman philosophy,

94

Science the Handmaid of Practice: Imperial Rome

Stoicism, and is not unconnected with the Roman passion for Rhetoric. In general we may say that Roman science appears at its strongest in the department of the general study of nature and at its weakest in pure mathematics. The success or failure of the Romans in any scientific field may be roughly gauged by its nearness to one or other of these disciplines. But Roman culture is so large a source of our own civilization that it is desirable to consider the Roman influence on the course of science in greater detail than the direct Roman contribution would itself warrant.

We have several works by Latins which deal with the implications of science in general. None involves any expert knowledge of natural phenomena, and they are concerned rather with the philosophical relations of science than with science itself. Of such works the most striking and widely read is LUCRETIUS (c. 95–55 B.C.), *On the Nature of Things*. The book is magnificent as literature and important as our best representation of Epicurean views (p. 54). It is, however, too much a work of propaganda to be of high scientific value. Moreover, it neither records first-hand observations nor does it even present a typically Roman attitude of mind.

The attention of the scientific reader of Lucretius will naturally be drawn to his atomic views. Following his master Epicurus, Lucretius explains the origin of the entire world as due to the interaction of atoms. This interaction, he believes, is without the intervention of any creative intelligence. Even mental phenomena are of atomic origin and there is no reality save 'atoms' and 'the void' (*inane*, p. 15). 'Nothing is ever begotten of nothing by divine will.' Everything springs from determinate units (*semina certa*). The genesis of all things is typified by the generation of organic beings. The species of plants and animals give us models for all processes and natural laws. This conception of generation has its converse. 'Things cannot then ever be turned to naught.' Such an attitude involves that ' indestructibility of matter' which, despite modern changes in our conceptions, is the historical foundation on which our chemical and physical knowledge has been built (pp. 283–4).

The resemblance of the Lucretian theory to modern atomic views is, however, more apparent than real. Not only are the atoms of Lucretius of different sizes and shapes, but also he knows nothing of definite laws by which they hold together as molecules. He has no inkling of chemical combination. He is without that 'doctrine of energy' that is so characteristic a feature in all modern physical theory. His work indeed had little direct influence on the development of the modern doctrine and probably was not widely read even in its own day. Epicurean thought was not favourable to scientific development. Moreover, the atomic view of matter was practically lost during the Middle Ages, and Aristotelian philosophy, which implied continuity of matter, was paramount for centuries.

Some have seen in Lucretius the beginnings of a theory of evolution. He certainly exhibits a 'ladder of nature' (p. 41) not unlike that of Aristotle. The earth produces first plants and then animals of ever higher type. 'Even as down and hair and bristles are first formed on the limbs of beasts . . . so the newborn earth raised up herbage and shrubs first, and thereafter the races of mortal things.' This idea of 'spontaneous generation' was inevitable until the realm of minute microscopic life could be explored (p. 245). It is thus no wonder that Lucretius follows Aristotle and all antiquity in assuring us that 'even now many animals spring forth from the earth, formed by rains and the heat of the sun'.

Did Lucretius take the matter further and did he have any conception of lower forms passing into higher forms? In a sense he did. Moreover, he invoked a process of 'survival of the fittest' for the formal exposition of which the world had to await the arrival of Darwin. But the Lucretian presentation of the manner in which the more perfect creatures reached their present state has no relation whatever to the historic geological record.

When we turn to the phenomena which Lucretius has chosen for special description we note that they are drawn from the magnificent, dramatic, or cataclysmic. His temper is far from the impartial spirit of science and there is nothing of the quietly scrupulous careful observer about him. Thunder and lightning, water-spout, volcano and thunderbolt, suffocating vapours and devastating pestilences—these are the themes he selects. There is no reason to give to Lucretius an important place among those who have helped or inspired the study of nature.

More characteristic of the Roman mind are the works of Varro (116–27 B.C.), of Pliny the elder (A.D. 23–79), and of Seneca (3 B.C.–A.D. 65).

VARRO, a country gentleman of the old Roman school, went to Athens and was influenced by Platonism, but developed definite Stoic leanings. He wrote an encyclopaedia of the sciences, and his works were the prototype of the numerous medieval treatises on the 'liberal arts'. He distinguished nine such disciplines, namely, grammar, dialectic, rhetoric, geometry, arithmetic, astronomy, music, medicine, and architecture. Of these the last two were not recognized by the later Latin writers who handed down the tradition to the Middle Ages. The number of liberal arts was thus reduced to seven (p. 127).

Varro tried to collect Latin learning and set it over against the Greek. He was in a good position to do this for he possessed the old Roman tradition and he had also received a good Greek education. He was employed by Caesar to arrange the great stores of Greek and Latin literature for the vast library which he intended to found. His work *On Farming* (*Res rusticae*) was written in his eightieth year. In it he records his own rich experience, but he has collected his material mainly from the writings of others. He thus exhibits the derivative tendency which is so disastrous a feature of Latin writers on scientific topics. He uses every opportunity to bring in etymology, rejoicing in artificial separations and divisions, so that the work gives much the impression conveyed by many treatises of medieval origin.

In the elder PLINY the Greek leaven has worked further than in Varro. Pliny had a literary education in Rome, where he took to studying plants. Coming under the influence of Seneca (p. 98) he turned to philosophy and rhetoric, and practised as an advocate. After military service in Germany, and having visited Gaul and Spain, he returned to Rome. There he completed his *Natural History*, dedicating it to the Emperor Titus. As prefect of the fleet he was stationed in the bay of Naples at the time of the eruption which overwhelmed Pompeii and Herculaneum in A.D. 79. He owed his death to his desire to observe that phenomenon more

The Failure of Inspiration

closely. His education, career, opinions, and character are all typical of the Italian tradition of his day.

The *Natural History* of Pliny was drawn from about 2,000 works —most of them now lost—by 146 Roman and 326 Greek authors. Its erudite, travelled, and industrious author exhibits an interest in natural phenomena that is quite uncontrolled by scientific or critical standards. The main thought that runs through the book is that nature serves man. Natural objects are hardly described as such but only in relation to man. All things have their 'uses'. 'Nature and the Earth', he says, 'fill us with admiration . . . as we contemplate the great variety of plants and find that they are created for the wants or enjoyments of mankind.' This world of wonder is, however, effectively without a God and works by rule— though it is a crazy rule which these disordered, credulous, wonder-loving volumes set before us.

Many of the matters on which Pliny expresses a judgement would have been impressed on him in the manifold life of Imperial Rome. Many of the animals he discusses were brought to the capital for the arena or for the kitchen from the farthest ends of the Earth. So too with plants. He describes a botanic garden kept by a Roman for the purpose of ascertaining the medical and allied properties of herbs. In descriptions of living creatures Pliny goes back to Aristotle and Theophrastus, but there is no systematic building of the subject and he is scientifically far inferior to his sources. Medical plants are treated in greatest detail, and he holds that all plants have their own special medical powers. The thought that nature exists for man constantly recurs. His philosophy, which accords in general with the Stoic scheme, is largely drowned and lost in his love of detail, and is often submerged in rhetoric. He presents a confused cosmology.

SENECA has gone over to the Greek more fully than either Varro or Pliny. A Spaniard by birth, he moved to Rome at an early age. There he came under Stoic influence and made his mark as an advocate and public servant. A member of one of the new provincial families, a brilliant rhetorician with a passion for philosophy, of which he was an eloquent but unsystematic exponent, a man whose undoubted balance and judgement had been earned in affairs rather than in action, with an interest in nature rather in its cosmical than in its detailed aspects, Seneca provides an interesting contrast to his contemporary Pliny.

Seneca's work is more philosophical and far more critical than that of Pliny. Yet his *Natural Questions,* even more than the *Natural History* of Pliny, is borrowed material. He, too, is a Stoic, but does not hesitate to criticize the opinions of that school. His subject is a general account of natural phenomena, but it is ill-arranged and imperfect. It deals chiefly with astronomy, meteorology, and with physical geography. He exhibits, like Lucretius, a special interest in the convulsions of nature. Moreover, Seneca was absorbed, like many Romans, by ethics, a moralist first and physicist afterwards. Thus physics—which for him meant a general description of the Universe—led to a knowledge of man's destiny and through that to a consideration of man's duty.

Seneca repeatedly tells of the moral to be derived from the phenomena investigated. The relation is often of the most distant and strained character. Thus, terminating his discussion of the phenomena of light, he asks, 'What were nature's purposes in providing material capable of receiving and reflecting images?' And he answers, 'To show us the Sun with his glare dulled, for eyes are too weak to gaze at him direct. Secondly, that we might investigate eclipses reflected in basins. Thirdly, mirrors were discovered in order that man might know himself.' [Abbreviated.]

Such a point of view appealed greatly to the medieval Church, by which Seneca was regarded as a Christian. He was included by St. Jerome among the 'ecclesiastical writers' and is frequently quoted by later Christian authors. But the ethical attitude to phenomena is inconsistent with the effective advancement of knowledge and has been one of the great enemies of science. In spite of the nobility of his sentiments, in spite of his lip-service to the advancement of learning, in spite of his faith in human destiny, Seneca could do nothing to stay the downfall of ancient wisdom.

2. Geography and Imperialism.

Just as the conquests of Alexander had opened up the East to science, so did the advance of Rome open up the West. Unfortunately the quality of the science had changed.

The Failure of Inspiration

A link between the Alexandrian and the Roman geographical standpoints is provided by the Arcadian POLYBIUS (204–122 B.C.), who had resided at Alexandria and later took service with the Roman army. He was present at the destruction of Carthage in 146 B.C., and was employed by the younger Scipio (185–129) to explore the coasts of Africa. He also visited Gaul and Spain. His descriptions, particularly of Spain, are very accurate, and he even attempts an estimation of the length of the Tagus. He has much valuable information about the Alps, and his knowledge of the geography of Italy was superior to that of any of his predecessors. Though an historian rather than a geographer, Polybius understood the necessity of constructing a correct map, and therefore gives much attention to the determination of distances and positions.

During the second and first centuries B.C., improved accounts of the Red, Black, and Mediterranean Seas, and the countries bounding them, began to be available for students. Determinations, even of points in India, were attempted. Mention should be made of the navigator EUDOXUS of Cyzicus (not to be confused with Eudoxus of Cnidus, p. 37). After exploring the Red Sea Eudoxus made at least two voyages southward along the African coast and brought back considerable new information.

The wars and military expeditions of the Romans yielded much further geographical knowledge. Thus STRABO of Amasia in Pontus (born c. 63 B.C.) had plenty of material when he began his general survey of the world. He was something of a traveller and had journeyed westward to the part of Etruria opposite Sardinia and southward from the Black Sea to the borders of Ethiopia. 'Perhaps not one of those who have written geographies', he says, 'has visited more places than I within these limits.' He travelled right through Egypt and made a considerable stay at Alexandria. Working for long at Rome, he was in a good position to receive authentic information. His mathematical qualifications were, however, inadequate and inferior to those of Eratosthenes (p. 70) on whom his work is based, though his circumstances gave him a greater knowledge of detail, especially for Europe.

Strabo opens by indicating the vast extension of knowledge as a result of the expansion of the Empire of Rome and that of her enemies on the east, the Parthians. Yet he is struck by the comparative smallness of the inhabited world. He makes the suggestion that there might be other continents still unknown. The length of the inhabited world from the Islands of the Blessed (that is the Canaries) to the Silk Land (that is China) was not more than about a third of the total circumference of the globe in the temperate zone. It was therefore possible that within the vacant space might be other lands inhabited by different races of men. In describing the inhabited world Strabo reduces its width from north to south to 30,000 stadia, an estimate below the 38,000 of Eratosthenes. The abbreviation is due to his scepticism as regards the northern regions. He rejects Thule, and disbelieves in any habitable land as far north as the Arctic Circle. Ireland, the most northerly of known territories, is 'barely habitable on account of the cold'. Southward, he considers the habitable world extends about 3,000 stadia beyond Meroe.

A feature of Strabo's work is his account of how a map of the world should be made. This, he points out, would not be difficult upon an actual globe, but such a globe would need to be very large for the insertion of details. He therefore considers the countries as though represented on a flat surface. Many of the distortions in Strabo's account are due to erroneous projection. His best accounts are of the countries bordering on the Mediterranean, where his map is distorted least. As he gets farther from the Mediterranean, his errors become greater. Even in the Mediterranean, however, he makes unexpected blunders. Thus the Pyrenees are represented as running north and south instead of east and west (cp. Fig. 46). With regard to the Caspian, Strabo shared the opinion of geographers since Herodotus that it was an inlet of the Northern Ocean (Figs. 28, 44). The north of Asia and the region east of Sogdiana was, he tells us, a mere blank to him. A vast chain of mountains extended, he thought, from east to west across Asia, bounding India on the north. From this range the Tigris and Euphrates took their rise in the west, the Indus and Ganges in the east. Thus the Himalayas are confused with the mountains of Asia Minor and with the Caucasus.

Among the very few native Romans who had a true conception of the nature of scientific inquiry was JULIUS CAESAR (102–44).

The Failure of Inspiration

He formulated the splendid scheme of a complete survey of the Empire. The government of the provinces, the demands of trade, and the distribution of the fleet all made the need evident. The death of Julius left the execution of this plan to his successor, Augustus. The survey was superintended by his son-in-law, VIPSANIUS AGRIPPA (died 12 B.C.) and finally completed after nearly thirty years' work in 20 B.C. It was rendered possible by the fact that the Empire was well furnished with roads, marked with milestones. There was a regular service of skilled surveyors, whose work, incorporated in the reports of provincial governors, was available at head-quarters. The vast chart prepared from these details was exhibited in a building especially erected for the purpose at Rome. In this map all other geographical elements were subordinated to indications for the marching of armies.

Geography in the limited sense, as distinct from cosmography, was a topic that might be expected to appeal to the practical and imperialistically minded Roman. He was, however, hardly in an intellectual position to appreciate geography, save in the form of a road-book or rough strategic chart. To general geography the Roman paid little attention. The only important Latin writer on the subject is the Spaniard POMPONIUS MELA (c. A.D. 40), who refers to Britain as about to be more fully explored by an expedition then in progress. This was the visit of the Emperor Claudius in A.D. 43.

Pomponius Mela clearly meant his work as an easy account of his subject. In his general description of the Earth he avoids mathematical topics in the true Roman manner, nor does he give distances or measurements. The world is a sphere, and the land upon it is surrounded on all sides by sea. Five zones may be distinguished. Of these the middle zone is as uninhabitable by reason of its heat as are the two extreme zones by reason of cold. We live in one of the two intermediate temperate zones while in the other dwell the 'Antichthones'. The land in our own hemisphere is completely surrounded by ocean, from which it receives four seas or gulfs, one at the north, the Caspian, two in the south, the Persian Gulf and the Red Sea, and the fourth to the west, the Mediterranean. The scheme is taken from Eratosthenes (p. 70), and it is clear that Pomponius Mela is a mere borrower from Greek sources (Fig. 44).

Science the Handmaid of Practice: Imperial Rome

Mela gives a general description of the three continents, Europe, Asia, and Africa. Between the three is the Mediterranean, which he speaks of as 'our sea'. He takes the river Tanis (*Don*), Lake Maeotis (*Sea of Azov*), and the Euxine Sea (*Black Sea*) as frontiers between Europe and Asia, while it is the Nile that divides Asia from Africa. Asia is as large as Europe and Africa together. These ideas were passed on to the earlier Middle Ages and are expressed

FIG. 44. The World according to Pomponius Mela.

in the world-maps of which the earliest is in a seventh-century codex of St. Isidore of Seville (560–636). The so-called OT map of the Middle Ages is well known (Fig. 45).

The haziness of the geographical ideas even of an intelligent Roman of Imperial times may be gathered from TACITUS (*c.* A.D. 55–120). He tells how, under Agricola, the Roman fleet rounded Britain and proved it to be an island, discovering at the same time the Orcades (*Orkney Islands*) and coming in sight of 'Thule' (? *Shetlands*). Yet Tacitus, like Caesar and the elder Pliny, believes that Spain lies to the west of Britain (Fig. 46). Like Strabo he describes the Pyrenees as running north and south (p. 101). He goes on to explain the phenomenon of the Midnight Sun—which he brings as far south as the north of Scotland—by telling us that 'the flat extremities of the Earth, casting a low shadow, do not

The Failure of Inspiration

throw the darkness up high, and the night does not reach to the sky and stars'. The statement implies the view that the Earth is a disk with flattened edges. This from a Roman gentleman who had access to the ideas of Aristotle, Hipparchus, Archimedes, and Eratosthenes.

As antiquity passes into the Middle Ages, geography as a science becomes yet further degraded and is represented by mere route-

FIG. 45. Conventional medieval OT map, as in Isidore of Seville.

books. Of these the best are the earliest, for the deterioration is progressive. We have a fairly complete register of the roads of the whole Empire, put together in its present form about A.D. 300. Both principal and cross-roads are indicated by lists of the towns and stations upon them, the distance from place to place being given in Roman miles. Of more limited scope are the pilgrim books, which mostly give the itinerary to and from Jerusalem. The earliest of these Christian works is by a lady, SYLVIA of Aquitaine (about 380). Of a somewhat similar character is the work of RUTILIUS NAMATIANUS of Toulouse, who wrote in 417 a versified account of a journey from Rome to Gaul. He was a pagan who fiercely attacked the monks—'men who dread the evils without being able to support the blessings of the human condition'. His work naturally delighted the heart of Gibbon, and is of interest as still exhibiting the faith that Rome is immortal. Of special note, as marking the passage to the Middle Ages, is the work of an anonymous geographer of Ravenna put together in the seventh century. It contains valuable information concerning Roman roads and towns and is still using sources employed five centuries earlier by Ptolemy.

3. *Imperial Organization of Medicine, Hygiene, and Public Health.*

The original native Roman medical system was that of a people of the lower culture and devoid of scientific elements. Interwoven with ideas that trespass on the domain of religion, it possessed that multitude of 'specialist deities' characteristic of the Roman cults. Thus Fever had three temples in Rome, and was supplicated as the goddess *Febris* and flatteringly addressed as 'Divine Fever', 'Holy Fever', 'Great Goddess Fever'. Foul odours were invoked in the name of *Mephitis*, to whom a temple was erected at a place where asphyxiating fumes emerged from the earth. Lassitude was implored as *Fessonia*. *Uterina* guarded the womb. *Lucina*, with her assistant goddesses, had charge of childbirth. Over the entire pantheon of disease and physiological function presided the *Dea Salus*, 'Goddess Health', who had a special temple on the Quirinal. She was the deity who took the public health under her supervision.

FIG. 46. Map of Western Europe from descriptions of Tacitus.

The entire external aspect of Roman medicine was gradually transformed by the advent of Greek science. The change, however, hardly penetrated below the upper classes. Thus many references in the *City of God* of St. Augustine (354–430) show the ancient beliefs still current in the Italy of his day. After the fall of the

Empire, they lingered among the·barbaric peoples that entered into its heritage. Nor are they yet extinct. Prescriptions and practices of Pliny (p. 108) and of his even more gullible successors may still be traced in European and in American folk-customs and folk-beliefs.

During the Republic, medical education had been a private matter. The direct relation of pupil and master exhibited by the magnificent *Hippocratic oath* was evidently that which prevailed under the early Empire. The initiate declared:

> ' I will reckon him who taught me this Art as dear to me as those who bore me. I will look upon his offspring as my own brethren and will teach them this Art, if they would learn it, without fee or stipulation. By precept, lecture, and every other mode of instruction, I will impart a knowledge of this art to my own sons, and to those of my teacher, and to disciples bound by a stipulation and an oath, according to the Law of Medicine, but to none other.' (See p. 27.)

Despite the ancient Greek dress in which this formula is cast, there is evidence that it is of Imperial date and of Roman rather than of Greek origin. The very form suggests the arrangements which were gradually made for medical instruction at Rome.

The first important teacher there was the Greek ASCLEPIADES of Bithynia (died *c.* 40 B.C.), a contemporary of Lucretius and like him an Epicurean (p. 54). He influenced deeply the course of later medical thought, ridiculed, and perhaps we should add misunderstood, the Hippocratic attitude of relying on the *vis medicatrix naturae*, 'the healing power of nature', which he regarded as a mere 'meditation on death', and urged that active measures were needed for the process of cure. He founded a regular school at Rome which continued after him.

At first the school was the mere personal following of the physician, who took his pupils and apprentices round with him on his visits. Later, such groups met to discuss questions of their art. Towards the end of the reign of Augustus (died A.D. 14) these societies constructed for themselves a meeting-place with a regular organization. Finally the emperors built colleges for the teaching of medicine. At first the professors received only the fees of pupils, but before the end of the first century they were given a

Science the Handmaid of Practice: Imperial Rome

salary at the public expense. The system was extended in the second and third centuries. Thus Rome became a centre of medical instruction. Moreover, subsidiary centres were established in other Italian towns. These provincial schools were largely training places for army surgeons.

A very weak point in the Roman medical curriculum was the absence of any practical study of anatomy. Considering the indifference to human life which the Romans exhibited, considering their brutality to slaves and the opportunities offered by gladiatorial combats, considering the value—obvious to us—of anatomical knowledge for surgical practice, and considering the organization of the military medical service of the Empire, it is highly significant that the knowledge of antiquity was thus allowed to lapse.

Had a great Roman military leader been questioned on this point he would probably have replied, 'Of course doctors want anatomy, but isn't Galen's anatomy good enough? Cannot they read that?' But he would have been wrong. It is not by reading that science is sustained. It is by contact with the object—by systematic observation and experiment. From these the Roman army doctor was cut off, and we see the result of his deprivation in the poverty of Roman science.

As regards the literature of medicine, the earliest scientific work in Latin bears the name of CELSUS and was prepared in Rome about A.D. 30. It is in many ways the most readable and well arranged of all ancient medical works. The ethical tone is high and the general line of treatment sensible and humane. The most interesting section is perhaps that on surgery, which gives an excellent account of what might be thought to be the modern operation for removing the tonsils. The dental practice includes the wiring of loose teeth and the use of the dental mirror. In view of the attractive character of the work it is disappointing to find that it is but a compilation from Greek sources. This fact also is significant of the status of science in Rome.

The remaining Latin medical writings of Imperial times are not of high scientific value. In this connexion we must recall Pliny (p. 97). A large section of his *Natural History* is devoted to medical matters. Yet he scorned medical science and the Greeks who practised it.

'Medicine, in spite of its lucrativeness,' he says, 'is the one Greek art that the serious Roman has so far refused to cultivate. Few of our fellow-citizens have been willing even to touch it, and if they do so they desert at once to the Greeks. . . . Unfortunately there is no law to punish ignorant physicians, nor is the capital punishment inflicted upon them. Yet they learn by our suffering, and experiment by putting us to death!'

The collection of Pliny that was to displace the works of the despised Greeks is a vast series of remedies chosen on the supposedly firm ground of 'experience'. Their selection is based on no theory, supported by no doctrine, founded on no experiment. Yet this drug book is the prototype of the medical output of the next fifteen hundred years. The cry of Pliny for 'experience' as against 'theory' has been plaintively echoed by the 'practical' man down the ages. Yet there are subjects and there are conditions in which the man without a theory may be the most unpractical of all. Medicine is such a subject; disease is such a condition.

When 'experience' is invoked by Pliny and by later writers, especially of the Middle Ages, we must beware against confusing it with the 'experience' of science. In scientific matters the essence of experience is that it be under control. Such experience is normally capable of repetition at will, as a chemical reaction, for instance, may be repeated. All true scientific experience, in fact, approaches the character of 'experiment'. Scientific experience is thus the result of a series of *observations provoquées*.[1]

A single example from Pliny will suffice to illustrate this distinction. 'The herb dittany', he says, 'has power to extract arrows. This was *proved* [note the word; it really means *tested*] by stags who had been struck by these missiles, which were loosened when they fed on this plant.' Had Pliny exhibited any desire to verify such a statement? Could he have verified it even if he had desired? The answer is not difficult. He had, in fact, taken his

---

[1] There are scientific experiences in which the mind comes to rest with conviction, even when not repeated. Thus an astronomical prediction, involving exact and detailed calculation, if confirmed in an exact and detailed way, may carry conviction as to the soundness of its principle even though verified by but a single observation.

'experience' from an interpolated and spurious passage of a work by Theophrastus (p. 51) and he omits to mention his source! Prepossession with the idea of the value of such experience led Pliny and the ages which followed him—as it leads men to this day—into innumerable absurdities. 'General experience' whether first hand or second hand is no substitute for exact scientific knowledge.

If in medicine itself the Roman achieved but little, in organization of medical service, and especially in the department of public health, his position is far more honourable. Several Roman writers on architecture give much attention to the orientation, position, and drainage of buildings, and from an early date sanitation and public health drew the attention of statesmen. Considering the dread of the neighbourhood of marshes on the part of these practical sanitarians and in view of modern knowledge of the mosquito-borne character of malaria, it is entertaining to find the use of the mosquito net (*conopeum*) ridiculed as effeminate by poets such as Horace and Juvenal.

Sanitation was a feature of Roman life. Rome was already provided with *cloacae* or subterranean sewers in the age of the Tarquins (6th cent. B.C.). The first construction of the *Cloaca maxima*, the main drain of Rome, parts of which are in use to this day, is referrable to that period.

The growth of hygienic ideas is seen in an interdict of as early as 450 B.C. against burial in the city. There is in this edict no reference to any physician. The same absence of professional intervention may be noted in the instructions issued to the city officers for cleansing the streets and for the distribution of water. Nor is any medical help or opinion invoked by the ancient law, attributed to Numa the first king of Rome, which directed the opening of the body of a woman who had died pregnant in the hope of extracting a live child. This is the so-called *Caesarian section* by which Caesar himself is said to have been brought into the world. The expression still has a surgical meaning.

The finest monument to the Roman care for the public health stands yet for all to see in the fourteen great aqueducts which supplied the city with 300,000,000 gallons of potable water daily.

The Failure of Inspiration

Few modern cities are better equipped. The distribution of water to individual houses was also well organized, and excellent specimens of Roman plumbing have survived (Fig. 47).

Under the early Empire a definite public medical service was constituted. Public physicians were appointed to the various towns and institutions. A statute of the Emperor Antoninus of about the year A.D. 160 regulates the appointment of these

FIG. 47. Mechanism of Roman double-action pump.

physicians, whose main duty was to attend the needs of the poor. In the code of the great law-giving Emperor Justinian (A.D. 533) there is an article urging such men to give this service cheerfully and to prefer it to the more subservient attendance on the wealthy. Their salaries were fixed but they were encouraged to undertake the training of pupils.

Linked with the public medical service is the hospital system. It arose out of the Roman genius for organization and is connected with the Roman military system. Among the Greeks private surgeries were well known. Larger institutions were connected with the temples to Aesculapius, the god of healing, but there is no evidence of scientific medical treatment in these places. Such a temple had been established on an island of the Tiber in Republican times. On this island of Aesculapius writes the historian Suetonius (c. A.D. 120) 'certain men exposed their sick and worn-out slaves because of the trouble of treating them. The Emperor Claudius (41–54), however, decreed that such slaves were free, and that, if they recovered, they should not return to the control of their masters'. Thus the island became a place of refuge for the sick poor. It was an early form of public hospital. The example was imitated, the facilities improved, and the service extended to free men.

The development of public hospitals naturally early affected military life. As the Roman frontiers spread ever wider, military hospitals were founded at important strategic points. Later there were constructed similar institutions for the numerous imperial officials and their families in the provincial towns. Motives of benevolence, too, gradually acquired weight, and finally public hospitals were founded in many localities. The idea naturally passed on to Christian times, and the pious foundation of hospitals for the sick and outcast in the Middle Ages is to be traced back to these Roman institutions.

The first charitable institution of this kind concerning the foundation of which we have clear information was established at Rome in the fourth century by a Christian lady named FABIOLA of whom we learn from St. Jerome. The plan of such a hospital projected at St. Gall in the early days of the ninth century has survived. It reminds us in many respects of the early Roman military hospitals. These medieval hospitals for the sick must naturally be distinguished from the even more numerous 'spitals' for travellers and pilgrims, the idea of which may perhaps be traced back to the rest-houses along the strategic roads of the Empire.

4. *Roman Mathematical, Physical, and Calendarial Science.*

As with all peoples, the first system of numeration adopted by the Romans was finger counting. From it developed methods of mechanical reckoning. The simplest was a board covered with sand, divided into columns by the finger, counters being used in calculation. Such counters had graven upon them figures of the hand in various positions to represent different numbers. These symbols are identical with those which remained in vogue till late medieval times.

The Failure of Inspiration

A more complicated apparatus was the true *abacus*. This began as a board with a series of grooves in which pebbles or *calculi* would be moved up and down, hence the verb *calculo* and the modern use of 'calculate'. In its more developed form the abacus consisted of an upper row of short rods and a longer row of long rods (Fig. 48). Each short rod had a single perforated bead running on it; each of the longer ones four such beads. The first rod on the right was marked for units, the next on its left for tens,

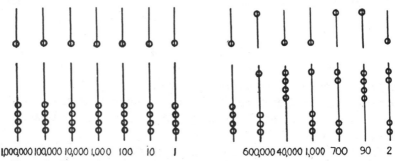

1,000,000 100,000 10,000 1,000 100 10 1   600,000 40,000 1,000 700 90 2

FIG. 48. Essentials of the Roman abacus, consisting of beads running on wires. On the left it is set for reckoning. On the right a total of 641,792 is represented. Without an abacal representation and in Roman figures this would need twenty-one elements, namely CCCCCXLIMVIICLXXXXII.

and so on up to a million. The mode of application of the abacus was more complicated than might be imagined.

The whole mathematical system of antiquity was handicapped by its inadequate notation. The system with which we are nowadays familiar, with nine separate integers and a zero, each of which has a local value, did not reach Europe until the Middle Ages. The Greeks used mostly geometrical methods where we should invoke the aid of algebra (p. 21), and their mathematical developments made little impression on the Romans. How slight was the mathematical knowledge absorbed by Latin scientific authors may be gathered from the *Geometrica* and the *Arithmetica* bearing the name of BOETHIUS (A.D. 480–524). Those elementary works ascribed to 'the last of the ancients' represent the mathematical legacy of antiquity to the earlier Middle Ages. It is interesting to note that Boethius divides mathematics into four sections, Arithmetic, Music, Geometry, and Astronomy, and that he is the first to describe these four disciplines as the quadrivium ('four pathways'). Even when Rome had world dominion, Cicero bemoaned that 'Greek mathematicians lead the field in pure geometry while we limit ourselves to reckoning and measuring'.

The Romans held that the art of surveying was at least as old as their city, and had been practised from the first by the priests.

FIG. 49. The Groma.

In Imperial times a regular school for surveyors was established. The chief instrument in general use was known as the *groma* (Fig. 49). It consisted of two sets of plumb-lines fixed at right angles and arranged to turn about a vertical pivot. One set was used for sighting and the other to determine the direction at right angles to the first. As both agricultural and town-planning were mainly on rectangular lines this instrument was of wide application. A dioptra (p. 82) was in use and also a very clumsy water-level.

Compasses and other instruments employed in mensuration recovered from Pompeii are well made, and the excellence of Roman masonry is a household word. Thus the inaccuracy of some Roman measurements is strange. For instance, $3\frac{1}{8}$ is given as the value of π by VITRUVIUS (*c.* A.D. 10), a competent architect who must often have had occasion to examine the drums of columns. A better result might have been expected from any schoolboy provided with a compass and a tape-measure, and $3\frac{1}{7}$ had already been suggested as an approximation by Archimedes.

Vitruvius gives a method of estimating the distance from an observer of an inaccessible point on the same level as himself, e.g. on the opposite bank of a river. A line is traced along the near bank, and is measured by rolling along it a *hodometer*, an instrument consisting of a wheel the length of the circumference of which is known and whose revolutions can be counted. This is in principle a 'taxicab'. From each end of the measured line a sight is taken by means of the dioptra (p. 82). Angles and base being thus available a triangle congruent to that formed by joining the point on the far bank to the extremities of the measured line, can be constructed on the near bank. The vertical height of this triangle as measured by the hodometer gives the breadth of the river.

Mechanical knowledge among the Romans always had a practical direction. Among the few devices of native Roman origin is perhaps the steelyard. This instrument is a device of considerable antiquity and may be traced back at least as far as the third century B.C. The principle of the pulley, too, was well known. An elaborate system of pulleys was adapted to cranes and to engines of war.

The inadequate theoretical basis of the physical conceptions of Latin writers is shown in various directions. Thus Pliny recounts a fable of the Remora, a fish of the Mediterranean which has a sucker on its head. 'This tiny fish can restrain all the forces of ocean. Winds may rage and storms may roar, yet the fish withstands their might and holds ships still by simply adhering to them!' Three centuries before, Archimedes had demanded 'a *fixed* place on which to stand that he might move the world' (p. 65). The full understanding of the works of Archimedes failed for the next millennium and a half. Yet his simpler practical devices, such as the water-screw, were familiar enough to the Romans.

Science the Handmaid of Practice: Imperial Rome

Applied mathematics underwent some development in early Imperial times. JULIUS CAESAR (102–44) himself was an astronomical author and wished to improve the Roman calendar which had fallen into great confusion. The early history of the Roman calendar is obscure. At an early date there emerged a lunar year of 355 days, which is almost exactly twelve lunations. Of this calendar Martius (the month of Mars) was the first month, Aprilis (probably for *aperilis* from *aperire*, 'to open'), Maius (perhaps related to *major*), and Junius (which may be related to *junior* and *juvenis*) were named in connexion with the opening, growth, and ripening of vegetation. The following six months, Quinctilis, Sextilis, September, October, November, and December were given merely the numerical names from fifth to tenth which the last four still bear. Januarius was named from the god *Janus*, and Februarius, the last month, was the season of ritual purification (*februare*, 'to purify' or 'expiate').

To obtain some relation of this lunar reckoning to the solar year a cycle of four years had been invented of which the first year contained 355 days, the second 377, the third 355, and the fourth 378. The cycle thus covered 1,465 days, and the average year was of $\frac{1465}{4}=366\frac{1}{4}$ days. So variable a year had little value for agricultural purposes. The farmer had thus still to rely on the rising and setting of certain constellations for timing his operations. The year was variously modified at different periods, but until the reforms of Julius Caesar no adequate correspondence to solar events was attained.

In place of this system Julius Caesar, acting upon the advice of an Alexandrian mathematician, substituted a solar year of 365 days and abandoned any attempt to adapt the years or months to the lengths of the lunations. In every fourth year one day was interpolated, thus introducing the system of leap years. This reform was probably a reproduction of an Alexandrian calendar enacted in 238 B.C. and had perhaps been designed at a yet earlier date by the Greek astronomer Eudoxus (p. 37). In 44 B.C., the second year of the Julian Calendar, one of the months, *Quinctilis*, was named *Iulius*—our July—in honour of its founder.

115

The Failure of Inspiration

In 8 B.C. another month, *Sextilis*, was called *Augustus* after his successor. The Julian Calendar, the year of which began in the month of March, remained in general use until reformed by Pope Gregory XIII in 1582.

5. *Roman Astronomy and Astrology.*

The Romans did not deal with astronomical matters until late, and then only for practical purposes such as the calendar, seamanship, or agriculture. Popular astronomy is represented in Latin by certain metrical writings bearing the name of AVIENUS (c. A.D. 380). These, which were popular in the early Middle Ages, are adapted from various Greek works. To one of the Greek sources of Avienus, namely ARATUS of Soli (271–213 B.C.), peculiar interest is attached. St. Jerome tells us that when, in *Acts*, St. Paul says 'In Him we live, and move, and have our being; as certain even of your own poets have said, *For we are also his offspring*' (*Acts* xvii. 28), he is quoting the *Phaenomena* of Aratus. The words 'for we are also his offspring' are in fact to be found in the opening invocation to Zeus in Aratus, and in a slightly different form in a work of the poet Cleanthes (c. 250 B.C., p. 54) and in an expanded form in Avienus. Aratus was a native of Cilicia, St. Paul's native province. Both Aratus and Cleanthes were claimed by the Stoics, who, with the Epicureans, were opposing the apostle at Athens (*Acts* xvii. 18).

Though backward in astronomy, the Romans had early developed a good knowledge of such elementary developments as the sundial, which was known to them in the third century B.C., and the results of which were early applied to calendarial reckoning. Full directions for the construction of sundials are given by the architect VITRUVIUS (c. A.D. 10, p. 114) who tells of a number of different forms in use in his time. Some of these, he says, were invented by various Greeks, of whom Aristarchus (p. 59) and Eudoxus (p. 37) are the best known. The construction of these various forms implies command of considerable mechanical skill and some efficiency in the making and recording of elementary astronomical observations. Sundials suitable for use by travellers were also not uncommon. Vitruvius describes also a water-clock of an extremely simple and effective type.

Science the Handmaid of Practice: Imperial Rome

The difference in the length of day in different latitudes was well known to the Romans. From the fact that the longest day in Alexandria was 14 hours, in Italy 15, and in Britain 17, Pliny deduces that lands close to the Pole must have a 24-hours' day in the summer and a 24-hours' night in winter.

Many passages in Pliny reflect a contest concerning the form of the Earth, reminding us of earlier disputes of the same order (pp. 21, 103). He opens his work with a description of the general structure of the universe and discusses the spherical form of the Earth:

'Science and the opinion of the mob', says Pliny, 'are in direct opposition. According to the former the whole sphere of the Earth is inhabited by men whose feet point towards each other while all have the heavens above their heads. But the mob ask how men on the antipodes do not fall off; as though that did not present the opposite query why they should not wonder at *our* not falling off. Usually, however, the crowd objects if one urges that water also tends to be spherical. *Yet nothing is more obvious, since hanging drops always form little spheres.*'

To the Moon and fixed stars the Romans had already, in Pliny's time, begun to attribute an influence on human affairs. 'Who does not know', he asks, 'that when the Dog Star rises it exercises influence on the widest stretch of Earth?' The influence of the Dog Star is an idea that may be traced back in Greek literature at least as far as Hesiod (8th cent. B.C.) and has given us our modern superstition of the 'dog days'. The Moon's influence on tides was recognized, and it was thought that besides influencing the outer world, the *macrocosm*, the Moon had influence also on the body of man, the *microcosm* (p. 37). With the waxing of the Moon it was believed that the muscles became bigger and blood increased. This theory gave rise to the practice of periodical blood-letting which took so prominent a place in early monastic life.

The supposed influence of the heavenly bodies on the Earth and on the life of man is a topic that leads to judicial astrology (p. 151). A knowledge of that subject became under the Empire a professional possession, illegal and prohibited, but often tolerated and invoked even by emperors. Astrology was beginning to spread in Rome in the first century of the Christian era.

117

The Failure of Inspiration

'There are those', Pliny tells us, 'who assign [all human events] to the influence of the stars, and to the laws of their nativity. They suppose that God, once for all, issues his decrees and never after intervenes. This opinion begins to gain ground, and both the learned and the vulgar are accepting it.'

The art was of foreign origin. The credit of its invention is always ascribed to 'Chaldeans', but the main channel of transmission was Greek.

'As for the branch of astronomy which concerns the influences of the twelve signs of the zodiac, the five Planets, and the Sun and Moon on man's life', says Vitruvius, 'we must leave it to the calculations of the Chaldeans to whom belongs the art of casting nativities, which enables them to declare the past and future.'

The original meaning of the zodiacal figures is disputed, but they were certainly in very ancient use in Mesopotamia (Fig. 50) whence came the methods of dividing time and the divisions of the heavenly sphere based on them. Against these Chaldeans Cicero directed his dialogue *On Divination*. He misunderstood

FIG. 50. Babylonian boundary stone showing a seated deity above whose head are the heavenly bodies. The Zodiacal sign of the Scorpion is exhibited. The inscription records a donation to the temple of Marduk in Babylon. Second millennium B.C.

the basis of astrology and marshalled ancient and fallacious arguments against it. Yet even Cicero accepted some astrological doctrine, and in his *Dream of Scipio* he spoke of the planet Jupiter as helpful and Mars as harmful. To the early Christian writers astrology was even more abhorrent, for it seemed to them to be the negation of that doctrine of free will that was so dear to

118

them. The fathers Tertullian (*c.* 155–*c.* 222), Lactantius (*c.* 260–
c. 340), and Augustine (354–430) all inveigh against it. With the
spread of Christianity in the West and the disappearance of the
Stoic philosophy, astrology passed into the background, to return
with the Arabian revival and the rise of the Universities.

At an early date there arose a large literature on the subject.
Nevertheless, astrology seems on the whole to have been rather
less cultivated in Rome itself than the general state of society and
the wide spread of the Stoic philosophy might perhaps suggest.
Lovers sought to learn of astrologers a lucky day for a wedding,
travellers inquired what was the best day for starting on a journey,
and builders asked the correct date for laying a foundation stone.
All these may easily be paralleled by instances among the empty-
headed in our own time and country. But Galen (130–200), who
practised among the well-to-do and educated, assures us that they
only bothered about astrology for forecasting legacies—and again
a parallel might be drawn.

But astrology must not be considered only as a superstition and
an occupation for empty heads and idle hands. The astrological
system of antiquity was, in essence, a formal presentation of those
beliefs concerning the nature and working of our mundane sphere
which had been fostered by a scientific astronomy and cosmology.
Faith in it was part of the Stoic creed. In the mechanism of the
world there was no room for those anthropomorphic gods, the belief
in whom was still encouraged by the priests and held by the multitude. The spread of belief in that mechanism had led at last to
a complete breach between the official faith and the opinions of the
educated classes. The idea of the interdependence of all parts of
the universe produced in time a new form of religion. The world
itself must be divine. 'Deity,' says Pliny, 'only means nature.'
From such a view to the monotheism of Virgil, in which the world
as a whole is regarded as the artistic product of an external god,
is perhaps no great step. Roman Stoicism, however, failed to take
that step, and assumed among later Latin writers a fatalistic and
pessimistic mood. 'God, if God there be, is outside the world and
could not be expected to care for it', says Pliny. The idea
of immortality seems to him but the 'childish babble' of those
who are possessed by the fear of death, as Lucretius had once

The Failure of Inspiration

maintained. After death, so Pliny would have us believe, man is as he was before he was born—and this he tells us as he plunges into his magic-ridden pages!

Once and once only in these Latin scientific writings have we a clear note of real hope. It is significant that that note is sounded in connexion with a statement of a belief in the *progress* of knowledge, an echo of the Greek thought of the fifth and fourth centuries B.C. It is significant, too, that the note is sounded by one who approached, nearer perhaps than any other pagan Latin philosopher, to the idea of the divine immanence. In his natural questions Seneca wrote:

'How many heavenly bodies revolve unseen by human eye! . . . How many discoveries are reserved for the ages to come when our memory shall be no more, for this world of ours contains matter for investigation for all generations. . . . God hath not revealed all things to man and hath entrusted us with but a fragment of His mighty work. But He who directeth all things, who hath established the foundation of the world, and clothed Himself with Creation, is greater and better than that which He hath wrought. Hidden from our eyes, He can only be reached by the spirit. . . . On entering a temple we assume all signs of reverence. How much more reverent then should we be before the heavenly bodies, the stars, the very nature of God!'

But the science of antiquity as exhibited elsewhere in Latin writings contains very little of this belief in man's destiny, this hope for human knowledge. The world in which the Imperial Roman lived was a finite world bound by the firmament and limited by a flaming rampart. His fathers had thought that great space peopled by *numina*, 'divinities', that needed to be propitiated. The new scientific dispensation—the *lex naturae* of the world that had so many parallels with the *jus gentium* of the Empire—had now taken the place of those awesome beings.

In the inevitableness of the action of that law Lucretius the Epicurean might find comfort from the unknown terror. Yet for the Stoic it must have remained a limited, fixed, rigid, and cruel law. His vision, we must remember, was very different from that given by the spacious claim of modern science which explores into ever wider and wider regions of space and time and thought. It was an iron, nerveless, tyrannical universe which science had

Science the Handmaid of Practice: Imperial Rome

raised and in which the Roman thinker must have felt himself fettered, imprisoned, crushed. The Roman had forsaken his early gods, that crowd of strangely vague yet personal beings whose ceremonial propitiation in every event and circumstance had filled his fathers' lives. He had had before him an alternative of the oriental cults whose gods were but mad magicians—a religion unworthy of a philosopher—and the new religion of science whose god, he now saw, worked by a mechanical rule. He had abandoned the faith of his fathers, had flung himself into the arms of what he believed to be a lovelier goddess, and lo! he found himself embracing a machine! His soul recoiled and he fled into Christianity. A determinate view of the world induced that essential pessimism which clouds much of the thought of later antiquity. It was reaction against this pessimism which led to the great spiritual changes in the midst of which antiquity went up in flames and smoke.

6. *The Passage from Pagan to Christian Thought.*

We have gained a general view of the course of ancient thought in relation to science. Four stages may be distinguished:

(*a*) During the rise of Greek thought, philosophy is based on natural science. It neglects ethics and ignores popular religion (Chapter I). Here was the emergence of Mental Coherence.

(*b*) Plato and Aristotle seek to adjust the rival claims of ethics and science, while giving preference to the former. Popular religion is repudiated (Chapter II). This is the Great Adventure.

(*c*) Alexandrian thought develops separate departments for science, ethics, and religion. The age of the 'specialist' has begun. The Alexandrian period terminates with definite scientific deterioration (Chapter III). Intellectual Nerve is failing.

(*d*) Under the Empire the prevalent schools of thought, Stoicism and Epicureanism, are indifferent to science, which deteriorates further (Chapter IV). Great emphasis is laid on Ethics. Scientific inspiration has waned to nothing.

We must now consider somewhat more deeply certain aspects in this final stage of ancient thought in so far as it is related to the material world. Stoicism, in the first two Christian centuries, divided the thinking world with Epicureanism and certain less

The Failure of Inspiration

important philosophical sects. The Stoic philosophy assumed that man's life in all its details is controlled by an interplay of forces. Both the nature and the behaviour of these were, in theory, completely knowable. The same assumptions were made by Epicureanism, save that different forces were held to control man's fate. The Stoic invoked the action of the spheres and astrology. The Epicurean invoked the play of atoms. Both schemes were *determinate*. In this they differed from the new and rising school of Neoplatonism, the *indeterminacy* of which fitted better the doctrine of free will on which Christianity came to insist. Atomism being opposed by the authority of both Aristotle and Plato and by Stoicism and Neoplatonism alike, Epicureanism fell into the background. All philosophical sects became ultimately absorbed into Neoplatonism, the history of which it is necessary to trace.

Alexandria of the third century of the Christian era presented an extraordinary mixture of religions, philosophies, and sects. The old scientific school was in decay. Christian, Jewish, and pagan elements jostled each other. The cults of ancient Egypt, of Greece, of Rome, and of the Orient appealed to the devout and the superstitious. The decayed schools of Aristotle and Plato had still conservative followers. There were also those who called themselves Stoics and Epicureans. A common factor among these various elements was contempt for science.

It must be remembered that the science of those days differed from that of ours in that it had introduced no obvious and extensive amelioration of man's earthly lot. Nature had not been harnessed as we have harnessed her. Science was a way of looking at the world rather than a way of dealing with the world. And as a way of looking at the world—a way of life—positive knowledge that is, science was a failure. The world was a thing that men could neither enjoy nor master nor study. A new light was sought and found. In its glare the old wisdom became foolishness and the old foolishness wisdom. Weary of questioning, men embraced at last and gladly the promises of faith. The faith that was immediately most successful was that which included within itself the experiences of the largest number of educated men. This was the syncretic system known as *Neoplatonism*.

The syncretic tendency exhibited itself very early in Alexandria.

Science the Handmaid of Practice: Imperial Rome

Philo, who was about twenty years older than Christ, developed a system that used the Jewish scriptures in the light provided by Plato and Aristotle, and with some admixture of mysticism. He introduced the doctrine of the *logos*, and his tendency is away from observational science. Following Philo in the first, second, and third centuries were writers of 'Neopythagorean' and 'Hermetic' leanings whose views and tenets were as syncretic as Philo's. They need not delay us. It would be possible to consider the earliest Christian writers as members of this syncretic group.

Early in the third century there arose in Alexandria one AMMONIUS SACCAS—that is the 'sack-carrier' or 'porter'—(died 245), whose personal influence was destined to be fatal for science. Born a Christian he apostatized and opened a school of philosophy which became known as the *Neoplatonic*. The teaching of his school was secret, after the Pythagorean model (p. 17). His pupils, however, were not averse to writing; and the greatest of them, PLOTINUS (204–70), himself a Roman, carried Neoplatonism to Rome and thence to the pagan world at large.

We are not here concerned with any general consideration of Neoplatonism and but little with a further discussion of its numerous sources. These included Aristotle and Plato and their successors and various religious cults, together with the philosophical sects such as Stoicism. There is, however, a certain doctrine of great historic importance which demands some notice here. It is a doctrine shared by Neoplatonism and Stoicism. Both philosophies set off the Universe, the great world, the *macrocosm*, against Man, the little world, the *microcosm* (p. 117). The one was a reflection of the other. Broadly speaking, the Neoplatonist would have said that the Universe had been made for Man who is the essential reality; the Stoic that Man has been made for the Universe. The Neoplatonic view was victorious. The view of the macrocosm and microcosm as elaborated by Neoplatonism was not unacceptable to Christianity.

Neoplatonism developed a characteristic metaphysic derived mainly from Plato but in part also from Stoicism whence it drew its ethics. The Platonic 'Idea' was greatly emphasized and almost personified. The Idea, as expressed by form, governs matter just as the soul governs the body. But matter may at times break away from the Idea and then the world of matter becomes a world of strife and discord. Idea is in the end identifiable with form. Matter, destitute of form or idea, is evil; with form it is at best neutral. It must be the soul's aspiration to free itself from such dangers. Then and then only it can hope for ecstatic union with the Divine.

During the fourth century Neoplatonism flourished. Associating itself with the theologies of various sects, it was a serious rival to Christianity. Its hopes rose high when Julian the Apostate became Emperor (361–3), but they fell again even before the end of his short reign to sink still lower with the victory of Christianity in the age of Valentinian (364–75) and Theodosius (379–95). Christianity in its spread absorbed, with the masses, their superstitions, their magic, and their theurgy. Neoplatonism, on the other hand, at first saturated with these elements, became at last purged of them, though passing thereby out of touch with the spirit of the age. Towards the end of the fourth century the head of the Neoplatonist school at Alexandria was Hypatia (379–415). Her murder ended the effectiveness of the Neoplatonic school as such. She influenced Christian thought directly through her pupils, the most famous of whom, Synesius of Cyrene (373–414), became a very free-thinking bishop.

The passage of Neoplatonic doctrine into Christianity was in the main the work of ST. AUGUSTINE (354–430). After a youth and young manhood spent in devotion to Manichean studies he turned, at last, to study the exact sciences. In 383 he came to Rome whence he moved in 384 to Milan. There he became acquainted with Neoplatonic teachers. In 386 he became converted to Christianity. His great literary activity, begun in 393, ended only with his life.

We have it from Augustine himself that his debt to Neoplatonism was very great. In all his cardinal doctrines—God, matter, the relation of God to the world, freedom, and evil—Augustine borrowed freely from Neoplatonism. Through him we may regard Neoplatonism, itself the final stage of Greek thought, as passing in its final stage into Christianity. Through St. Augustine, above all men, early Christianity acquired its distaste for a consideration of phenomena. 'Go not out of doors', said

the great Father of the Church. 'Return into thyself. In the
inner man dwells truth.' For a thousand years men responsible
for the thought of the Western world did not go out of doors.
It was through St. Augustine that certain Neoplatonic doctrines,
notably that of the macrocosm and microcosm, passed to the
Latin West, where they awaited the Arabist revival (p. 150)
for their fuller development. In a somewhat similar way such
traditions lingered for centuries in the Byzantine East until, with
the great outburst of Islam, they were caught up and elaborated
by the Arabic culture (pp. 139–41). Stamped with specific Islamic
characters the same doctrines were sent forth a second time to
Christian Europe in the process of translation from the Arabic
(pp. 150–3).

V. THE FAILURE OF KNOWLEDGE

The Middle Ages (about A.D. 400–1400): Theology, Queen of the Sciences

1. *The Dark Age* (400–1000).

WE now enter the last and longest phase of the Great Failure. With the decline and fall of the Empire the decay of philosophy was as pronounced as the decline of science. Neoplatonism gives place to the great philosophical and religious movement known as Christianity. The standpoint of its early champions, the Church Fathers, Tertullian (155–222), Lactantius (260–340), and, above all, St. Jerome (340–420) and St. Augustine (354–430), is outside the department with which we deal, but it was assuredly not conducive to the exact study and record of phenomena. Nevertheless, the Middle Ages, under the influence of the Church, developed a characteristic attitude towards nature.

For our purposes we may place the limits of the medieval period between about the years 400 and 1400. This millennium is divided unequally by an event of the highest importance for the history of the human intellect. From about 900 to 1200 there was a remarkable development of intellectual activity in Islam. The movement reacted with great effect on Latin Europe through works which reached it, chiefly in the twelfth and thirteenth centuries, in Latin translations from the Arabic. This intellectual event divides the medieval period in the Latin West into two parts, an earlier *Dark Age* which terminates in the twelfth century, and a later *Age of Arabian Influence* which expressed itself characteristically in Scholasticism. As we pass from one period to the other, the general outline of beliefs as to the nature of the external world changes relatively little, but their presentation is vastly altered and the whole doctrinal scheme of the material world assumes a formal rationality.

During the closing centuries of the classical decline, the body of literature destined to pass down to subsequent ages had been delimited and translated into Latin, the only language common

126

The Middle Ages. Theology: Queen of the Sciences

to the learned West. We must briefly discuss this legacy that antiquity passed on to the Dark Age.

Of the works of Plato, the *Timaeus* fitted well the views of the Neoplatonic thinkers of the late Empire and fitted not ill to Christian belief. A Latin commentary on the *Timaeus*, prepared in the third century, presents a doctrine held throughout the entire Middle Ages as to the nature of the universe and of man. This book became one of the most influential of all the works of antiquity, and especially it conveyed the central dogma of medieval science, the doctrine of the macrocosm and microcosm. This conception, that the nature and structure of the universe foreshadows the nature and structure of man, is basic for the understanding of medieval science.

Of the writings of Aristotle there survived only the logical works translated in the sixth century by BOETHIUS (480–524). These determined the main extra-theological interest for many centuries. Boethius had purposed to translate all of Aristotle and it is a world-misfortune that he did not live to prepare versions of those works that display Aristotle's powers of observation. Had a translation of his biological treatises reached the earlier Middle Ages, the whole history of thought might have been different. Boethius repaired the omission, to some small extent by compiling elementary mathematical treatises based on Greek sources. Thanks to them we can at least say that during the long degradation of the human intellect, mathematics, the science last to sink with the fall of Greek thought, did not come quite so low as the other departments of knowledge.

A somewhat similar service to that of Boethius was rendered by MACROBIUS (395–423) and by MARTIANUS CAPELLA (*c.* 500). The latter, especially, provided the Dark Age with a complete though very elementary encyclopaedia of the seven 'liberal arts', namely the 'trivium', grammar, dialectic, rhetoric, and the 'quadrivium' (p. 112) geometry, arithmetic, astronomy, and music. This classification of studies dates back to Varro (p. 97) and was retained throughout the Middle Ages. The section on Astronomy has a short passage containing a suggestion that Mars and Venus may circle the sun, perhaps derived from Aristarchus (p. 59). The passage, however, is without relation to the text as a whole, and the cosmology of Capella and of Macrobius is similar to that of the *Timaeus*. It may be described as Neoplatonic.

In addition to the little cosmography, mathematics, and astronomy that could be gleaned from such writings as these, the Dark Age inherited a group of scientific and medical works from the period of classical decline. By far the most widely read was the *Natural History* of Pliny (p. 98). Very curious and characteristic is a group of medical pseudepigrapha bearing such names as Dioscorides, Hippocrates, Apuleius, and others. These extremely popular works were translated into Latin between the fourth and sixth centuries. They provided much of the medical equipment of the Dark Age.

Such material, then, was the basis of the medieval scientific heritage. Traces of it are encountered in works of CASSIODORUS (490–585), perhaps the earliest general writer who bears the authentiç medieval stamp. The scientific heritage is, however, much more fully displayed by Bishop Isidore of Seville (560–636) who produced a cyclopaedia of all the sciences in the form of an 'Etymology' or explanation of the terms proper to each. For many centuries this was very widely read. The works of the series of writers, the Spaniard ISIDORE (560–636), the Englishmen BEDE (673–735) and ALCUIN (735–804), and the German RABANUS MAURUS (776–856), who borrow successively each from his predecessor and all from Pliny, contain between them almost the entire corpus of the natural knowledge of the Dark Age.

It must be remembered that the Dark Age presented no coherent philosophical system, and men were capable of holding beliefs inconsistent with each other. The world was but God's footstool, and all its phenomena were far less worthy of study than were the things of religion. In the view of many patristic writers, the study of the stars was likely to lead to indifference to Him that sitteth above the heavens. This is the general attitude of the fourth and fifth centuries, set forth for instance by St. Augustine, who speaks of 'those imposters the mathematicians (i.e. astrologers) . . . who use no sacrifice, nor pray to any spirit for their divinations, which arts Christian and true piety consistently rejects and condemns'.

By the sixth and seventh centuries the Church had come to

some sort of terms with astrology. Thus St. Isidore regards astrology as, in part at least, a legitimate science. He distinguishes, however, between *natural* and *superstitious* astrology. The latter is 'the science practised by the *mathematici* who read prophecies in the heavens, and place the twelve constellations (of the Zodiac) as rulers over the members of man's body and soul, and predict the nativities and dispositions of men by the courses of the stars'. Nevertheless, St. Isidore accepts many of the conclusions of astrology. He advises physicians to study it, and he ascribes to the moon an influence over plant and animal life and control over the humours of man, while he accepts without question the influence of the Dog Star and of the comets. He is followed by the other Dark Age writers on natural knowledge, who accept successively more and more astrological doctrine.

A certain 'revival of learning' under Charlemagne, centred round about the year 800, is very important for its literary activity and certainly did much to preserve such few scientific texts as were available. This movement is greatly emphasized by general historians, but it cannot be considered in the light of a scientific awakening. Perhaps only one figure in the Dark Age is worth our attention here. It is that of GERBERT who became Pope as Sylvester II (died 1003). His merit is to have introduced the abacus (p. 112) which had disappeared with the Roman decline. Its use lingered among the Byzantines whence it reached Arabic-speaking Spain in Gerbert's time. Gerbert had studied in Spain and there, perhaps, learnt of it. He also visited the court of Otto I (913–73) in Southern Italy (970). From wheresoever he derived his abacus there can be no doubt that the details of his arithmetic, like the numerals that he used, he drew from the works of Boethius (p. 127). The immediate future of learning lay not in the West but in the East.

2. *Science in the Orient* (750–1200).

During the Dark Age the intellectual level of the Greek world stood higher than that of the Latin. Science, it is true, was as dead in the one as in the other, but in the Byzantine Empire there was still some activity in the preservation and multiplication of copies of the works of antiquity. The classical dialect was not

The Failure of Knowledge

wholly unknown to the educated class. Despite intense theological preoccupation, classical learning was still occasionally cultivated. A few scholars still glossed the works of Aristotle and Plato.

The Byzantine Empire included many Syriac-speaking subjects. The Syriac language had, from the third century, replaced Greek in Western Asia. There in the fifth century the heretical Nestorian Church had been established. The Nestorians, bitterly persecuted by the Byzantines, emigrated to Mesopotamia. Yet later, they moved to south-west Persia where, from the sixth century onward, they long exhibited great activity especially at their capital Gondisapur. Literature in Syriac became very extensive. It included translations of the works of Aristotle, Plato, Euclid, Archimedes, Hero, Ptolemy, Galen, and Hippocrates.

It was in the seventh century that the Arabs first entered into the heritage of the ancient civilizations of Byzantium and Persia. From their desert home they brought no intellectual contributions save their religion, their music, and their language. Moreover, in the Byzantine and Persian Empires, Greek science was at a low ebb save among the Syriac-speaking Nestorians. Thus the Nestorian metropolis, Gondisapur, became the scientific centre of the new Islamic Empire. From Gondisapur during the Umayyad period (661–749), learned men and especially physicians came to Damascus, the capital. They were mostly Nestorian Christians, or Jews bearing Arabic names.

The rise of the Abbaside Caliphs (750) inaugurated the epoch of greatest power, splendour, and prosperity of Islamic rule, but Islamic thought was still in the absorptive period. The most important agents in the transmission of Greek learning through Syriac into Arabic were members of the great family of Nestorian scholars that bore the name of BUKHT-YISHU ('Jesus hath delivered'). This family produced no less than· seven generations of distinguished scholars, the last of whom lived into the second half of the eleventh century. It was the skill of the physicians of this family that instigated the Caliphs to propagate Greek medical knowledge in their realm.

During the century 750–850 the old Syriac versions were revised and others added. The translators, mostly Nestorians of the Bukht-Yishu family or their pupils, had a command of the

The Middle Ages: Theology, Queen of the Sciences

Greek, Syriac, and Arabic languages and often also of Persian. Most of them wrote first in Syriac. The venerable Yuhanna Ibn Masawiah, the JOHN MESUE of the Latins (d. 857), one of the Bukht Yishu, and medical adviser to Harun ar-Rashid, the fifth Abbaside Caliph, produced, however, many works in Arabic. As time went on Arabic began to replace Syriac for scientific and medical works. Just as 750 to 850 was the century of translation into Syriac, so 850 to 950 was the century of translation into Arabic.

The seventh Abbaside Caliph, Al-Mamun (813–33), created in Bagdad a regular school for translation. It was equipped with a library. HONAIN IBN ISHAQ (809–77), a particularly gifted philosophical and erudite Nestorian, was the dominating figure of this school. He passed his life in Bagdad, serving nine caliphs and exhibiting phenomenal intellectual activity. He translated into Arabic almost the whole immense corpus of Galenic writings. His predilection for the scholastic turn in Galen's theories contributed much to give Galen his supreme position in the Middle Ages in the Orient, and indirectly also in the Occident. He began the translation of Ptolemy's *Almagest* and of works of Aristotle. Honain and his pupils rendered also a number of astronomical and mathematical works into Arabic as well as the Hippocratic writings. Many of these translations passed ultimately into medieval Latin.

Bagdad now rapidly replaced Gondisapur as the centre of learning. The Caliphs and their grandees furnished the necessary means to allow the Christian scholars to travel in search of Greek manuscripts and to bring them to Bagdad for translation. It was at Bagdad that most of the Aristotelian writings were first made accessible in Arabic, together with works on botany, mineralogy, and mechanics, as well as many Greek alchemical works. There was also an active intake of ideas and of texts from Indian and Persian sources. It seems likely that many alchemical methods were of Persian origin, while there was a strong mathematical influence, expressed especially in the system of numeration, exercised by Indian civilization.

The general course of thought may be considered separately for Eastern and Western Islam. Of these the East is more important for the positive sciences, which we can consider under the headings (i) *Alchemy*, (ii) *Medicine* (p. 133) (iii) *Mathematics and*

The Failure of Knowledge

Astronomy (p. 134), and (iv) *Physics* (p. 135).[1] For Western Islam see pp. 138–40.

(i) *Alchemy in Eastern Islam.*

Of original scientific writers using the Arabic language one of the earliest was GEBER (*c.* 850),[2] a pagan Syrian. Geber was the father of Arabic alchemy and through it of modern chemistry. In discussing his work we must rid ourselves of the conception of alchemy as a bundle of fantastic superstitions. The word 'alchemy' is usually said to be derived from the Egyptian *kem-it*, 'the black', or from the Greek *chyma* (molten metal), but in any event it comes to us through Arabic. The fundamental premises of alchemy are of Alexandrian origin (p. 93) and may be set forth thus:

(*a*) All matter consists of the same ingredients, the four elements, in various mixtures.

(*b*) Gold is the 'noblest' and 'purest' of all metals, silver next to it.

(*c*) Transmutation of one metal into another is possible, by an alteration in the admixture of the elements.

(*d*) Transmutation of 'base' into 'noble' metal can be achieved by means of a certain precious substance often called the fifth element, or *quintessence*. (The earliest alchemical documents call the process 'tincturing' the base metal, and in fact describe an alloy.)

These conceptions, absurd though they seem to us, are no more so than those of many eminent chemists of as late as the eighteenth century. In fact they had the great merit of provoking experiment. It is a misfortune that at Alexandria, where alchemy specially flourished, mystical tendencies, largely of Neoplatonic origin, overlaid the experimental factor. Alchemy, which for Geber was a matter of experimental research, thus tended with his successors to superstitious practice, passing into fraudulent deception.

On the practical side, Geber described improved methods for

[1] There was also considerable geographical activity. As, however, it contributed little to the general current of Western science, we omit its discussion.

[2] The date of Geber is much in dispute. Recent evidence points to the ninth century.

132

evaporation, filtration, sublimation, melting, distillation, and crystallization. He prepared many chemical substances, e.g. cinnabar (sulphide of mercury), arsenious oxide, and others. He knew how to obtain nearly pure vitriols, alums, alkalis, sal-ammoniac, and saltpetre, and how to produce so-called 'liver' and 'milk' of sulphur by heating sulphur with an alkali. He prepared fairly pure mercury oxide and sublimate, as well as acetates of lead and other metals, sometimes crystallized. He understood the preparation of crude sulphuric and nitric acids and of the mixture of acids called 'aqua regia' and of the solubility of gold and silver in it. Several technical terms have passed from Geber's Arabic writings through Latin into the European languages (see p. 147).

After Geber there is a great number of alchemical writers, many of whose works found their way into Latin. Except for Rhazes (see below), the quality of their work is commonly much below that of the great original and is frequently cursèd by that wilful obscurity that sometimes usurps the name of 'mysticism'.

(ii) *Medicine in Eastern Islam.*

The first original Arabic writer on medicine was the Persian known to the Latin West as RHAZES (865–925). He was undoubtedly one of the great physicians of all time. He studied in Bagdad under a disciple of Honain (p. 131) who was acquainted with Greek, Persian, and Indian medicine. His erudition was all-embracing, and his scientific output remarkable, amounting to more than two hundred works, half of which were medical. In his youth Rhazes practised as an alchemist but later, when his reputation attracted pupils and patients from all parts of western Asia, he devoted himself to medicine.

The greatest medical work of Rhazes, and one of the most extensive ever written, is his 'Comprehensive Book' known to medieval Europe as the *Liber continens* (p. 149). It gathers into one huge corpus the whole of Greek, Syriac, and early Arabic medical knowledge and incorporates also the life experience of Rhazes himself. Rhazes was the first to describe small-pox and measles adequately. His account of them is a medical classic.

Besides medicine, Rhazes left writings on theology, philosophy,

The Failure of Knowledge

mathematics, astronomy, and alchemy. His great *Book of the Art* (*of Alchemy*) is dependent partly on his predecessor Geber. Rhazes excels Geber in his exact classification of substances, and in his clear descriptions of chemical processes and apparatus. These are always devoid of 'mystical' elements. While Geber and the other Arabian alchemists divide mineral substances into 'bodies' (gold, silver, &c.), 'souls' (sulphur, arsenic, &c.), and 'spirits' (mercury and sal-ammoniac), Rhazes classified alchemical substances as animal, vegetable, or mineral, a conception which passed from him into a commonplace of modern speech.

A prominent contemporary of Rhazes was the writer known to the Latins as ISAAC JUDAEUS (855–955). This Egyptian Jew became physician to the Fatimid rulers of Kairouan in Tunisia. His works were among the first to be translated from Arabic into Latin (p. 143). That *On Fevers* was one of the best medical works available in the Middle Ages.

AVICENNA (980–1037) of Bokhara was one of the greatest thinkers of the Islamic world. He was less remarkable as a physician than as a philosopher, but his influence on medieval Europe was chiefly through his gigantic *Canon of Medicine*. It is the culmination and masterpiece of Arabic systematization and has been perhaps more studied than any medical work ever written. The classification adopted in it is excessively complex, and is in part responsible for the mania for subdivision which afflicted Western Scholasticism. Avicenna wrote also on alchemy. The early Arabic literature of medicine is very extensive.

(iii) *Mathematics and Astronomy in Eastern Islam.*

Of all the peoples of antiquity there was none except the Greeks that attained so high a standard in mathematics as the Hindus. Just as the Greeks developed geometry, the Hindus developed arithmetic and algebra. It is extremely difficult to fix the dates or even the chronological sequence of the Indian mathematical works. The Arabs, however, had much commerce with India and there can be no doubt that by the ninth century Hindu science was available in Arabic. Thus Arabic algebra and arithmetic are essentially Indian.

The most influential mathematical work produced in Arabic

was the *Arithmetic* of the Persian AL-KWARIZMI (*c*. 830). In it is used our so-called 'Arabic' numerical notation, in which the digits depend on their position for their value. The method is, in fact, of Indian origin. The *Algebra* of Al-Kwarizmi is the first work in which that word appears in the mathematical sense. 'Algebra' means 'restoration', that is to say the transposing of negative terms of an equation to the opposite side. The word is used also in Arabic surgery for treatment of fractures, the word there meaning 'restoration' of a broken bone to its correct position. Al-Kwarizmi also prepared astronomical tables.

The mathematics of Al-Kwarizmi shows little originality, and in general the achievement of the Arabs in the department of pure mathematics is below the Greeks in geometry and below the Hindus in algebra. On the other hand, they exhibited great skill in applying their mathematics to physical and to a less extent to astronomical problems.

Astronomy and astrology were constant preoccupations of the Arabic-speaking world. Very early works on the subject were the compendia by the Bagdad Jewish writer, MESSAHALA (770–820), whose name means 'What God will'.

The Caliph Al-Mamun (813–33) built a fine observatory at Bagdad (829) where observations were long recorded. The greatest of all the Arabic astronomers AL-BATTANI, Albategnius of the Latins (d. 929), observed chiefly at his home Raqqa (Aracte) in Asia Minor, but also at Bagdad. He worked over the observations of Ptolemy in a searching and exact manner. He thus obtained more accurate values for the obliquity of the ecliptic and the precession of the equinoxes (pp. 76–77). His improved tables of the sun and the moon contained his great discovery that the direction of the sun's excentric (p. 78), as recorded by Ptolemy, was changing. Expressed in the terms of more modern astronomical conceptions, this is to say that the earth is moving in a varying ellipse (p. 265). Al-Battani drew up his observations in tabular form.

A popular elementary writer on astronomy was ALFARGANI of Transoxiana (d. *c*. 850). He worked at Bagdad and served the Caliph Al-Mamun and his followers. His work deeply influenced the Latin West.

The Failure of Knowledge

(iv) *Physics in Eastern Islam.*

Among the Arabic writers on Physics ALKINDI (813–80) of Basra and Bagdad was the earliest. No less than 265 works have been ascribed to this 'first philosopher of the Arabs'. Of these at least fifteen are on meteorology, several are on specific weight, and others on tides. His best work is on optics, and deals with the reflection of light.

In the ninth century the technical arts were rapidly developing in Mesopotamia and Egypt, where irrigation works and canals for water-supply and communications were created. Theoretical mechanics roused much interest, and many books were written on such topics as raising water, on water-wheels, on balances, and on water-clocks. The earliest appeared about 860 as the *Book of Artifices* by the brother mathematicians Muhammed, Ahmed, and Hasan, sons of Musa ben Shakir, who were themselves patrons of translators. They describe one hundred technical devices, of which some twenty are of practical value, among them being vessels for warm and cold water, wells with a fixed level, and water-clocks. Most, however, are mere scientific toys like those of Hero (p. 81).

The tenth and early eleventh centuries were the golden age of Arabic literature. It was also remarkable for its wealth of technical knowledge. Optics especially was developed to its highest degree. ALHAZEN (965–1038) of Basra was the greatest exponent of this science. He entered the service of the Fatimid Caliph al-Hakim (996–1020) at Cairo. In his main work, the *Treasury of Optics*, Alhazen opposes the theory of Euclid and Ptolemy and others among the ancients that the eye sends out visual rays to the object of vision. It is, he thinks, rather the form of the perceived object that passes into the eye and is transmuted by its 'transparent body', that is the lens. He discusses the propagation of light and colours, optic illusions and reflection, with experiments for testing the angles of reflection and of incidence. His name is still associated with the so-called 'Alhazen's problem'. 'In a convex mirror, spherical, conical, or cylindrical, to find the point at which a ray coming from one given position will be reflected to another given position.' It leads to an equation of the fourth degree which Alhazen solved by the use of an hyperbola. Alhazen examines also the refraction of light-rays through transparent media (air, water). In detailing his experiments with spherical segments he comes very near to the theoretical exposition of magnifying lenses which was made centuries later (p. 195). Alhazen regards light as a kind of fire that is reflected at the spheric limit of the atmosphere. His calculation of the height of this atmosphere gives about ten English miles. He treats also of the rainbow, the halo, and the reflection from spherical and parabolic mirrors. He constructed such mirrors of metal on the basis of most elaborate calculations. His fundamental study *On the Burning-sphere* represents real scientific advance, and exhibits a profound and accurate conception of the nature of focusing, magnifying, and inversion of the image, and of formation of rings and colours by experiments. The work is far beyond anything of its kind produced by the Greeks. Alhazen records in it the semi-lunar shape of the image of the sun during eclipses on a wall opposite a fine hole made in the window-shutters. This is the first mention of the *camera obscura*.

Among the most characteristic products of Arabic thought is a group of writings on what we may call scientific theory and classification. An early exponent of these was the Turkish philosopher ALFARABI (d. *c.* 951), the author of the most important oriental work on the theory of music. His treatise on the classification of the sciences was very influential.

The Persian ALBIRUNI (973–1048), physician, astronomer, mathematician, physicist, geographer, and historian, is perhaps the most prominent figure in the phalanx of the versatile scholars of the Golden Age. His *Chronology of Ancient Nations* is an important historical document. Most of his mathematical work and many others of his writings await publication. In physics his greatest achievement is the very exact determination of the specific weight of eighteen precious stones and metals. His method was, in effect, that of the bath of Archimedes (p. 65).

In the tenth and eleventh centuries several secret or at least esoteric sects professing the atomic nature of matter established themselves in Mesopotamia. Certain of them professed an Epicurean attitude to the world and to Creation which was opposed to the orthodox Aristotelianism of Moslem theologians. A struggle

The Failure of Knowledge

ensued comparable to that of later date in Europe. In the end the unorthodox atomists were vanquished.

Among the secret societies mention may be made of the BRETHREN OF PURITY, a philosophical sect founded in Mesopotamia about 980. They combined to produce an encyclopaedia of fifty-two treatises, seventeen of which deal with natural science, mainly on Greek lines. They contain discussions on the formation of minerals, on earthquakes, tides, meteorological phenomena, and on the elements, all brought into relation with the celestial spheres and bodies. The work of the Brethren, although burnt as heretical by the orthodox in Bagdad, spread as far as Spain where it influenced philosophic and scientific thought.

In Western Islam the scientific tradition was established later than in the East. It first appears in Spain, during the glorious reigns of the Caliphs Abd Ar-Rahman III and Al-Hakam II of Cordova, in the person of HASDAI BEN SHAPRUT (d. *c.* 990), a Jew who was at once minister, court physician, and patron of science. He translated into Arabic, with the help of a Byzantine monk, a splendid manuscript of Dioscorides (p. 89) sent, as a diplomatic present, to his sovereign from Constantine VI of Byzantium. The Moslem known to the Latins as ABULCASIS (d. *c.* 1013) was likewise court physician in Córdova. His name is associated with a great medical handbook in thirty sections, the last of which deals with surgery, an art which had till then been neglected by Islamic authors.

A library and academy was founded at Cordova in 970, and similar establishments sprang up at Toledo and elsewhere. Astronomy was specially studied. The chief astronomer of the Moslem Spain was known to the Latins as ARZACHEL. He was a Cordovan but worked at Toledo. He drew up so-called *Toledan tables* which attained a high degree of accuracy (1080). One of the last significant men of science of Moslem Spain was Al-Bitrugi of Seville, known to the Latins as ALPETRAGIUS. He wrote a popular text-book of astronomy (*c.* 1180). The work contains an attempt to replace the Ptolemaic by a strictly concentric planetary system and is important for having provided suggestions to Copernicus (p. 179).

138

The Middle Ages: Theology, Queen of the Sciences

In the twelfth century a great change came over Islamic thought. Under the influence of the religious teacher Al-Ghazzali (d. 1111), tolerance gave place to persecution of studies thought to 'lead to loss of belief in the origin of the world and in the Creator'. Outstanding and independent works become rarer. Among the scientific writers an increasing proportion of Jews is to be observed, because they were relatively free from the restraints of orthodox Islam. Of these the most eminent was the court physician, philosopher, and religious teacher, MAIMONIDES (1135–1204). Born in Spain, he spent most of his active life in Cairo under the great Saladin and his sons. In his medical works he even ventured to criticize the opinions of Galen. As a court official he wrote hygienic treatises for the Sultan which are good typical specimens of the medical literature of Islam. His cosmological views are of great importance and influenced St. Thomas Aquinas and through him the whole thought of Catholic Europe. His *Guide for the Perplexed* is perhaps the most readable treatise on general philosophy produced by the Middle Ages, whether Arabic, Byzantine, or Latin. It has the crowning merit, most unusual for the period, of relative brevity.

The latest and the greatest exponent of Islamic philosophy was the Spaniard AVERROES (1126–98). He was born at Cordova, son and grandson of a legal officer. He himself held the office of judge, but also studied and practised medicine. His very voluminous philosophical writings earned the enmity of orthodox Moslem theologians, some of whom regarded him as having become a Jew. In fact, no writer exerted greater influence than Averroes on later medieval Jewish thought. His writings were burned by royal decree, and most of the latter part of his life was passed in disgrace.

Averroes has certainly been one of the most influential of all thinkers. He placed his thought in the form of a long series of commentaries on the works of Aristotle, whom he exalted above all other men. Nevertheless, his teaching was basically, though unconsciously, modified by Neoplatonism, notably in his conception of the human soul as part of the Divine world soul. His most discussed doctrine was that the world is eternal. This some of his interpreters represented as a denial of creation. Nevertheless,

The Failure of Knowledge

Averroes did accept the idea of creation, though not of an entire universe out of nothing as demanded by the current theology of Islam, Christianity, and Judaism alike.

Averroes believed, not in a single act of creation, but in a continuous creation, renewed every instant in a constantly changing world, always taking its new form from that which has existed previously. This is true philosophic evolutionism. For Averroes the world, though eternal, is subject to a *Mover* constantly producing it and, like it, eternal. This Mover can be realized by observation of the eternal celestial bodies whose perfected existence is conditioned by their movement. Thereby may be distinguished two forms of eternity, that with cause and that without cause. Only the *Prime Mover* is eternal and without cause. All the rest of the universe has a cause or, as we should say nowadays, is 'subject to evolution'.

Averroes, like all medieval thinkers, pictured the universe as finite in space. For a formal denial of that doctrine we have to look forward to Nicholas of Cusa (p. 171) and Giordano Bruno (p. 185).

With the thirteenth century there sets in a very definite deterioration in the quality of Arabic science. The future lay with the Latin West on which Arabic thought was now setting its stamp.

Perhaps the most significant of all Moslem influences on the West has been the philosophy transmitted through Averroes and chiefly by Jewish agents. By his doctrine of the eternity of the world, his denial of Creation in time, and his conception of the unity of the soul or intellect, Averroes split Western thought from top to bottom. Orthodox Catholic philosophy of the Middle Ages may be regarded as an organized attempt to refute his views. The fact that this seemed necessary tells of the gravity of the opposition. His influence may be traced in many medieval heresies, in the works of Nicholas of Cusa (p. 171), and of several Renaissance thinkers, in the standpoint of Copernicus (p. 179), in the thought of Giordano Bruno (p. 185), and beyond. It may seem strange that a professedly faithful exponent of Aristotle should have initiated that movement which led to the final overthrow of Aristotelian cosmology in the Insurgent Century (Ch. VII). It must be remembered, however, that Averroes, like the other Arabic philosophers, saw Aristotle through Neoplatonic spectacles; though he was himself unconscious of the fact. The Neoplatonic tinge became, moreover, intensified in the Latin versions and commentaries on his works.

3. *Oriental Penetration of Occident* (1000–1300).

The eleventh century and those that follow brought the West into relation with the wisdom of the East. In these centuries the relation of East and West with which we are nowadays familiar is reversed.

In our time most Oriental peoples value Western civilization and accord it the sincerest form of flattery. The Oriental recognizes that with the Occident are science and learning, power and organization, and business enterprise. But the admitted superiority of the West does not extend to the sphere of religion. The Oriental who nowadays gladly accepts the Occidental as his judge, his physician, or his teacher, repudiates, and perhaps despises, his religion and his philosophy.

In the Europe of the eleventh and twelfth centuries it was far other. The Westerner knew full well that Islam held the learning and science of antiquity. Moslem proficiency in arms and administration had been sufficiently proved—the Occidental belief in them is enshrined in our Semitic words 'arsenal' and 'admiral', 'tariff', 'douane', and 'average'. There was a longing, too, for the intellectual treasures of the East, but the same fear and repugnance to its religion that the East now feels for West. And the Western experienced obstacles in obtaining the desired Oriental learning analogous to those now encountered by the Eastern in the Occident.

We may consider Arabic influence on Western Europe in two stages, an earlier indirect stage, 'the Age of Rumours', and a later direct stage, 'The Age of Translations'.

(i) *The Age of Arabian Rumours* (1000–1100).

The first definitely Oriental influence that we can discern as affecting ideas about nature is of the character of infiltration rather than direct translation. Thus GERBERT, who died in 1003

The Failure of Knowledge

as Pope Sylvester II had studied in north-east Spain, beyond the Moslem zone. He described an abacus (p. 112) that was almost certainly of Arabic origin though he used for it counters bearing numerals similar to those of Boethius (p. 129). He also instigated a translation from Arabic of a work on the astrolabe. He was clearly in touch with some sort of Arabic learning.

Similarly with HERMAN THE CRIPPLE (1013–54) who spent his life at the Benedictine Abbey of Reichenau in Switzerland. He wrote certain mathematical and astrological works which were extensively used in the following century. Herman was unable to read Arabic, and could not travel by reason of his infirmity. Yet his writings display much Oriental influence, which must have been conveyed to him by wandering scholars. Similar evidence of Arabic infiltration is exhibited in lapidaries and herbals of the eleventh and twelfth centuries.

By the mid-eleventh century, Arabic learning was thus beginning to trickle through to the West. It was derived ultimately, as we have seen, from Greek sources (p. 130). There was, however, just one channel by which the original Greek wisdom might still reach Europe, though in a much debased form. Communication between the West and the Byzantine East was very restricted in the Dark Age, but a Greek tradition still lingered in south Italy and Sicily. These remained for centuries under the nominal suzerainty of Byzantium, and the dialects of the 'many-tongued isle' still bear traces of the Greek spoken there, as in Calabria and Apulia, until late medieval times. But Saracens had begun their conquest of Sicily in the eighth century, and did not loosen their hold until the Norman attack of the eleventh. The Semitic language of the Saracens left the same impress on the island as did their art and architecture. Thus between the tenth and thirteenth centuries the 'Sicilies' were a source of both Greek and Arabic science.

One seat of learning in the southern Italian area felt especially the influence of both Greek and Arabic culture. Salerno, on the Gulf of Naples, had been a medical centre as far back as the ninth century. There was a Greek-speaking element in the town and some traces of ancient Greek medicine lingered there as in other parts of south Italy after the downfall of the Western Empire.

142

There were, moreover, a number of Jews in the town and many of these had affiliations with the Orient. Such learning as was found at Salerno was galvanized into life by Saracenic energy. From about 1050 onwards medical works were produced at Salerno. It is easy to understand why some of them contain Semitic words, and why others present unexpected and strangely altered Greek terms.

A very important carrying agent of the Arabic learning was CONSTANTINE THE AFRICAN (1017–87), a native of Carthage. He reached Salerno about 1070 and some years later acted as secretary to the Norman conqueror of that city. Later he retired to a monastery and spent the rest of his life turning current Arabic medical and scientific works into Latin.

Constantine's sources are mainly Jewish writers of North African origin and Arabic language, among them Isaac Judaeus (p. 134). In his desire for self-exaltation Constantine often conceals the names of the authors from whom he borrows, or he gives them inaccurately. His knowledge of both the languages which he was treating was far from thorough. Yet his versions were very influential, and remained current in the West long after they had been replaced by the better workmanship of students of the type of Gerard of Cremona (p. 148). With Constantine is linked ALPHANUS, Archbishop of Salerno (d. 1085), who was himself the first medical translator direct from the Greek, and who turned a Neoplatonic physiological work of the fourth century into Latin.

(ii) *The Mechanism of Translation.*

The earliest Oriental influences that reached the West had thus been brought by foreign agents or carriers, but the desire for knowledge could not be satisfied thus. The movement that was to give rise to the universities was shaping itself during the twelfth century. The Western student was beginning to become more curious and more desirous of going to the well-springs of Eastern wisdom.

Language was his main difficulty. The idiom of Arabic was utterly different from the speech of the peoples of Europe. Moreover, its grammar had not been reduced to rule in any Latin work, nor could teachers be easily procured. The only way to learn the

The Failure of Knowledge

language was to go to an Arabic-speaking country. This was a dangerous and difficult adventure, involving hardship, secrecy, and perhaps abjuration of faith. Moreover, a knowledge adequate for rendering scientific treatises into Latin meant a stay of years, since some understanding of the subject-matter as well as the technical vocabulary was needed. There is good evidence that such knowledge was very rarely attained by western Christians, and probably never until the later twelfth century.

At the period during which Western science began to draw from Moslem sources there were only two areas of contact of the rival civilizations: Spain and 'the Sicilies'. The conditions in the two were somewhat similar. In the tenth century the Iberian peninsula was Moslem save for Leon, Navarre, and Aragon, small kingdoms of the French march. In that northern area the grip of Islam had soonest relaxed, and this territory remained religiously and linguistically a part of the Latin West. The Moslem south was ruled from Cordova, which became a very Islamic stronghold. At the more northern Toledo the townsfolk while speaking an Arabic patois, were chiefly Christian, though with a large Jewish element. In 1085, Alphonso VI of Leon, aided by the Cid, conquered the town. A large Arabic-speaking population remained. It was at Toledo that most of the work of transmission took place (Figs. 51 and 52).

The question is often asked why in the Middle Ages, the prevailing tendency was to translate works from the Arabic rather than from the Greek, and why this tendency affected even works originally written in Greek. The reasons may be set forth thus:

(a) Between 1000 and 1300 Moslem learning was better organized, more original, more vital in every way than Byzantine learning.

(b) Byzantine Greek is far distant from the classical tongue. The language of Aristotle was incomprehensible to the monastic guardians of his manuscripts. On the other hand, classical Arabic was intelligible to every well-educated man —Moslem or other—who spoke and wrote Arabic.

(c) The whole trend of Byzantine learning was to theology and away from philosophy and science.

144

The Middle Ages: Theology, Queen of the Sciences

(*d*) The channels of trade with the West were rather with Islam than with the Byzantine Empire.

(*e*) In the Middle Ages languages were learned by speaking and not from grammars. Spoken Arabic was more accessible than spoken Greek.

FIG. 51. The recession of Islam in Spain. The figures after the names of towns are the dates of their reconquest by Christendom.

(*f*) Latin Christendom made little progress in occupying Byzantine territory. On the other hand, from 1085, when Toledo fell, Islam was in retreat in the West. It was thus easier to find a skilled Arabic than a skilled Greek teacher.

(*g*) Jewish help could be obtained for Arabic, but seldom for Greek.

The process of translation from Arabic, especially in Spain, was frequently carried on by the intervention of Jewish students. Many of the translated works were themselves by Jews. The tenth, eleventh, and twelfth centuries, a time of low degradation of the Latin intellect, was the best period of Jewish learning in Spain. Arabic was the natural linguistic medium of these learned

145

Jews, among whom were Solomon ibn Gabirol (1021–58?) of Sara-gossa, who was disguised in scholastic writings as AVICEBRON, and Moses ben Maimon (1135–1204) of Cordova, more familiarly known as MAIMONIDES (p. 139). The writings of these two authors together with the Jewish version of AVERROES were the most philosophically influential of those rendered into Latin from Arabic

FIG. 52. Italy in the first half of the 13th century.

during the Middle Ages. Their works helped to mould Western scholasticism.

Despite the activity of the translators, medieval Latin was not yet equipped with an adequate supply of technical terms. The meanings of some of these in the Arabic were imperfectly known to the translators themselves. Such words were therefore often simply carried over, transliterated from their Arabic or Hebrew form. The early versions are full of Semitic expressions. Thus of chemical substances we have *realgar* (red sulphide of arsenic), *tutia* (zinc oxide), *alkali*, *antimony*, *zircon*, and of chemical apparatus *alembic* for the upper, and *aludel* for the lower part of a distillation vessel. A new chemical substance unknown to the Greeks which appears for the first time in the works of Geber is *sal-ammoniac*. The *ammoniacon* of the Greeks was rock-salt, and it seems that the transference of the old names to a new salt was effected by the Syrians. Of pharmaceutical terms, we have a number of Persian terms that have passed through Arabic, such as *zedoary, alcohol, sherbet, camphor, lemon,* and *syrup*, while more purely Arabic are *alizarin, borax, elixir, natron, talc,* and *tartar*. In astronomy there are numerous Arabic star names as *Aldebaran, Altair, Betelgeyse, Rigel, Vega,* some astronomical terms as *nadir, zenith, azimuth, azure,* a few instrumental designations as *alidade* and *theodolite,* and at least one word which has passed into common language, *almanac*. To these may be added the mathematical terms *zero, cipher, sine, root, algebra* (p. 135), *algorism* (see below). Music was also deeply affected, as witness *lute, guitar, shawm, rebeck*. There was a complete Arabic-Latin anatomical vocabulary of which almost the sole remains is *nucha*, though the titles of the *cephalic, basilic,* and *saphenous* veins have passed through Arabic. The modern botanical vocabulary provides us with many plant-names of Arabic origin such as *artichoke, coffee, lilac, musk, ribes* and *sumach* or names that have passed through Arabic as *jasmine, mezereon, saffron, sesame,* and *taraxacum*.

(iii) *The Translators.*

Among the pioneer Western translators from Arabic to Latin was ADELARD OF BATH (*c.* 1090–*c.* 1150), who journeyed both to Spain and the Sicilies. His services to mathematics were very distinguished. He began early with a treatise on the abacus. Then he turned to Arabic mathematics and translated into Latin the *Arithmetic* of Al-Kwarizmi involving the use of the 'Arabic', i.e. Indian, numerals (p. 135), which he thus introduced to the West. Al-Kwarizmi has, through him, left his name in *algorism*, the old word for arithmetic. Moreover, Adelard also rendered Euclid from the Arabic and so made the Alexandrian mathematician known for the first time to the Latins. He wrote a popular dialogue, *Natural Questions*, which is a sort of compendium of Arabic science.

A generation later than Adelard was ROBERT OF CHESTER (*c.* 1110–*c.* 1160), who lived long in northern Spain (1141–7). He was the first to translate the Koran (1143). Among his scientific renderings was the first alchemical text to appear in Latin (1144). His translation of the *Algebra* of Al-Kwarizmi (p. 135) introduced the subject to the Latins (1145). Later he returned to England and settled in London (1147). There he produced astronomical tables for the longitude of London (1149–50) based on Albategnius (p. 135) and for the latitude of London based on Al-Kwarizmi and Adelard.

Contemporary with Robert and perhaps stimulated by him, were certain native translators who worked at Toledo. One of these was DOMENIGO GONZALEZ (fl. 1140), a Christian who rendered into Latin from Arabic the *Physics* and other works of Aristotle. Another, JOHN OF SEVILLE (fl. 1139–55), a converted Jew, was very active and translated among many other works a pseudo-Aristotelian treatise which greatly influenced Roger Bacon, as well as astronomical and astrological works of Al-battani, Alfarabi, Alfargani, Al-Kwarizmi, Alkindi, and Messahala.

The greatest and most typical of all the translators from the Arabic was GERARD OF CREMONA (1114–87), who spent many years at Toledo and obtained a thorough knowledge of Arabic from a native Christian teacher. He is credited with having translated into Latin no less than ninety-two complete Arabic works. Many of them are of very great length, among them the *Almagest* of Ptolemy (p. 84) on which Georg Purbach (p. 171) began his work in the fifteenth century, and the enormous *Canon* of Avicenna (p. 134), perhaps the most widely read medical treatise ever penned. Latin editions of Avicenna continued to be issued right down to the middle of the seventeenth century. The *Canon* is still in current use in the East.

Among the other achievements of Gerard are translations from the Arabic of Archimedes *On the Quadrature of the Circle* (p. 67), of an optical work of Apollonius (p. 69), of many of the works of

The Middle Ages: Theology, Queen of the Sciences

Aristotle both spurious and genuine, of Euclid's *Elements*, of many medical works of Galen, Hippocrates, Isaac Judaeus (p. 134), Rhazes (p. 133), and Albucasis (p. 138), of alchemical works of Geber (p. 132), of mathematical and astronomical works by Alkindi (p. 136), Alfargani (p. 135), Alhazen (p. 136), Alfarabi (p. 137), Messahala (p. 135), and others. He also translated certain important Neoplatonic works.

The Sicilian group was less active. Among its products was the *Optics* of Ptolemy (p. 83), translated about 1160 by the Sicilian admiral EUGENIUS OF PALERMO. He rendered it from the Arabic, though he had an effective knowledge of Greek. The great astronomical and mathematical system of Ptolemy known to the Middle Ages as the *Almagest* (p. 84) was also first translated into Latin from the Greek in Sicily in 1163, some twelve years before it was rendered from the Arabic by Gerard at Toledo (p. 148). This version from the Greek gained no currency and only that from the Arabic was available until the fifteenth century.

The last important medieval translator from the Arabic was of Sicilian origin. He was the Jew MOSES FARACHI (d. 1285), a student at Salerno, and his works were among the latest of influence that issued from that ancient seat of learning. His great achievement was the translation for his master Charles of Anjou (1220–85), King of the Sicilies, of the enormous *Liber continens* of Rhazes (p. 133), a standard medical work of the Middle Ages.

Special consideration among the translators may be given to MICHAEL THE SCOT (*c*. 1175–*c*. 1235) because we have more picturesque details of him than of the others. He had a career similar to Adelard. He visited Toledo and afterwards northern Italy, staying at Padua, Bologna (1220), and Rome (1224–7). He ended his days in the south in the service of the 'Stupor Mundi', Frederick II. He rendered into Latin from Arabic the astronomy of Alpetragius (p. 138), a number of Averroan commentaries, and the biological works of Aristotle. His pseudo-Aristotelian compendium, the *Secrets of Nature*, from a number of Greek, Arabic, and Hebrew sources, contains a section on generation that is still reprinted in the European vernaculars. Michael also produced a great treatise on astrology.

Michael's activity was significant for several reasons. His version of Alpetragius contained the first attack on traditional astronomy. His translations of Averroes were among the first works of that heresiarch available to the Latins. His version of Aristotelian biology gave Aristotle's own scientific observations for the first time to the West. His work on astrology was the first major treatise on the subject available in Latin. Michael certainly had Jewish and Moslem help and was long associated with the arch-enemy of the papacy, Frederick II. Thus it is no great wonder that in the popular imagination his name became associated with sorcery and black magic. This was the fate of other translators from the Arabic. The vulgar attitude towards such men is faithfully reflected in Sir Walter Scott's *Lay of the Last Minstrel* where a monk tells us that

> Paynim countries I have trod,
> And fought beneath the Cross of God.
>
> . . . .
>
> In those far climes it was my lot
> To meet the wondrous Michael Scott;
> A wizard of such dreaded fame,
> That when, in Salamanca's cave,
> Him listed his magic wand to wave,
> The bells would ring in Notre Dame!
> Some of his skill he taught to me;
> And, warrior, I could say to thee,
> The words that cleft Eildon hills in three,
> And bridled the Tweed with a curb of stone:
> But to speak them were a deadly sin,
> And for having but thought them my heart within,
> A treble penance must be done.
>
> When Michael lay on his dying bed,
> His conscience was awakened;
> He bethought him of his sinful deed,
> And he gave me a sign to come with speed.
> I was in Spain when the morning rose,
> But I stood by his bed ere evening close.

4. *Scholasticism and Science* (1200–1400).

The view of the material universe conveyed by Arabic science to Latin Christendom was new in tone and presentation rather than in kind. The thought of the Latins in their Dark Age on material things was Neoplatonic, with the *Timaeus* as text-book and the theory of macrocosm and microcosm as key. With the advent of Arabic thought the outlines of this vision were sharpened, and details were elaborated from the Arabian commentators on the Aristotelian corpus.

Thus Aristotle's views or supposed views as to the structure of the universe formed the framework on which the whole of medieval science, from the thirteenth century onward, came to be built. Aristotle conceived the stars as beings whose nature and substance were purer and nobler than that of aught in the spheres below. This was a point of departure from which the influence of the heavenly bodies over human destinies might be developed. Changes undergone by bodies on the earth—all the phenomena of our life—were held to be paralleled and controlled by movements in the heavens above.

The theory carried the matter farther. Taking its clue from the Aristotelian conception of 'perfection' of the circle among geometrical figures (p. 46), it distinguished the perfect, regular, circular motion of the fixed stars from the imperfect, irregular, linear motion of the planets., The fixed stars, moving regularly in a circle, controlled the ordered course of nature, the events that proceeded in recurring, manifest, and unalterable rounds, such as winter and summer, night and day, growth and decay. The planets, on the other hand, erratic or at least errant in their movements, governed the more variable and less easily ascertainable events in the world around and within us, the happenings that make life the uncertain, hopeful, dangerous, happy thing it is. It was to the ascertainment of the factors governing this kaleidoscope of life that astrology set itself.

Thus the general outline was fixed, death in the end was sure, and, to the believing Christian, life after it. But there was a zone between the sure and the unsure that might be predicted and perhaps avoided, or, if not avoided, its worst consequences abated. It was to this process of insurance that the astrologer set himself, and his task remained the same throughout the Middle Ages. In this hope, *savoir afin de prévoir*, the medieval astrologer was at one with the modern man of science. The matter is summarized by Chaucer (1340–1400):

The Failure of Knowledge

Paraventure in thilke large book,
Which that men clepe the heven, y-written was
With sterres, whan that he his birthe took,
That he for love sholde han his deeth, allas!
For in the sterres, clerer than is glas,
Is written, God wot, whoso coude it rede,
The deeth of every man, withouten drede.

. . . .

. . . But mennes wittes ben so dulle
That no wight kan wel rede it atte fulle.

(The Man of Lawes Tale, ll. 190–6 and 202–3.)

With the advent of the Arabian learning, astrology had become, in fact, the central intellectual interest. It retained this position until the triumph of the experimental method in the seventeenth century.

Especial attention had always been paid to the zodiacal signs (p. 118) and to the planets. Each zodiacal sign was held to govern some region of the body, and each planet to influence a special organ. The supposed relations of zodiacal signs, planets, and bodily parts and organs, in relation to the advent of disease and calamity, had been set forth in many texts of late antiquity. This belief, conveyed to the Dark Age, but much corrupted and attenuated during its course, was reinforced and developed in the West by translations from the Arabic during the scholastic period which followed.

Doctrine of this type, once received into Europe, was stamped with the special form of Western thought. Now, it was characteristic of the scholastic thinker that, like the early Greek philosopher and unlike his predecessor of the Dark Age, he sought always a complete scheme of things. He was not content to separate, as we do, one department of knowledge or one class of phenomena, and consider it in and by itself. Still less would he have held it a virtue to become a 'specialist', to limit his outlook to one department with the object of increasing the sum of knowledge in it, and in it alone. His universe, it must be remembered, so far as it was material, was limited. Its frontier was the sphere of the fixed stars. Of the structure and nature of all within this sphere he had been provided with a definite scheme. The task of medieval science was to elaborate that scheme in connexion with the moral world. This was first especially undertaken by mystical writers working under the stimulus of the new Arabian influence. Such authors as HUGH OF ST. VICTOR (1095–1141), who drew on the earlier and more vague Arabian rumours, BERNARD SYLVESTER (*c.* 1150) of Chartres, who relied on Herman the Cripple (1013–54, p. 142), and ST. HILDEGARD (1099–1180) of Bingen, who was influenced by Bernard Sylvester and by other Arabicized writings, all produced most elaborate mystical schemes based on the doctrine of the macrocosm and microcosm. These schemes took into account the form of the world and of man as derived from Arabian sources, and read into each relationship a spiritual meaning.

For such an attitude of mind there could be no ultimate distinction between physical events, moral truths, and spiritual experiences. In their fusion of the internal and the external universe, these mystics have much in common with the mystics of all ages. The culmination of the process is reached with Dante (1265–1321).

There were other typical currents of medieval thought that were susceptible of more systematic development. It was the age of the foundation of universities and of religious orders. Among these new orders were two that specially influenced the universities, the Dominicans or Black Friars founded at Toulouse in 1215 by the austere and orthodox Dominic (1170–1221), and the Franciscans or Grey Friars founded in 1209 by the gentle and loving Francis of Assisi. The name of Dominic is associated with the terrible extermination of the Albigenses, and the Dominicans, whose title was paraphrased as *Domini canes*, 'hounds of the Lord', set themselves to the strengthening of the doctrine of the Church and to the extirpation of error. The activity of the Inquisition was one of the less edifying interests of the 'hounds' of whom Torquemada was a specially unamiable representative. The work of the Franciscans led up more clearly to the scientific revival. During the thirteenth century these two orders provided most of the great university teachers, who occupied themselves in marshalling the new knowledge and making it more accessible. Alexander of Hales (d. 1245), Robert Grosseteste (d. 1253), and

The Failure of Knowledge

Roger Bacon (d. 1270) were Franciscans, Albertus Magnus (1206–80) and St. Thomas Aquinas (1227–74) were Dominicans. A foremost influence in the revival was the recovery of the writings of Aristotle. It was the interpretation of these works by a few great thinkers that gave to Scholasticism its essential character. The first scholastic to be acquainted with the whole works of Aristotle was ALEXANDER OF HALES. ALBERT was the first who reduced the whole philosophy of Aristotle to systematic order with constant reference to the Arabian commentators, while ST. THOMAS AQUINAS remodelled the Aristotelian philosophy in accordance with the requirements of ecclesiastical doctrine. As time went on, the works of Aristotle, at first represented in translation from Arabic, became partially accessible in renderings direct from the Greek. A very important translator from the Greek was the Dominican WILLIAM OF MOERBEKE (d. 1286), who was in close contact with St. Thomas.

It is remarkable that the process of codifying the new knowledge derived from the Arabic, involving as it did a rapid development in the whole mental life, did not early give rise to a more passionate and more conscious faith in the reality and value of progress in knowledge. The test of such faith, so far as nature is concerned, must be the direct appeal to nature. Yet there is very little evidence of direct observation of nature in the great physical encyclopaedias of the thirteenth century, such as those of the Dominican VINCENT OF BEAUVAIS (1190–1264), or of the Franciscan BARTHOLOMEW THE ENGLISHMAN (c. 1260). The explanation is that the medieval mind was obsessed with the idea of the world as mortal, destructible, finite, and therefore completely knowable in both space and in time and as being, at once, both fully knowable and not worth knowing. Hear St. Augustine:

'Men seek out the hidden powers of nature, which to know profits not and wherein men desire nothing but knowledge. With the same perverted aim they seek after magic arts. . . . As for me, I care not to know the courses of the stars, while all sacrilegious mysteries I hate' (*Confessions*, x. 35). 'Even if the causes of the movements of bodies were known to us, none would be important except such as influence our health. But since, being ignorant of these, we seek physicians, is it not clear that we should rest content to be ignorant of the mysteries of the heavens and the earth' (*De fide*, 16).

Thus medicine is the one science that St. Augustine would allow. Is it wonder that medicine had deteriorated into mere traditional drug lists until the Arabian revival? In the Latin West during the Middle Ages the motive for detailed *research*, in our modern sense of the word, was absent.

One great Islamic philosopher there was, Averroes (p. 139), who took another view of the universe, denying it to be finite, at least in time. His works were available in Latin, but the great ecclesiastics set their faces against him, though he was widely and illicitly read. His theories were adopted mainly by Jews and by Latins with heretical leanings.

5. *Main Personalities of Scholastic Science (Thirteenth century).*

The medieval world thus knew nothing of that infinite sea of experience on which the man of science nowadays launches his bark in adventurous exploration. Medieval science tended to the encyclopaedic form. The task of the writer of the encyclopaedia was to set forth such a survey of the universe as would be in accord with spiritual truth rather than to reveal new truths or new relations. The framework on which this scheme was built was Aristotle, largely as conveyed by commentaries upon his works. Yet it affords a reflection on the incompleteness of all philosophical systems that the great teacher and systematist, ALBERTUS MAGNUS (1206–80), who perhaps more than any other man was responsible for the scholastic world-system, was among the very few medieval writers who were real observers of nature. It is, after all, in the very essence of the human animal to love the world around it and to watch its creatures. 'Throw out nature with a pitchfork and back she comes again.' Albertus, scholastic of the scholastics, drowned in erudition and the most learned man of his time, has left us evidence in his great works on natural history that the scientific spirit was beginning to awake. As an independent observer he is not altogether contemptible, and this element in him marks the new dawn which we trace more clearly in his successors.

Contemporary with the Dominicans, Albert (1206–80) and St.

The Failure of Knowledge

Thomas Aquinas (1227–74), were several Franciscan writers who form the earliest group with whom the advancement of knowledge was a permanent interest. These men were the first consciously forward-looking thinkers since antiquity. The most arresting of them was ROBERT GROSSETESTE (*c.* 1175–1253), Bishop of Lincoln. Grosseteste determined the main direction of physical interests during the thirteenth century. He knew something of the action of mirrors and of the nature of lenses. It would appear that he had actually experimented with lenses, and many of the optical ideas of Roger Bacon were taken from his master. The main Arabian source of Grosseteste was a Latin translation of the mathematical work of Alhazen (p. 136). The great Bishop of Lincoln was an enthusiastic advocate of the study of Greek and Hebrew and an important forerunner of the Revival of Learning.

An important writer was the Pole WITELO (fl. 1270), an acute mathematical investigator and writer who worked in northern Italy and wrote a commentary on Alhazen. The Franciscan Roger Bacon was largely dependent on Witelo for his optical views. Another optical writer dependent on Alhazen was the English Franciscan JOHN OF PECKHAM (*c.* 1220–92), who became Archbishop of Canterbury. His works exhibit some mathematical skill, and one of them continued to be printed in the seventeenth century after the appearance of the writings of Kepler and Galileo!

The greatest figure in medieval scientific thought is unquestionably ROGER BACON (1214–94). He was a Franciscan who taught at Paris and Oxford. He was essentially an encyclopaedist who realized better than most the urgent need for the enlargement of the basis of knowledge, especially in connexion with accurate knowledge of language and the collection and collation of scientific data. In setting forth these needs he made an appeal, verbose, diffuse, yet definite, for the encouragement of the experimental spirit. He was not himself an experimenter or mathematician, but he clearly saw that without experimentation and without mathematics, natural philosophy is but verbiage.

Perhaps Bacon's greatest claim on our attention is that he recognized the usefulness of natural knowledge, foreseeing man's control of nature set forth more clearly, three and a half centuries later, by his great namesake Francis (p. 226). Vaguely, too, he foresaw a number of important modern scientific ventures, flying, the use of explosives, circumnavigation of the globe, mechanical propulsion, &c. A single anticipation of this kind would hardly deserve mention, but the convergence of so many

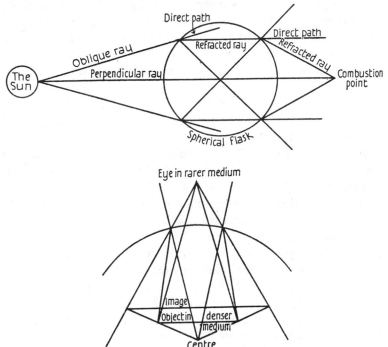

FIGS. 53 and 54. Roger Bacon's diagrams of the paths of rays through a spherical glass and through a plano-convex lens.

in one head is impressive. Specially noteworthy—not so much for their originality as for their clarity—are Bacon's excursions into optics. He understood the nature of refraction and grasped its implications for curved surfaces. He thus attained to an approximately accurate view of the path of the rays in a burning-glass and he had more than an inkling of the mode of action of convex lenses. He seems to have been the first to suggest the use of lenses for spectacles and, perhaps, from hinting at the combination of lenses can be regarded as the progenitor of optical apparatus.

The Failure of Knowledge

Despite all this Bacon must not be considered as a man born out of his time. On the contrary, he was in many ways very typical of the scholastic movement and an important link in the chain of scholastic scientific development. In especial, so far from seeing any opposition between science and religion, he regarded the advancement of science as important for the support of religion. That he was frequently in trouble with his superiors there can be no doubt, but to suggest that these differences were caused by his scientific views is not only to go beyond the facts but beyond all probability.

During the century after Bacon, though his other works were still at times studied in the schools, it happened that for a variety of reasons mathematics and philosophy, in which he was chiefly interested, fell into abeyance. In this interval the chief advances were made by medical men of whom the last half of the thirteenth and the first half of the fourteenth century exhibit an especially brilliant group. Bologna and Montpellier were the centres at which this progress was made.

Bologna had possessed a medical school since the twelfth century, and had inherited the learning of Salerno. At Bologna surgery may be said to have been born again with ROGER OF SALERNO (c. 1220) and his successor and faithful follower ROLAND OF PARMA (c. 1250), who link the new 'Arabic' medical movement with the old that had survived in southern Italy (p. 142). At Bologna, above all, the later thirteenth century saw established a regular tradition of anatomization. This was expounded by MONDINO DA LUZZI (1276–1328), whose work, despite its practical character, was based on translations from the Arabic text of Avicenna. The *Anatomy* of Mondino became the general text-book of the subject in the later Middle Ages. By the fourteenth century the practice of dissection of the human body had become well recognized in several universities.

At the end of the thirteenth century the ancient foundation of the medical school of Montpellier was coming to the fore. The Catalan ARNALD OF VILLANOVA (c. 1240–1311), one of the most remarkable personalities of medieval medicine, taught there.

158

The Middle Ages: Theology, Queen of the Sciences

Arnald was not only the earliest modern exponent of the Hippocratic method of observing and carefully recording actual cases of disease, but he also deeply influenced alchemy. That study was effectively of Arabian origin so far as the Western world is concerned (p. 132). It begins in 1144 with the translation into Latin by Robert of Chester (p. 148) of an important alchemical work by MORIENUS ROMANUS, a contemporary Arabic Christian of Jerusalem who derived it from an earlier Arabic source. Alchemy had taken its rise with a real effort to understand the properties of metals, prompted by the hope of transmuting the baser into the more precious (p. 132). Like other medieval studies, it became linked with astrology. Thus the 'seven metals' were each controlled or influenced by one of the 'seven planets' much in the same way as were the organs of the human body (p. 152).

Of such ideas, Arnald was a prolific exponent. He had direct access to both Arabic and Hebrew and had personal relations with both Moslems and Jews. A student at Naples and Salerno, a traveller in Italy, Sicily, France, and Spain, he served as medical adviser to the Papal Curia both at Rome and Avignon, and was employed as ambassador on more than one special mission. Arnald influenced politics no less than learning and ended his adventurous life at sea.

Astronomy—which cannot at this stage be distinguished from astrology—was certainly the main scientific interest of the scholastic age. The practical results of scholastic astronomical activity are, however, pitifully meagre. Western knowledge of astronomy was largely based on the activity of King ALFONSO THE WISE (1223–84) of Castile. He collected at Toledo a considerable body of scholars, mostly Jews, who calculated a set of astronomical tables (1252). These *Alfonsine tables* spread rapidly through Europe. They contain few new ideas, but several numerical data, notably the length of the year, were calculated with very remarkable accuracy. Alfonso is also responsible for a vast encyclopaedia of astronomical knowledge compiled by a similar group from Arabic sources.

The standard astronomical text-book of the scholastic period was by the Yorkshireman JOHN HOLYWOOD (Sacrobosco, died 1250) who was long a teacher at Paris. The work was universally

159

popular, exists in numerous manuscripts, and was translated into most European vernaculars. It contains, however, no new or original element and is put together from translations of the works of Albattani and Alfargani. Holywood wrote also a book on arithmetic, or rather 'algorism' (p. 147). It was extremely popular and did more to introduce the Arabic notation than any other. Both the astronomy and arithmetic were very frequently printed.

Apart from the *Alfonsine tables* the best astronomical work of the period is that of the French Jew LEVI BEN GERSON (1288–1344). His great astronomical treatise is essentially an attempt to demonstrate the falseness of the prevalent *homocentric* theory (p. 152). It is, in a sense, a return to Hipparchus (p. 76) and a predecessor of Copernicus (p. 179). It was written in Hebrew, but part of it, under the title *The Instrument that reveals Secrets*, was translated into Latin in 1342 by order of Pope Clement VI. The instrument is 'Jacob's staff'. This well-known surveying implement was invented by Levi's countryman and co-religionist JACOB BEN MAKIR (died 1308).

After medicine, alchemy, and astronomy, the practical sciences in which the West exhibited activity in the Middle Ages were botany and optics. Botany was always studied in connexion with medicine. No advance was made in the use of drugs save what was borrowed from the Arabs. There is, however, some indication of a revived interest in nature in the graphic representation of plants. Numerous optical texts exhibit a certain advance in ideas. Nevertheless, neither any of the Latin texts nor even all of them together are equal in value to the great work of Alhazen (p. 136) that itself became available in Latin about the middle of the thirteenth century.

In pure mathematics the original achievement of the scholastic age was small. There was, however, a borrowed element that was to prove of high significance. At the end of the twelfth century a merchant LEONARDO OF PISA (*c.* 1170–*c.* 1245) travelled for commercial purposes in the East and especially in Barbary. There he learnt of that use of Indian numerals in which the value of a digit depends on its place in a series. It is the ordinary method of numeration that we now employ. In 1202 Leonardo produced his famous *Book of the Abacus*, in which he advocates this system with great skill. It is the first book by a Latin Christian that employed this system and is the essential source of our modern system, which, however, was extremely slow of general adoption. Other works of Leonardo were much more original, but being before their time had less influence. He was undoubtedly a mathematician of extraordinary ability, but his positive contributions are as nothing compared to his importance as the carrier of the new method. A much more popular work than Leonardo's that employs the 'Arabic' numerals was the *Algorismus* of John Holywood (died 1250). It appeared about 1240, was very frequently printed in the fifteenth and sixteenth centuries, and was still being reprinted in the seventeenth century.

It is one of the puzzles of history that the great improvement represented by the 'Arabic' as against the Latin system of numerical notation took three centuries to gain general acceptance. The scholastic age was over before the modern system came into general use.

Many attempts have been made to rehabilitate the intellectual achievement of the Middle Ages. So far as science is concerned they have been unsuccessful. There is no reason to reverse the decision that in this domain the period is one of intellectual degradation.

VI. THE REVIVAL OF LEARNING

The Rise of Humanism (1250-1600). The Attempted
Return to Antiquity

1. *Humanism.*

THE advent of Catholic philosophy is one of the most impressive events in the whole history of thought. This great effort to rationalize Christianity is closely linked with the recovery of the Aristotelian texts.

Until the thirteenth century the only works of Aristotle available were those on logic. These had been turned into Latin from Greek by Boethius in later antiquity (p. 127). Early in the thirteenth century versions from the Arabic associated with the commentaries of the Moslem philosopher Averroes (p. 139) began to circulate. The centre of the intellectual world at that period was the University of Paris. There the reading of these Averroan interpretations of Aristotle met with ecclesiastical opposition. This was, however, lifted by the middle of the thirteenth century, perhaps because versions less coloured by the Averroan outlook had become available. The architect of Catholic philosophy, the Dominican St. Thomas Aquinas (p. 154), was able to work largely on versions of Aristotle prepared directly from the Greek by William of Moerbeke (p. 154) and others. These medieval Greek versions remained in use until the end of the fifteenth century.

This summary of the knowledge of Aristotle in the thirteenth century requires some explanation. Before the days of printing a work seldom replaced completely another that was actually in circulation. Manuscripts were too expensive to jettison. Libraries were small, scholars conservative and uncritical, catalogues inadequate. The better or newer versions did not commonly drive out the worse or older. The two generally continued in use, some at one centre, some at another, and often both at the same seat of learning.

So far as science is concerned the versions of Aristotle, and the Aristotelian commentaries and interpretations in most common use, long continued to be those from the Arabic and *not* those from

162

the Greek. The hold of the Arabic-Latin versions began to be somewhat shaken by two important events; (*a*) the rise of humanism and (*b*) the advent of printing. But though these versions of Arabic origin had now competitors they were by no means discredited. Indeed 'Arabist' versions retained their supremacy through the fifteenth and sixteenth centuries. Even in the seventeenth century they were still in use in universities where the old Aristotelian philosophy flourished. The true and ultimate Arabist defeat was not the work of the Greek scholars who ranged themselves under the banner of 'Humanism'. It was rather the men of science, adherents of the new 'Experimental Way', who swept away the whole medieval philosophic approach—whether based on Greek or Arabic or Aristotle or Averroes. Their triumph was not fully apparent till the eighteenth century. There are backward centres where it is not complete even now.

In the nineteenth century scholarship itself was transformed by the experimental method. Adepts in that method came at last to study modern critical versions of the Aristotelian corpus. Then, and in the fullness of days, the scientific powers of the great teacher came to be properly appreciated. The beauty and symmetry of his mind appeared as never before and are not likely again to escape the historian of science.

We turn now to consider the small beginnings of a true appreciation of ancient science. The process is wrapped up with the advent of the versions of scientific works prepared from the Greek. One of the first to appreciate these was the heretical PETER OF ABANO (1250–1318). He had a knowledge of Greek, acquired at Constantinople, and he translated works from that language. He professed medicine at Paris and later at Padua in the generation after that in which the newly won Aristotelian works on physics had entered the curriculum. He earned a reputation as a magician, and only his natural death saved him from an unnatural one at the hands of the Inquisition.

The best-known work of Peter, the *Conciliator*, expresses his mediation between the new Greek and the old Arabist school. It shows traces, too, of wider contacts, for from it we learn that he had met the great traveller Marco Polo (*c.* 1254–1324). He was much less conservative than most medieval writers on scientific themes. Among Peter's views most worth record may be mentioned his statements that the air has weight, that the brain is the source of the nerves, and the heart the source of the blood-vessels —novel ideas in his time. He made a remarkably accurate measure of the length of the year as 365 days, 6 hours, 4 minutes.

With the fourteenth century appeared a great movement the hand of which is still heavy on our own day. The ancient classics began to be recovered and Greek began to be studied. Historians have perhaps linked the 'humanistic' movement too intimately with a knowledge of the Greek language. Instances of familiarity with that language in the West can be adduced far back into the Dark Age (e.g. John Scot Erigena, *c.* 850), while many of the greatest of the humanists, including PETRARCH himself (1304–74), were without any facility in Greek. It is worth noting, too, as linking humanism with the Middle Ages, that Petrarch's epistolary style was still moulded on St. Augustine rather than on Cicero.

The backward-looking habit, strong in man from his nature and strengthened by Christian teaching, was yet further enforced by the humanists. From Petrarch onward the humanist was brooding on the past that had been Greece and Rome. Seeking to penetrate the dark shadows of what was now recognized as a 'Middle Age', the humanist tried hard to discern the antiquity that was beyond. And as he strained his eyes another vision, a reflection perhaps of himself, came sometimes to him. In the cloud-land of the past he caught or thought he caught a glimpse of what was to come—nay of what was in the act of becoming. And then again the vision would be clouded over by that terrible erudition, which, in the absence of general ideas, has been and is one of the enemies of science.

Even in the thirteenth century Roger Bacon and a few isolated souls had had this double vision, but for a whole school to possess it was something new. In his *Book of Memorable Things* Petrarch says outright, 'Here stand I as though on a frontier between two peoples, looking both to the past and to the future.' While studying the classics some of these men were also forging new intellectual weapons by developing those national vernaculars that have made possible modern literature, modern philosophy, and modern science. It is no mere coincidence that Boccaccio (1313–75), friend and contemporary of Petrarch, should have been at once the first modern literary man to study Greek and the first great master of Italian prose.

Italy was the birthplace and nursery of humanism. We would emphasize that save for reference to the one supreme poet in their own tongue, DANTE (1265–1321), the backward gaze of the Italian humanist is always fixed on the more distant classical past, not on the nearer period that came to be regarded as an abyss across which he sought to reach back to the thought of antiquity. To him the abyss seemed real enough and dark enough. It stood for the period during which the sweet Greek literature had been forgotten. Even in this new age it could be understood by few except in Latin dress, and the work of translation and interpretation remained a specialist's occupation. To the end of the fifteenth century an effective knowledge of Greek continued to be rare even among the learned. Some of the most important philosophical teachers even of the sixteenth century were still quite without it.

The great influence of the masterpieces of Greece, therefore, was then as now something indirect, often conveyed through translators and special interpreters; something esoteric, the full intricacy of which was shared only by a few adepts; a subtle thing that influenced men's way of thinking rather than the actual content of their thought. The mere capacity for translation from the Greek goes back very far. It was not simply the discovery of the actual Greek language which brought about the revival of letters. How, then, can we account for the change of heart that came over the world when humanism was born? Or is that change of heart but an illusion, a difference of degree rather than of kind, in a world where everything is in a state of becoming?

Some answer to this absorbing question we may glean by comparing the earlier Greek works which came to the West with those of later advent. The general character of the earlier translations was determined by the outlook of a world becoming ever more deeply Arabicized. Islam, the inheritor of antiquity, entered into the enjoyment of its legacy with great spirit, but with a taste already fixed. The ancient literary and artistic works were debarred to the Moslem scholar. Homer and Hesiod, Sophocles and Euripides, Greek Art and Greek Architecture, were chapters as closed and forbidden to Islam as to early Christian Europe. It was the philosophical, the scientific, the mathematical, the medical works that made an appeal. The bulk and number of these gave sufficient material for thought and gave an illusory impression of completeness with which Islam long rested content.

It was these very works, to which the world of Islam clung, that were the first to be rendered into Latin from the Arabic. The Latin taste being thus determined, the mere knowledge of Greek wrought little change. It was works similar to those already rendered accessible from the Arabic that were the first to be turned into Latin direct from the Greek, for, in fact, Byzantine literary taste was not very different from Arabic taste. The texts were merely improved by direct access to the tongue in which they had been written, but they were still the same philosophical, medical, mathematical texts.

Such material—and it is bulky and intricate enough—represents the Western access to Greek wisdom before the fourteenth century. It does not lack quantity—it lacks life. They err who think the discovery of the humanists was the Greek language—here the humanists were but followers where others had been pioneers. It is something much deeper and more fundamental which they have handed on, something the nature of which they hardly knew and the meaning of which they missed—and perhaps still miss.

The humanists discovered the literary works of antiquity. In them they became absorbed to the exclusion of all else. Their eagerness passed into a literary vogue, and cast the blight of a purely literary education on the modern world. The barren striving after form as distinct from substance, the miserable imitativeness that is an insult to its model, these features, exhibited typically in the literature of the late Empire, were repeated by the humanists as they have been often repeated in modern times. They still remain the curse of our educational system. The importance of the humanist is not that he gave us the knowledge of a language, nor that he gave us an insight into the life of antiquity. What the humanist really gave was a something which, added to the heritage already there, made possible a completer reconstruction of the Greek spirit. That reconstruction, indeed, he was himself never able to make. It was the succeeding generations that made it for themselves. With that reconstruction Greece lived again, the modern world was ushered in, and modern science, art, literature, and philosophy were born. It is an illuminating reflection, not without bearing on our present state, that both the medieval heritage of Greek science and the Renaissance heritage of Greek literature proved barren by themselves. It was not until the one fertilized the other that there was vital growth.

Modern thought, modern science, modern art, modern letters are offspring of that union. Let us put from our minds the time-worn fallacy that they are the virgin births of one of these elements alone. Men accomplished alike in the arts and in the sciences, Leonardo (p. 172), Vesalius (p. 177), Galileo (p. 195), are more truly the heirs of Plato and Aristotle than are the men who spent their lives in editing the works of these giants of old. It is literature, art, and science, not classical scholarship, that has inherited the legacy of ancient wisdom.

2. *Recovery of the Ancient Scientific Classics.*

An event of primary importance for the history of science as for that of all branches of culture was the introduction of the art of printing into Europe about the middle of the fifteenth century. There are certain aspects of early printing in connexion with science to which attention must be directed.

(*a*) We now use the printed page to express our views on current matters. In science we mark a discovery by its first publication, while both publishers and readers demand of a new book that it should contain at least something new. But in the early days of printing the press was not thus employed. The Bible and other sacred writings were the first to be printed. Then followed the works of medieval authors of theological authority. Next medieval treatises on law and especially ecclesiastical law, occupied the press, and were followed by medieval medical texts. The writings of classical antiquity came later. Only a very small proportion of early printed books are by contemporary writers. Almost all are either medieval or ancient texts or compilations therefrom. The custom of using the printed page to record one's own views or experiences crept but slowly into practice.

(*b*) In the process of recovery of the classical originals the attention of scholars was first directed to works of literary merit. Scientific treatises appealed to a much smaller audience and, moreover, few scholars were adequately equipped to deal with them. Thus the revival of classical science came later than the revival of other sections of classical literature.

(*c*) There has arisen a curious misconception of the importance of the classical literature in the fifteenth and sixteenth centuries. The great influence of the Revival of Learning on the subsequent history of thought and of education has distorted our view of fifteenth-century cultural interests. Even in the great days of humanism, in the later fifteenth century, Greek was a very rare accomplishment. Those who had any facility in it were extremely few, even among the best-educated class. Right through the sixteenth century and even into the seventeenth century, the overwhelming mass of published philosophical and scientific literature was still of the medieval type.

(*d*) The publication of Greek scientific writings had little influence unless or until such works began to appear in Latin or vernacular translations. The humanists seldom had adequate scientific equipment and the men of science seldom had adequate linguistic equipment.

Bearing these matters in mind, it is interesting to follow the chronological course of the appearance in print of the classical scientific works of antiquity. For the progress of science at the time it was the printing of these works rather than the discovery of their manuscript texts that was of chief significance.

The earliest scientific classics to be printed were naturally those of the Latins. The first was the *Natural History* of Pliny, which appeared at Venice as early as 1469. But Pliny, it must be remembered, was in no sense 'recovered'. On the contrary he had never ceased to be read throughout the Middle Ages (p.128). The work was in fact so familiar that the Venetian printer did not think it worth while to attach the name of an editor to his work. Throughout the sixteenth century Pliny was as popular as during the Middle Ages and was very frequently reprinted. In 1601 his *Natural*

Rise of Humanism. *Attempted Return to Antiquity*

History was translated by Philemon Holland into English and was the second work of ancient science to appear in that language, the first being Euclid (1570, p. 170).

Following on Pliny were editions of Varro (p. 97, Rome, 1471), of a collection of agricultural writers (Venice, 1472), and of the poem of Manilius (Nuremberg, 1472). These were all of practical application. Manilius is interesting as the earliest classical scientific treatise to appear outside Italy. It was printed at the private press of Regiomontanus (p. 171). The interest in it is explained by its astrological content, for astrology had become part of the University curriculum. Lucretius followed in 1473 (Brescia). But Lucretius, as we have seen (p. 95), is not properly speaking a scientific writer. Celsus (p. 107), again of immediate practical value, followed some years later (Florence, 1478). The medical work of Celsus was thus the first technical scientific classical work to appear. It had been unknown in the Middle Ages and was a real discovery. It began at once to influence the practice of medicine. The architectural writers, Vitruvius, Frontinus, and Vegetius (Rome, 1486–7)—again practical works—complete the short list of early printed ancient Latin science.

The Greek writings that deal with the true abstract sciences are both more numerous and have a more complex history. We may first note how backward was the treatment of the Aristotelian scientific corpus. For the most part the Renaissance reader was content with the medieval Latin versions mainly from Arabic (p. 162). The first 'modern' translation and the first important scientific book to be printed was the Latin version by Theodore Gaza (1400–78) of the three great Aristotelian biological treatises (Venice, 1476).

Actual Greek type was hardly used before 1476, and it was near the end of the fifteenth century before the scholar-printer Aldo Manuzio (1449–1515) produced an adequate edition of the Greek text of Aristotle and Theophrastus (Venice, 1495–98). He added to his services by issuing the Greek text of Dioscorides (p. 89, 1499), of Pollux, a classical scholar who determined Renaissance anatomical nomenclature (1502), and of Strabo (p. 100, 1516). Aldo's successors in the 'Aldine' firm were responsible for the first Greek editions of Galen (1525) and Hippocrates (1526).

The Revival of Learning

Very influential for the whole course of Renaissance science were the editions of Euclid. He first appeared in Latin dress at Venice in 1482. Editions continued to flow from the press throughout the sixteenth century. The first edition in Greek appeared at Basel in 1533 and the first in English in London in 1570.

A work that had a large share in fixing the geographical ideas of the Renaissance was the *Geographia* of Ptolemy, which first appeared in Latin at Vicenza in 1475 and in Greek at Basel in 1533. The maps illustrating early editions of Ptolemy are most interesting (p. 88). Even more influential was the *Almagest*, which was first printed in Latin at Basel in 1538 and very frequently at later dates. The works of Ptolemy in Renaissance versions are common in comparison to those of early Greek mathematicians and astronomers. Thus a collection of Archimedes was not made until 1544 (Basel) and was not reprinted till the seventeenth century.

The most frequently printed of the ancient scientific works at this time were undoubtedly the medical. Hippocrates, Dioscorides, Galen, and others appeared in scores of editions in Greek, Latin, and the vernacular throughout the sixteenth century. They were very generally studied, and in conjunction with the Arabic medical writers Rhazes, Mesue, Avicenna, and Albucasis they came to provide the basis of the actual medical practice of the age.

3. Scientific Atmosphere of the Early Renaissance.

The humanists as a class exhibited little sympathy with the scientific outlook. Their interests were literary; their peculiar aversion was the Arabist tendency of the age that they were leaving behind. Arabism expressed itself rather in comment than in development of ancient scientific and philosophical themes. In the movement typified by Roger Bacon in the thirteenth century a new element had entered (p. 156). That movement had fallen into the background after Roger's death. It had not entirely died, but it had remained as the seldom expressed faith of a small band of philosophically minded recluses. At last faith in the appeal to nature found more open expression. With the fifteenth century, discontent with the entire medieval scientific scheme becomes more generally obvious. The idea that it may be possible to adjust theory by experiment again comes to the fore.

Rise of Humanism. Attempted Return to Antiquity

The earliest open suggestion is made by a man of lofty philosophic genius and penetrating scholarship, the Rhinelander NICOLAS OF CUSA (1401–64), who became a cardinal and made a fruitless attempt to reform the calendar. Nicolas was groping towards a philosophical basis for the experimental method. He records a careful experiment on a growing plant—afterwards pirated by the seventeenth-century writer van Helmont (see p. 231)—proving that it absorbs something of weight from the air. This is the first biological experiment of modern times, and incidentally the first experimental proof that air has weight. Nicolas wrote a book on the employment of the balance in physical experimentation. In more than one of his works he showed that he knew how to apply the experimental method in detail, and he suggests in outline many investigations which were not taken in hand until the time of Galileo, 150 years later. His theoretical views led him to a belief that the Earth is moving and the universe infinite, though he attained to no formal astronomic theory. He certainly influenced Bruno (p. 185) and gave philosophical assent to the proposition that the universe is boundless in both space and time.

The tradition of the combination of scholarship and observation that Nicolas had practised was carried on by several astronomers in the second half of the fifteenth century. For much of this we are indebted to the far-sightedness of another cardinal who, though born long before Nicolas, died long after. This was JOHANNES BESSARION (1389–1472), a Greek by birth, who was equally anxious to aid the progress of astronomical knowledge and to diffuse Greek literature in the West. Bessarion's friendship, extended to two German students in Italy, Purbach and Regiomontanus, made possible their work which formed the foundation of that of Copernicus.

GEORG PURBACH (1423–61) followed with great avidity the study of Ptolemy's *Almagest* (p. 148). He died prematurely and had only translations from the Arabic on which to base his work. He improved on his original, however, by calculating a table for every ten minutes, using sines instead of chords.

Johannes Müller (1436–76) of Königsberg (='king's mountain'), usually known from his birthplace in Bavaria as REGIOMONTANUS, lived hardly longer than Purbach. He had, however, the good

The Revival of Learning

fortune to have Greek originals on which to work. He completed his predecessor's digest of Ptolemy. He also produced the first systematic treatise on trigonometry[1] and a table of sines for every minute and of tangents for every degree. His astronomical tables were used by Columbus. Regiomontanus died at Rome, whither he had been summoned by the Pope to aid in the long-contemplated reform of the Calendar. This, in the event, was deferred for more than a century. The important works of Regiomontanus were only published after his death.

The Renaissance of Letters was contemporary with the Renaissance of Art, which had its reaction upon scientific thought. The great painters had begun to study nature more closely. Antonio Pollaiuolo (1428–98) and Andrea del Verrocchio (1435–99), among others, made careful investigations of surface anatomy, while the exquisite figures of plants in the pictures of Sandro Botticelli (1444–1510) mark him out as a very accurate observer. There was, however, one artist of the time who takes a quite peculiar place among students of nature. LEONARDO DA VINCI (1452–1519) stands for many as the turning-point of the Renaissance into modern times. The ingenuity of his ideas, the marvellous rapidity of his insight, the sureness of his intuitions, the exactness of his observations, the extreme versatility of his extraordinary genius, made earlier students place him in an isolated and almost superhuman position. His very limitations increase the apparent gulf which separates him from other men, and hamper us in our comprehension of him. To understand his scientific work and its fate we must recognize his defects.

Leonardo's great limitations were literary and linguistic. He hardly acquired even an elementary knowledge of Latin, and he exhibited no power of literary expression. The vernacular that he employs is that of a Florentine shopkeeper of the lower class. He created no great phrase or saying. His sentences are often ungrammatical and frequently unfinished. In a literary sense he was incoherent. The very rush of his ideas obstructed the channels for their expression. He might have said, with Petrarch,

[1] It was not printed till 1533, that is 57 years after its author's death.

Rise of Humanism. *Attempted Return to Antiquity*

E l'amor di saper che m'ha sì acceso
Che l'opera è retardata dal desio.

My love of knowledge so inflamed me
That my work was retarded by my very desire.

Among the great artists he was notorious for the smallness of his output and for the extreme slowness with which he worked. Did his art consume the major part of his energy and his thoughts? His private papers contain evidence not only of a unique scientific insight but of an industry which is almost incredible. He covers the whole field of science from mathematics to physiology, and there is nothing that he touches which he does not illuminate. Thus he presents us with a model of a flying machine and suggestions for a helicopter and a parachute and, interested in the problem of flight, he analyses the nature of the flight of birds in a way that has only been surpassed during the last few years. He designed a parabolic compass on a principle adopted only late in the seventeenth century. He hints at a heliocentric view of the world. He has drawings of quick-firing and breech-loading guns. He makes many ingenious suggestions for engineering apparatus. He has mastered the theoretical principles of perspective. He sets forth some of the homologies of the vertebrate skeleton. He has passages which suggest the laws of motion. His anatomical and embryological studies were not passed in certain respects for hundreds of years.

Marvellous as were the attainments and achievement of Leonardo, he does not occupy a completely isolated position. Others of his age rival him both in versatility and penetration. Thus his German contemporary ALBRECHT DÜRER of Nüremberg (1471–1528), apart from his achievement as an artist, studied the details of human anatomy, made a profound and painstaking investigation of the proportions of the human body at different ages and in the two sexes, was an exceedingly close observer of the habits and growth of animals and plants, conducted experiments in optics, perspective, and the properties of sound, had a remarkable command of the mathematics of his day, and, in his great drawing, *Melancholia*, set forth in allegorical form the changes in thought and attitude with which the age was instinct. Dürer worked long in Italy. He is a German, but all that he does and says is touched by the spirit of the Italian renaissance.

Dürer's work was done under strong Italian influence. This is less true of the Swiss writer Aureolus Philippus Theophrastus Bombastus von Hohenheim, more compendiously known as PARACELSUS (1493–1541). He was a person of violent, boastful, and repellent temper, whose iconoclasm, garrulous and often incoherent though it was, probably did something to deter men from the worship of the old idols. His symbolic act of burning the works both of the Greek Galen and of the Arab Avicenna, as an introduction to his lecture course at Basel, was meant to typify the position of the independent investigator. A writer of excessive obscurity, an obscurity of language and of form as well as of thought, very few claim the privilege of penetrating to his full meaning. It is unfortunate that these few have developed a vagueness of expression and an obscurity of style that rival those of their original. There is, however, a general agreement among the saner Paracelsists that their hero did in a vague sense foreshadow the 'New Instauration'. His aim was to see the world in the 'light of nature'. That light of his is dimmed for us because of his extreme gullibility in some matters, his violence and self-contradiction in others, and his involved and mystical presentment in all. 'Nature' included for him the influence of the stars upon the lives of men and many other relationships then generally credited and now universally discredited. He believed still in a relation of microcosm and macrocosm—as in a residual sense we all do—but his free modification of that theory may have helped to pave the way for its rejection in the generation which followed.

It is not easy to ascribe any positive scientific contribution to Paracelsus. He did, however, give currency to one important modification of Aristotelian doctrine whereby alchemy was deflected into a direction which led to chemistry. He held that, apart from the 'four elements' of Greek philosophy, there were certain proximate principles that gave matter its distinguishing characteristics. The principles were three in number. Unfortunately the names that he selected for these, 'mercury', 'sulphur', and 'salt', were already in use for definite substances. Thus confusion was worse confounded. By *mercury* he means the nature, principles, or characteristics which are common to the metals; by *sulphur* he means the power of combustibility and the essence of changeability and chemical impurity; and by *salt* the principle of fixity and of resistance to fire. This was an advance in the sense that these principles relate to experience and do not demand that nature must of necessity be simple and accord to some rigid scheme.

A much more coherent author than Paracelsus was the German mining engineer GEORG AGRICOLA (1490–1555) 'the father of mineralogy'. In his work *Concerning Metals* ('De re metallica') of 1546 he summarizes from experience the metallurgical knowledge of his day. In an admirable series of illustrations and descriptions he sets out for us the whole technology of mining. It is difficult to say how much of the book is original, but there are a number of devices and of processes that are mentioned by him for the first time. In other works he laid the foundations of physical geography and also devoted considerable attention to fossils, which he regarded as remains of extinct organisms.

The period of activity of Paracelsus represents the beginnings of the modern study of mathematics. The best exponent of the subject was the unprincipled genius JEROME CARDAN (1501–76), whose name is still remembered in 'Cardan's rule' for the solution of cubic equations and the 'Cardan shaft' of the motor-car. Cardan's rule was, in fact, shamelessly pirated from another who had imparted it to him under a pledge of secrecy. Nevertheless, the appearance of Cardan's work on algebra (1545) undoubtedly marks for mathematics the end of the Middle Ages and the openings of a new era.

4. Revival of Direct Study of Nature.

The sixteenth century brought with it a combination of circumstances particularly favourable to certain types of observational activity. The printed page had grown familiar. Books were becoming commoner and were now the recognized means for the conveyance of new knowledge. The many new and strange forms that explorers were bringing back to Europe were drawing attention to the beauty and variety of living things. The medicine of the age laid special emphasis on vegetable drugs, so that physicians

The Revival of Learning

were accustomed to distinguish a large variety of native and foreign plants. The artists also had paid much attention to plants, and several had devoted themselves to the study of their habits and habitats. Lastly, the arts of the woodcut and the copper engraving had been perfected, and there was a number of craftsmen capable of producing admirable illustrations of living things and especially of plants. Thus books began to appear in which plants were portrayed with lively skill. No better botanical figures have ever been produced than some issued from the presses of the sixteenth century.

The special development of plant portraiture began in Germany, the home of printing, where that art had reached a very high standard. OTTO BRUNFELS of Mainz (1489–1534) was the first to produce a work on plants, the figures of which rely wholly on observation (Strasbourg, 1530). The drawings are firm, sure, faithful. It is very interesting to compare them with those of a good modern text-book. The text, however, is befogged by an error from which botanists took long to free themselves. Brunfels identifies his plants—gathered in the Rhineland—with those of Dioscorides, who worked in eastern Mediterranean lands. The equation was impossible and confusion results.

A younger German botanist was JEROME BOCK of Heiderbach (1498–1554), who escaped some of the errors of Brunfels. Bock's careful descriptions of plants and of their mode of occurrence (Strasbourg, 1539) are the first of the kind since Greek times. Only by collating a large number of such descriptions did botanists outgrow the habit of comparing all their plants with those of the ancients.

The most remarkable of the early German botanists was LEONARD FUCHS (1501–66). His botanical work (Basel, 1542), intended as a guide to the collector of medical plants, is a landmark in the history of natural knowledge. Fuchs had a good acquaintance with the Greek and Latin classics, and was, withal, an excellent observer, so that his identifications are supported by adequate knowledge. His woodcuts are of extraordinary beauty and truth. They established a tradition of plant illustration traceable to the present day. Fuchs enjoys a verdant immortality in the beautiful group of American plants, the 'Fuchsias'.

Fuchs arranged his plants alphabetically. He gives us nothing

Rise of Humanism. *Attempted Return to Antiquity*

of classification, hardly anything that can be called plant geography, little concerning the essential nature of plants or of their relation to other living things. His book is, in fact, a 'herbal' pure and simple. Yet by its close observation of details and by their accurate record on the printed page, it may claim a place among the pioneer works of modern science. Fuchs includes in his work an admirable glossary of botanical terms.

Modern plant study thus became effective with this happy combination of humanistic learning, Renaissance art, and the perfected craft of printing. The same is no less true of the study of the animal body. The real father of modern anatomy was the Fleming, ANDREAS VESALIUS (1514–64), whose work brings out this combination admirably.

Even as a boy Vesalius was always observing Nature and dissecting the bodies of animals. He studied first at Louvain in his native Belgium and afterwards at Paris. Both universities were extremely conservative. Anatomical instruction was still medieval and pinned to the texts of Galen. Vesalius was highly successful as student and teacher there, and he became very learned in Galen. Fortunately for himself and for the world he quarrelled with his superiors and decided to seek his fortune elsewhere. He determined on Italy, was appointed professor at Padua (1537), and immediately introduced sweeping reforms.

In the old days of Mondino (p. 158) the professor had dissected on his own account. The successors of Mondino abandoned this difficult and tiring process. They were content to *read* their lectures from the text of Galen, while a *demonstrator* (Latin *demonstro*, 'I point out') indicated the parts to the students. Hence our modern academic titles *Reader* or *Lecturer* (*lego*, 'I read') and *Demonstrator*. The basic reform of Vesalius was to do away with demonstrators and other intermediaries between himself and the object—'to put his own hand to the business', as he called it. His drive was irresistible. In five years he had completed and printed the masterpiece on which his fame is based, and he was still only twenty-eight. He did no further important work. Vesalius's *On the Fabric of the Human Body* (Basel, 1543) is both the first great modern work of science and a foundation-stone of modern biology.

The Revival of Learning

The book opens with a description of the bones and joints, the general classification of which is from Galen. The first bone considered is the skull. It is astonishing to find here an examination in the modern manner of the different shapes of human skulls. Anthropologists to-day attach great importance to these. Skulls are systematically measured and individuals and races classed as broad-headed, long-headed, round-headed. This is exactly what Vesalius does. He follows this matter up by comparing the skull of man with that of certain animals, notably the dog.

Of all the subjects of which Vesalius treats, he is most successful with the muscles. In certain respects his representations of these are actually superior to most modern anatomical figures. Vesalius, with an artist's eye, has succeeded in representing the muscles with their normal degree of contraction.[1] In other words, he has represented living anatomy. This is a more difficult task, and one involving more real knowledge, than any presentation of the details of dead structures. For this reason naturalists still return to these figures of Vesalius and have something to learn from them, although they were prepared 400 years ago.

The account by Vesalius of the structure of the heart has a special interest. The workings of the heart and blood system had always been a puzzle. The current solution was that of Galen, which depends on the supposed existence of pores in the septum between the ventricles (p. 91). Vesalius generally follows the physiological view of Galen. When, however, he comes to the septum between the ventricles he is mystified. He tells us that

> 'The septum is formed from the very densest substance of the heart. It abounds on both sides with pits. Of these none, so far as the senses can perceive, penetrate from the right to the left ventricle. We wonder at the art of the Creator which causes blood to pass from right to left ventricle through invisible pores.'

Thus he was not satisfied with Galen's view. Twelve years later he brought out a second edition of his great book. He has again examined the pits on the septum. This time he says:

> 'Although sometimes these pits are conspicuous, yet none, so far

[1] In fact most of the drawings are not by Vesalius himself, but there can be no doubt that he supervised them in every detail and determined the poses.

as the senses can perceive, passes from right to left ventricle . . . not long ago I would not have dared to turn aside even a hair's breadth from Galen. But the septum of the heart is as thick, dense, and compact as the rest of the heart. I do not see, therefore, how even the smallest particle can be transferred from the right to the left ventricle through the septum.'

This attitude to Galen makes it evident that we are on the eve of a scientific revolution. Men are no longer satisfied with the traditions of the ancients. In this Vesalius was not alone. He was but the first of a whole line of Paduan anatomists that leads on continuously to the great biological awakening of the seventeenth century.

5. *Astronomical Observation and Hypothesis in the Sixteenth Century.*

The astronomy of the earlier sixteenth century exhibits certain activities that mark it off with some definiteness from that of the Middle Ages. The work of Regiomontanus (p. 171) was widely known and was in large part responsible for this.

Leonardo da Vinci (p. 172) about 1510 explained correctly the dim illumination of the surface of the Moon, when the bright part is but a narrow crescent—'the new Moon in the arms of the old'—as due to *earthshine*. It is light reflected from the Earth. His younger contemporary JEROME FRACASTOR of Verona (1483–1543), an able writer on the revived atomism of Lucretius and founder of the modern view of infection, made contributions both to astronomical theory and practice. He observed that the tails of comets are always turned from the Sun. This fact throws light on the nature of these bodies. The French physician JEAN FERNEL (1497–1558) made a calculation of the size of the Earth (1528) accurate within 1 per cent. The fame of all these writers has, however, been wholly overshadowed by that of Copernicus (1473–1543).

The Pole, NICOLAS COPERNICUS (1473–1543), despite the vast change that was introduced in his name into men's ideas, was himself more in the line of such comparatively conservative scholars as Regiomontanus than the more revolutionary Leonardo, Fracastor, or Fernel. He was a student rather than an observer, and

179

he continued to attend university courses until over thirty years of age. He studied at several Italian universities, giving attention to classics, mathematics, astronomy, medicine, law, and theology. It was in Italy that he first discussed the Pythagorean theory with which his name has become associated. Copernicus had skill in painting which suggests that he had that type of visualizing imagination frequently associated with scientific power. He was not at all active as a practical astronomer. He had, it is true, taken a few observations of eclipses and oppositions of planets, but for the most part his results were obtained in the study.

Copernicus tells us that he was induced to seek a new theory of the heavenly bodies by finding that mathematicians differed among themselves on this subject. He had considered the various motions of the heavenly bodies according to the old system, and concluded that some essential factor had been missed. He found his hint in the traditions that had survived of the thought of Philolaus the Pythagorean (p. 21) and of Aristarchus (p. 59).

'Occasioned by this', he says, 'I decided to try whether, on the assumption of some motion of the Earth, better explanations of the revolutions of the heavenly spheres might not be found. Thus assuming the motions which I attribute to the Earth . . . I have found that when the motions of the other planets are referred to the circulation of the Earth and are computed for the revolution of each star, not only do the phenomena necessarily follow therefrom, but also that the order and magnitude of the stars and of all their orbits and the heaven itself are so connected that in no part can anything be transposed without confusion to the rest and to the whole universe.' (Copernicus, Introduction to *De Revolutionibus*).

The new or rather renovated scheme of Copernicus retained much of the ancient theory. It still assumed that the universe is spherical and finite, terminating in the sphere of the fixed stars. It still assumed that the movements of the celestial bodies are always circular and always with uniform volocities. It still invoked epicycles. It still demanded the excentric (p. 77). In fact Milton's description of the Ptolemaic world fits not ill with the attempt to 'save the phenomena' by means of a system of circles and spheres that was made by Copernicus. In *Paradise Lost* the Archangel Raphael tells that

180

Heaven
Is as the Book of God before thee set
Wherein to read his wondrous works, and learn
His seasons, hours, or days, or months, or years.
This to attain, whether Heaven move or Earth
Imports not. . . .

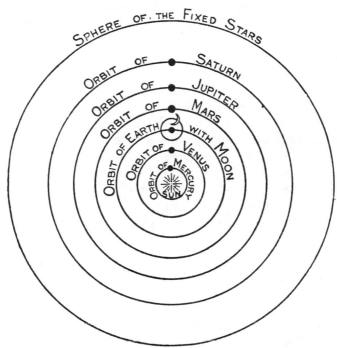

FIG. 55. The Copernican World-System.

and then goes on to the 'conjectures' of those who would
model Heaven
And calculate the stars: how they will wield
The mighty frame; how build, unbuild, contrive
To save appearances; how gird the sphere
With Centric and Eccentric scribbled o'er,
Cycle and Epicycle, Orb in Orb.
(*Paradise Lost*, viii. 70–84.)

The simplicity of the Copernican system—which has been inferred
rather from his famous diagram (Fig. 55) than from his book
itself—is really more apparent than real. Thus while he reduced

the number of circles demanded to explain celestial movements, he still invoked no less than thirty-four.

The immediate influence of the teaching of Copernicus on contemporary thought was, in fact, much less than might be supposed. Notices of it, for a generation or more, are surprisingly few and not always unfriendly. Religion was the main interest of the day. Religion is, by its nature, extremely consèrvative, and any scientific advance of the first magnitude usually disturbs its professors. Nevertheless Christian doctrine, guided by St. Thomas Aquinas, had adapted itself to the Aristotelian system (pp. 150–1). During the Middle Ages the doctrine of a spherical Earth had normally been taught in the schools. A spherical Earth is neither more in accord nor less in accord with the Biblical account than is a world system of which the Sun rather than the Earth is the centre. Christian doctrine accommodated itself to the one; it might have accommodated itself to the other. There were, however, certain extraneous circumstances that intervened in determining the reception of the Copernican system.

One group of these had relation with current religious teaching which was greatly disturbed by Giordano Bruno (p. 185).

A second group had to do rather with the contemporary view of the nature of the physical universe. It was an age that believed in astrology, and astrology had become part of the university curriculum (p. 169).

Now astrology was based on the doctrine that the outer spheres of the universe influenced the inner sphere (Figs. 20, 40). This conception coloured all departments of thought and imbedded itself so deeply in speech that many expressions still current are based on it. 'The scheme was conceived under an evil star', 'His fortune is in the ascendant', 'The seventh heaven of delight', 'He has gone to a higher sphere', 'The British sphere of influence', 'Canst thou bind the sweet influences of the Pleiades' (*Job* xxxviii. 31), 'He has the influenza' are such cases. All the conceptions on which these phrases were originally based—and they covered a large part of life—were disturbed by the Copernican view. Remove the Earth from her central position among the spheres and the whole astrological system becomes unworkable. It is too much to expect such disturbance to be accepted calmly.

Rise of Humanism. Attempted Return to Antiquity

The Dane TYCHO BRAHE (1546–1601) was born three years after the death of Copernicus. Unlike Copernicus he was, before everything, a patient and accurate observer. He was provided by his sovereign with a magnificent observatory which was the scene, during twenty-one years, of Tycho's labours in the systematic collection of astronomical observations for the correction of cosmic

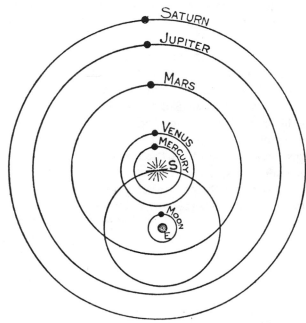

FIG. 56. Tycho's World-System.

theories. The records of Brahe were much the most extensive and accurate that had been made up to their time. Brahe's actual scientific achievements and contributions may be summarized:

(a) He set forth a planetary system with the Earth central to the orbits of Moon and Sun and central also to the fixed stars. The Sun revolves round the Earth in twenty-four hours carrying all the planets with it. Of the planets, Mercury and Venus have orbits smaller than that of the Sun while the other three have orbits that encircle the Earth (Fig. 56). Mathematically this system works out as identical with that of Copernicus (1588).

(b) Examining a comet he was able to determine its parallax and thus proved that it was farther off than the Moon. It was thus outside the sphere of the 'elementary' world (see Fig. 20). This was equivalent to introducing the principle of change into the changeless spheres and therefore contrary to Aristotelian principles (1577).

(c) He suggested that the movement of a comet might be 'not exactly circular but somewhat oblong'. This is the first suggestion that a celestial body might move in a path other than circular (1577).

(d) He described very accurately perturbations in the Moon's motion (1599). These had to await explanation by subsequent generations and new astronomical systems.

(e) His numerous observations on the planets enabled Kepler to reveal the true nature of their orbits.

Tycho's attempt to represent the structure of the Universe as according to the ideal form of the circle was the last great effort of the Pythagorean spirit save for that of his pupil Kepler. The insurgent century sought for direct evidence as to the nature of the world. The new science concerned itself neither with ideal forms nor with the theory of knowledge nor with the nature of reality nor with the principles of investigation, but with the evidences of the senses.

VII. THE INSURGENT CENTURY (1600-1700)

Downfall of Aristotle. New Attempts at Synthesis

1. *Doctrine of the Infinite Universe.*

COPERNICUS worked in Poland, the eastern march of European civilization. It was at the western limit of Europe, in England, where the spirit of the great intellectual revival had not yet obtained full hold, that his message was first translated into philosophic form.

In 1583 there came to London GIORDANO BRUNO (1547-1600), a native of Nola near Naples and a renegade monk. He was in his thirty-seventh year and had already sojourned as a teacher at the Universities of Lyons, Toulouse, Montpellier, and Paris. At each of these centres of learning his restless and turbulent spirit had combined with an aloofness from the affairs of men to make him unwelcome. Throughout his life he showed a lofty indifference to the dictates of common sense that cannot fail to command our respect—at a distance. He was accustomed to make a precarious livelihood by lecturing on a barren logical system which he had partly invented. It was intimately linked with an absurd principle of mnemonics which he had partly borrowed. So wayward a genius was predestined to tragedy.

It was during his visit to England that Bruno at length developed philosophical coherence. Even then his period of illumination lasted but a few months. In 1584 he published in London, though with the false impress of Venice, three tiny Italian works, *The Ash-Wednesday Supper* (*Cena de le Ceneri*), *On Cause, Principle, and Unity*, and *On the Infinite Universe and its Worlds*. These booklets contain wellnigh the whole of his effective philosophy, which was based in essence on Nicolas of Cusa (p. 171) and in form on Copernicus. Essential parts of the thought of Bruno are the doctrines that not only does the Earth move round the Sun but that the Sun itself moves, that there is no such thing as a point absolutely at rest, that the stars are at vast but various distances from the solar system and are themselves centres of comparable systems, that the universe, being itself infinite, can provide no

185

criterion of fixity, and that our planetary system is in no sense the centre of the universe.

'The Nolan maintains', he says, 'that the world is infinite. Therefore there is no body that can be said to be in the centre of the world or at the frontier thereof or between two of its frontiers. Bodies can only be said to have certain relations to other bodies or to frontiers that are chosen arbitrarily. Thus the motions of natural bodies are far from being simple circles with a single centre.' (*Cena de le Ceneri*.)

'All these innumerable worlds which we see in the universe are not contained therein as in a vessel but rather are comprehended or conserved by the efficient cause which moves them. Moreover, as the common soul is within the whole to which it gives being and at the same time is individual and yet is in all and every part, so the *essence* of the Universe is one in the infinite and in whatsoever thing you take as a member thereof. . . .' (*On the Infinite Universe*.)

In such a universe where may Paradise and Purgatory be placed? And is not the 'common soul', which uniformly permeates it, a memory of Neoplatonism which in turn took the idea from Stoicism (pp. 123–4)? Bruno's vision of an 'infinite Universe', endless both in time and space, whose soul abides uniformly in every part, differs utterly from the 'created Universe' of medieval Christian philosophy, the Creator of which must, of His nature, be separate from that which He has created. The universe of medieval Christian philosophy was necessarily centred in Man, for into Man alone, among created mundane things, the Divine Spirit had entered. Small wonder that the Church was disturbed by Bruno's thought. His revolution was incomparably greater than any dreamed of by that academic and conservative mathematician, Copernicus.

The issues involved were not at first generally recognized. Some who were profoundly stirred by the pagan character of Bruno's thought fixed on the irrelevant detail of the Earth moving round the Sun as contrary to scripture. This idea Bruno had certainly taken from Copernicus whose work was not, as yet, prohibited. But Bruno's vision had far deeper implications than a mathematical readjustment of the current world scheme. A finite universe, spherical or not, with or without the Earth as its centre, can be conceived as 'created'; an infinite universe cannot be so contemplated. Creation is fundamental to Christianity—at least to the Christianity of that age—and it need not surprise us that the Christianity of that age struck at Bruno. In 1600 he was burned at the stake, having passed seven years in the prisons of the Inquisition. His philosophical writings were suppressed, but their seed had been sown.

Bruno perished miserably without the hope or thought that he had a disciple. And yet his view was soon to displace that of medieval Christianity. Before he had been dead for thirty years, the world was, for the man of science, no longer a diagrammatic scheme which required investigation only as regards its details. It had become a world without bounds and therefore of infinite possibilities. And yet it was a world whose parts were uniformly related by mathematical rules, the physical bases of which were in process of discovery.

It was of course true then, as it is of course true now, that the view of universal law did not and does not occupy the whole mind of all men of science. Most men of science reserved, and still reserve, some department of experience in which they forbid full play to their vision of universal law. But when and where they give rein to that mood, then and there it is bound to displace the mood of faith, nor can the medieval compromise (p. 154) stand against it. Thus the three little tracts of Bruno printed in London in 1584 mark the real change from medieval to modern thought and especially to modern scientific thought. The change was long in coming, longer for some topics than for others, longer in some minds than in others. But the coming of that change was inevitable once these three tracts had got abroad. Every attempt was made to suppress them, but they had done their work. Bruno was right when he said at his trial 'Perchance you who condemn are in greater fear than I who am condemned'.

In summary we may express Bruno's thought thus:

(*a*) *There are other worlds than ours.* The Universe is made up of many worlds comparable to that in which we live. Our world is not the centre of the Universe.

(*b*) *The Universe is infinite in space and time.* This implies that conceptions of fixity of points in space or time or of movement in either space or time must be relative to other points arbitrarily regarded as fixed.

(c) The Universe is permeated throughout by a common soul. This carries the implication of a uniformity in all the workings of the parts of the universe and is easily adapted to the conception of uniform natural laws.

It is important to remember that Bruno's views were not based on experiment or observation. His contribution was a philosophy, not a scientific method or system, and was, in fact, a development of the thought of Nicolas of Cusa (p. 171). The doctrine of relativity in space, in motion, in thought, promulgated by the calm spirit of Nicolas became in the passionate Bruno an ardent and soul-absorbing faith.

It is not easy to trace in detail the progress of the dissemination of the ideas of Bruno. His life was obscure, the propagation of this thought furtive, his influence secret, indirect, unacknowledged. Yet his ideas crop up where they might not be expected. We will consider one such case.

In 1600 there appeared in London a Latin work *On the Magnet and on Magnetic Bodies and concerning that great magnet, the Earth, a New Physiology*, by WILLIAM GILBERT (1546–1603), personal physician to Queen Elizabeth, a man in authority, respectable and respected. His book is the first major original contribution to science that was published in England. It earned the admiration of Francis Bacon and of Galileo. Gilbert's work has medieval elements, still unpurged, but its main section sets forth his investigations of the properties of the magnet in thoroughly modern form. This section is entirely experimental in outlook, and opens the new era of physical research. It records numerous experiments and is illustrated by clear diagrams. The properties of the lodestone and of the magnet, the direction that the compass assumes in relation to the poles of the earth, its variation, its inclination and its declination are systematically treated.

The last section of the book is devoted to an exposition of the system of the universe. The universe of Gilbert is that of Bruno, whose name, however, he does not mention. Gilbert must have met Bruno at Elizabeth's court, probably in the company of Sir Philip Sidney (1554–86). It is to be remembered that

Downfall of Aristotle. New Attempts at Synthesis

this book of Gilbert was published in the year that Bruno was burned and that Bruno's views and mode of expression were well-nigh as inacceptable in Protestant England as in Catholic Italy. The work *On the Magnet* was the only publication issued by Gilbert. Long after his death, however, another work by him *On our Sublunary World, a New Philosophy*, was seen through the press by a surviving brother (1651). It expounds in detail, quoting Bruno, the idea that the 'fixed stars' are at differing distances from our planetary system and that these stars are the centres of other planetary systems.

The seventeenth century opened lurid with the fires that formed Giordano's shroud. That hideous event was the herald of a period that has no rival for the number and importance of its scientific discoveries. A glance at the mass of fundamental scientific work of the seventeenth century shows the major departments of science becoming clearly differentiated. The acceptance of observation and experiment, as the only methods of eliciting the laws of nature, reaches an ever-widening circle. The very first scientific generation of the century saw the development of a mathematical technique that became the instrument of the new discoveries.

2. Mathematics becomes the Instrument of Physical Investigation.

The improvement in the means of mathematical expression was a main condition for the development of exact conceptions of a new cosmology and physics. These were an intellectual necessity to replace the tottering Aristotelian scheme that Bruno had attacked. The insurgent century found a qualitative world based on abstract values. It bequeathed a quantitative world based on concrete impressions. The senses came to reign on that Olympus, where Platonic 'Ideas' had once held divine court. Mathematics was the mercurial messenger of the new gods.

A beginning had been made in the later sixteenth century. Thus the French lawyer FRANÇOIS VIÈTE (1540–1603) was among the first to employ letters to represent numbers. He applied algebra to geometry in such a way as to lay a foundation for analytical trigonometry (1591). At about the same time was introduced the decimal scheme for representing fractions (1586)

189

The Insurgent Century

by the Fleming SIMON STEVIN (1548–1620). This ingenious man preceded Galileo in experimenting on the relative rate of fall of bodies of different weight (1586). His name is also associated with the method of resolution of forces, with the distinction of stable and unstable equilibrium, with the law of equilibrium on inclined planes (Fig. 57), and with the 'hydrostatic paradox', that is that downward pressure of a liquid on the base of its containing vessel is independent of its shape and size and depends only on

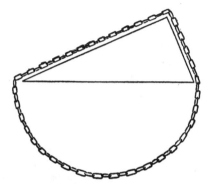

FIG. 57. *Stevin's proof of conditions of equilibrium on inclined planes.* Around the vertical angle of an upright triangle, of which the opposite side is horizontal, hang a ring-chain. It will be in equilibrium, for if not it would be in perpetual motion. Remove the suspended loop. Equilibrium remains. Therefore weights on planes inclined to each other are in equilibrium if they are proportional (as are those of the pieces of chain) to the lengths of the planes as cut by the horizontal.

the depth of the contained vessel and area of the base. Stevin was able also to calculate the pressure on any given portion of the side of the containing vessel. He laid the essential foundations for the whole science of hydrostatics (1586).

By the use of an improved form of Stevin's decimal notation calculation was much facilitated. Contemporary astronomical activity, however, still carried with it an endless task of computation. No technical advance was more needed than some further alleviation of this deadening burden. Thus the invention of logarithms by Napier was greeted with enthusiasm.

JOHN NAPIER (1550–1617), Laird of Merchiston in Scotland, began his investigations (1573) with an attempt to systematize algebraic knowledge. In his earliest work he says that in considering imaginary roots he discovered a general rule for roots of all degrees. He conceived the principles of logarithms in 1594. The next twenty years were spent in developing the theory and computing the 'canon' or table of logarithms itself. While thus engaged he invented the modern notation of fractions. His *Description of the Marvellous Canon of Logarithms* appeared in Latin at Edinburgh in 1614. An extension of Napier's long effort to do away with the tediousness of calculation was his *Rabdologia* (1617). It contains the description of 'Napier's bones', devices

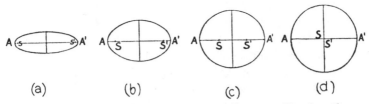

(a)   (b)   (c)   (d)

FIG. 58. The Circle as special case of the Ellipse. Keeping the major axis AA′ of constant length, construct a series of ellipses with the two foci SS′ successively closer, until they coincide. The process is also indicated in Fig. 26.

designed to simplify multiplication and division. These were in use for about a century and long attracted even more attention than his logarithms.

An advance, significant for the whole subsequent astronomical development, was made by Kepler (p. 200) in a commentary (1604) on the work of the thirteenth-century mathematician Witelo (p. 156). Kepler regarded conic sections as forming five species passing from the (1) line-pair, through (2) hyperbola, (3) parabola, and (4) ellipse, to (5) circle. In order to indicate the nature of this process Kepler designated as *foci* the fundamental points connected with these curves. The foci of the circle coalesce at the centre. Ellipse and hyperbola have two foci equidistant from the centre. The parabola has two foci, one within it and the other at an infinite distance on the axis, within or without the curve (Figs. 58, 26, 17).

Even more fundamental for future mathematical development were the ideas introduced by DESCARTES (p. 221). His analytical method appeared in his *Geometry* (1637). Its essential novelty is the introduction of the conception of motion into the geometric field. There is a well-known story that, lying late abed, as was his wont, and observing a fly hovering in the corner of his room, it occurred to him that its position in space could be defined at any moment by its distance from the three planes formed by the adjacent walls and ceiling. If two instead of three dimensions be considered, a point in a plane can be defined by its relation to two instead of to three 'Cartesian co-ordinates' as they have come to be called after him.

Thus Descartes saw a curve as described by a moving point, the point being the intersection of two moving lines which are always parallel to two fixed lines at right angles to one another. As the moving point describes its curve, its distances from the two fixed axes will vary in a manner characteristic of that particular curve. An equation between these distances can be formed which would express some property of the curve. The conception has had innumerable developments and has been adopted in every department of science. Its most familiar development is the 'graph'. Important parts of our mathematical notation are due to Descartes.

There is a basic conception in the mathematical attitude of Descartes that is far more significant than any technical addition that he made. His analytical procedure displayed the fundamental correspondences of number and form. Pythagoras and Plato perceived this correspondence. The Alexandrians tended to study the two in isolation. The separate development of algebra by the Hindus and of geometry by the Arabs, and the general trend of western mathematical studies in the Middle Ages and the fifteenth and sixteenth centuries concealed a most essential truth. Descartes, despite professed indifference to historical considerations, called men back to the old paths of Pythagoras and Plato on this most fundamental issue. It is probable that the application of algebraic methods to the geometric field is the greatest single step ever made in the progress of the exact sciences. Descartes himself insists on the unity of the study of mathematics.

'All sciences which have for their end investigations concerning order and measure are related to mathematics, it being of small import whether this measure be sought in numbers, forms, stars, sounds, or any other object; there ought therefore to be a general science, namely mathematics, which should explain all that can be known about order and measure, considered independently of any application to a particular subject. . . . A proof that it far surpasses in utility and importance the sciences which depend on it, is that it embraces at once all the objects to which these are devoted and a great many besides.' (*Rules for Direction of the Mind*, 1628.)

The expression of this Platonic view makes it evident that the reaction against the long reign of Aristotle has begun. (For Aristotle's own criticism on the point see p. 34–5.)

About the beginning of the seventeenth century were made the first decided advances since antiquity in synthetic geometry. In this department the leading name is that of Descartes' fellow countryman BLAISE PASCAL (1623–62). He also added much to mathematical theory especially in connexion with probability. Pascal invented one of the first arithmetical machines.

A versatile mathematician of the age was the learned Oxford Professor JOHN WALLIS (1616–1703). His first great mathematical work, *Arithmetica infinitorum* (1655), contains the germ of the differential calculus. Newton read it early and derived his binomial theorem from it. Wallis wrote the first mathematical work devoted to tides, in which he introduces the assumption 'that, for purposes of calculation, earth and moon can be treated as a single body concentrated at their centre of gravity.' His *Algebra* (1657) contains the idea of the interpretation of imaginary quantities in geometry which was fundamental for the development of analytic methods. Wallis introduced the symbol ∞ for infinity.

A younger man of mathematical genius was the polymath CHRISTIAN HUYGENS (1629–95). His mathematical skill early drew the attention of Descartes, who predicted his future eminence. Before he was twenty he did good work on the quadrature of the circle and conic sections (1651–4). He diffused his mathematical abilities in many departments, optics, astronomy, mechanics, theory of light, and elsewhere, and thus missed reaping some of the fame that his abilities justified.

Mathematical activity early influenced optics, a favourite subject for discussion by mathematicians even during the Middle

Ages (p. 160). The leading problem was the nature of the laws of refraction. Ptolemy (p. 83), Alhazen (p. 136), Witelo (p. 156), and their medieval followers were aware that light rays are bent or 'refracted' when passing from a rarer to a denser medium. In a commentary on Witelo, Kepler (p. 200) gives his measurements of the incident and the refracted rays in special cases, but failed to reach a general law (1604). This was successfully elicited (1621)

FIG. 59. Snell's law. Rays of light, passing from air into a denser medium, are bent toward the vertical to a definite amount. If AA' and BB' be two such rays $\dfrac{A'a'}{Aa} = \dfrac{B'b'}{Bb}$ at whatever angle they strike the surface. This ratio differs for different media and the less it is the higher is the 'refractive power' of the medium.

by the Hollander WILLIBRORD SNELL (1591–1626). Descartes placed the results in a more acceptable form and published them without acknowledging their source (1637).

The nature of the advance can be illustrated by a simple diagram. Rays of light AA' and BB' pass at different angles from air into water (Fig. 59). In all such cases they are bent toward the vertical and bent in such a degree that the ratio $\dfrac{A'a'}{Aa}$ is the same as the ratio $\dfrac{B'b'}{Bb}$. This ratio for water is $3:4$, which is said to be the *refractive index* of water. Each substance has its own characteristic refractive index. Different kinds of glass were, for instance, soon found to have different refractive indices.

The laws of refraction of light have been a most important factor in the construction of optical instruments. In that art the effective beginning was made by Galileo (1609). The optics of his compound systems of lenses were investigated by Kepler (p. 200), who first expressed in intelligible mathematical form the action of telescope and microscope (1611). Kepler's work led to further advance by Descartes (1637) who also produced a geometrical theory of the rainbow based on the law of refraction. The study of refraction occupied Huygens. His knowledge of the subject enabled him to improve lenses and to produce telescopes with much clearer definition than heretofore (1655).

If the development of optics was determined by mathematical advance the same is no less true of mechanics. In the latter case, however, the whole body of new teaching was in effect the work of one man, Galileo, to whom we now turn.

3. *Physico-Mathematical Synthesis.*

GALILEO GALILEI (1564–1642) lived a long life of unparalleled intellectual activity. Many of the products of his genius were of immediate practical application, many more involved profound modification of the current scientific opinions, yet others struck at the very basis of the general beliefs of the day.

The early training of Galileo had been along strictly scholastic and Aristotelian lines. In 1585 he began a systematic experimental investigation of the mechanical doctrines of Aristotle. By 1590 he had developed a number of objections to Aristotelian physical teaching. Notably he had accumulated his records of experiments on falling bodies. These were announced from his professorial chair and illustrated in 1591 from the leaning tower of Pisa. By that most famous of experiments he unmasked an Aristotelian error. Weights of 1 lb. and of 100 lb., dropped from the top of the tower, reached the earth together. How then was it possible to maintain with Aristotle that the rate of fall was a function of the weight of the falling object?[1]

For the rest of his life Galileo was constantly occupied with physical investigations. Of these the most famous resulted in his great astronomical discoveries (pp. 206–12). It is not, however, in his discoveries, numerous, fundamental, superb though they were, that we sense the full significance of Galileo in the history of thought. It is rather in his initiation of a new attitude toward the objective universe and in his construction of an enduring mathematico-physical scheme that would fit that attitude. More than to any other man, we owe to him the conception of our world in terms of interplay of *calculable* forces and *measurable* bodies. And moreover, to him more than to any other man we owe the experimental employment of that conception.

'Dynamics', says Lagrange (p. 266), 'is a science due entirely to the moderns. Galileo laid its foundations. Before him philosophers considered the forces which act on bodies in a state of equilibrium only. Although they attributed in a vague way the acceleration of falling bodies and the curvilinear movement of projectiles to the constant action of gravity, nobody had yet succeeded in determining the laws of these phenomena. Galileo made the first important steps, and thereby opened a way, new and immense, to the advancement of mechanics as a science.' (*Mécanique analytique*, 1788.)

He set forth his views in his great *Discourses concerning two new Sciences* (1638). The work at one step advanced the subject from the medieval to the modern stage. The two new sciences deal respectively with (*a*) 'Coherence and resistance to fracture' and (*b*) 'Uniform, accelerated, and violent or projectile motions'.

The first part of the work is mainly concerned with the resistance of solids to fracture, and the cause of their coherence. The value of this section lies in the incidental experiments and observations on motion through resisting media. The current belief that machines built on exactly similar designs, but on different scales, are of strength in proportion to their linear dimensions is discussed.

---

[1] The story of the weights dropped from the top of the leaning tower of Pisa is here told in its traditional form for which there is no satisfactory evidence. There is, however, adequate evidence that by the year 1590 he had attained to the attitude set out in the traditional account.

It is shown that the larger machine will equal the smaller in all respects, save that it will be less strong and less resistant to violent actions. The machines discussed include animal bodies (p. 214).

After explaining the strength of ropes of fibrous materials, Galileo turns to the cause of coherence of the parts of such things as stones and metals, which do not show a fibrous structure. What prevents a glass or metal rod, suspended from one end, from being broken by a pull at the other? The explanation suggested depends upon nature's so-called 'abhorrence of the vacuum' supposedly produced by the sudden separation of two flat surfaces. This idea is extended, and a cause of coherence is found by considering every body as composed of very minute particles, between any two of which is exerted a similar resistance to separation.

This line of reasoning leads to a very important experiment for measuring what is called the force of a vacuum. It occasions the remark that a pump will not work when the water had sunk 35 feet below the valve. This is sometimes told as if Galileo had said, jokingly, that nature's horror of a vacuum does not extend beyond 35 feet, but it is plain that the remark was made seriously. He held the conception of suction then current, for he compares the column of water to a rod of metal suspended from its upper end, which may be lengthened till it breaks with its own weight. It is strange that he failed to see how simply this phenomenon could be explained by the weight of the atmosphere, with which he was well acquainted. A fuller explanation had to await Torricelli (p. 232).

Aristotle's ideas on motion and especially that bodies fall with velocities proportional to their weights and inversely proportional to the densities of the media through which they fall, are examined. The end result is to substitute for Aristotle's assumption that law of the motion of falling bodies which is the foundation of modern dynamics.

The discussion of the strength of beams opens with a consideration of the resistance of solid bodies to fracture. This is very great in the case of a direct pull, but is less for a bending force. Thus, a rod of iron will bear a longitudinal pull of, say, 1,000 lb., while

50 lb. will break it if it be fastened by one end horizontally into a wall.

Galileo assumed, as the basis of his inquiry, that the forces of cohesion with which a beam resists a cross fracture in any section may all be considered as acting at the centre of gravity of the section, and that it breaks away at the lowest point. An elegant result deduced from this theory is that the form of a beam, to be equally strong in every part, should be that of a parabolic prism, the vertex of the parabola being the farthest removed from the point of support. As an approximation to this curve he recommends tracing the line in which a heavy flexible string hangs when supported from two nails.

The curvature of a beam under any system of strains is a subject into which, before the days of Newton, it was not possible to inquire, and even in the simpler problem considered by Galileo he makes assumptions which require justification. His theory of beams is erroneous in so far as it takes no account of the equilibrium which must exist between the forces of tension and those of compression over any cross-section.

The theorems and formulae deduced geometrically from the phenomena of uniform and accelerated motion lead to a more detailed statement of the principle of inertia. The definition of uniformly accelerated motion is given as that of a body which so moves that in equal intervals of time it receives equal increments of velocity.

There follows an application of the results. He examines the times of descent down inclined planes, assuming the velocity to be the same for the same height whatever the inclination. This he verified by careful experiments, although he was unable at the time to prove it mathematically.[1]

The next section plunges at once into the consideration of the properties of a body whose motion is compounded of two other motions, one uniform, and the other naturally accelerated. Such is the motion of a projectile. The law of the independence of the

---

[1] Viviani relates that, soon after he joined Galileo in 1639, he drew his master's attention to this. The same night, as Galileo lay sleepless in bed, he discovered the mathematical demonstration. It was introduced into the subsequent editions of the *Discourses*.

Downfall of Aristotle. New Attempts at Synthesis

horizontal and the vertical motions is here laid down. A body projected horizontally would—but for its weight and external impediments—continue to move in a straight line. Again, the effects of gravity acting by itself on the projected body would be entirely downwards. But gravity acting on the projected body can neither increase nor diminish the rate at which it travels horizontally. Therefore, whatever the path or the direction of motion at any moment, the distance travelled horizontally is a measure of the time that has elapsed since motion began. He proves that the path described has the geometrical properties of a parabola (Fig. 60).

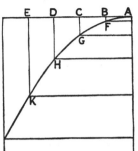

FIG. 60. AB, BC, CD ... represent equal forward displacements in equal increments of time of an horizontally ejected projectile. The distances of fall BF, CG, DH ... increase as the square of the time. The actual path AFGH is a parabola.

Galileo drew up a table giving the position and dimensions of the parabola described with any given direction of projection. He showed that the range on a horizontal plane is greatest when the angle of elevation is 45°. He was essentially applying the principles of the Differential or Fluxional Calculus. Had pure mathematics attracted him as strongly as its applications, he would have founded the Fluxional Calculus, which is the glory of Newton and of Leibnitz.

No sooner was the manuscript of these dialogues out of his hands in 1636 than Galileo began to occupy himself with new projects which he left unfinished at death. In them he approaches the laws of interdependence of force and motion which appear at the beginning of Newton's *Principia* (1687). But Galileo not only prepared the way for Newton: he supplied him with much of his materials. Thus, Newton's first law—*that a body will continue in a state of rest, or of uniform motion in a straight line, until compelled to change its state by some force impressed upon it*—is a generalization of Galileo's theory of uniform motion. Since all the motions that we see taking place on the surface of the earth soon come to an end, we are led to suppose that continuous movements, such, for instance, as those of the celestial bodies, can only be maintained by a perpetual consumption and a perpetual application of force, and hence it had been inferred that rest is the natural condition of things. We make, then, a great advance when we comprehend that a body is equally indifferent to motion or to rest, and that it perseveres equally in either state until disturbing forces are applied.

Newton's second law—*that every change of motion is proportional to the force that makes the change, and in the direction of that straight line in which the disturbing force is impressed*—is involved in Galileo's theory of projectiles. Before his time it was a commonly received axiom that a body could not be affected by more than one force at a time.

But now the establishment of this principle of the composition of forces supplied a conclusive answer to the most formidable of the arguments against the rotation of the earth. It is employed by Galileo in his *Dialogue of the Two Systems of the World* (1632, p. 211). The distinction between mass and weight was, however, not valued, and, consequently, Galileo failed to grasp the fact that acceleration might be made a means of measuring the magnitude of the force producing the motion, that is to say of the mass of the earth.

Of the third of Newton's laws of motion—that *action and reaction are always equal and opposite*—we find traces in many of Galileo's researches, as in his theory of the inclined plane, and in his definition of momentum. It is adumbrated in a little work on mechanics written by him in youth but published after his death. It is developed in his latest ideas on percussion.

4. *The Re-Formation of the Heavens.*

The first to apply mathematics as an empirical instrument in seeking the laws of celestial motion was the German JOHANNES KEPLER (1571–1630). He had strong mystical leanings, and a large proportion of his writings seem now unreadable foolishness, but there is a residuum of his works that is of the very highest scientific importance. His idea of the universe was, from the first, essentially Platonic and Pythagorean. He was convinced that the arrangement of the world and of its parts must correspond with some abstract conception of the beautiful and the harmonious

Downfall of Aristotle. *New Attempts at Synthesis*

and, further, must be expressible in numerical and geometric form. It was this belief that sustained him in his vast and almost incredible labours. He spent years of his life chained to the mere drudgery of computation without assistance and without any of the devices, such as logarithms or reckoning machines, that now lighten the computer's task. Nothing but a burning yet steady faith could make such drudgery endurable.

We gain an insight into the transition state between the old and the new in which Kepler worked when we recall that his professed occupation was largely astrological calculation. Nor was he cynically sceptical as to the claims of astrology, but sought in the events of his own life a verification of the theory of the influence of the heavenly bodies.

Kepler adopted the Copernican view from an early date. He turned his mind to the question of the number, size, and relation of the orbits of the planets. He was ever seeking a law binding together the members of the solar system. After trying various simple numerical relations, after attempting to fill the gaps by hypothetical planets and after discarding various other suggestions, he lit, at last, on a device which satisfied him (1596). There are only five possible regular solid figures (i.e. figures with equal sides and equal angles)—'Platonic bodies' as they were called (p. 22)—and there are only five intervals between the six planets that he recognized. As far as the calculations of Kepler extended at that time, the five regular solids could be fitted between the spheres of the planets so that each polyhedron was inscribed in the same sphere about which the next outer one was circumscribed (Fig. 61). Thus

Sphere of Saturn
Cube
Sphere of Jupiter
Tetrahedron
Sphere of Mars
Dodecahedron
Sphere of Earth
Icosahedron
Sphere of Venus
Octahedron
Sphere of Mercury

For the first time a unitary system had been actually introduced
in explanation of the structure of the universe. We may well smile

FIG. 61. From Kepler's *Mysterium Cosmographicum* (Tübingen 1596)[1]
illustrating supposed relationships between the five Platonic bodies (p. 22)
and the number and distances of the planets. The concentric figures are
inscribed within each other thus:

Outermost Sphere bearing sign of Saturn.
Cube.
Sphere bearing sign of Jupiter.
4-sided regular pyramid.
Sphere bearing sign of Mars.
12-sided regular body.
Sphere bearing sign of Earth.
20-sided regular body.
Sphere of Venus (hardly traceable).
8-sided regular body.
Sphere of Mercury.
Innermost Central body of the Sun.

at this instance of human presumption. Kepler soon found that he
had wrongly estimated the distance of the planets from their

[1] The reproduction is from the second edition of 1621 because the block
recut for it gives a little clearer result than that of the first edition.

centre! The basis of this unitary system was a miscalculation! It endured but a day. But to Kepler, who, like the medieval thinkers, held that the universe was designed on a moral plan, these new mathematical relationships—false as we know they are—came as a confirmation of what he conceived to be the divine purpose. The regular solids, he observed, were of two classes: primary (cube, tetrahedron, dodecahedron) and secondary (icosahedron, and octahedron), differing in various ways. What more fitting than that the earth, the residence of man 'created in God's image', be placed between the two kinds of solids? The scheme was confirmatory of many of the main tenets of his Pythagorean faith!

That Kepler sought so persistently for a simple mathematical scheme of the material world, and that, having found one, he regarded it as fitting his scheme of the moral world, suggests certain reflections on the workings of the mind itself. Whatever reality may be, we seem to be so made that we aspire towards an interpretation of the universe that shall hold together in a complete and reasonable scheme. The fact that we thus aspire does not in the least prove that such a scheme corresponds to reality. Nevertheless, all great religions attempt to provide such an interpretation. All become skilfully 'rationalized'.

It is because science disturbs part of this already carefully rationalized field that religion resents its intrusion. The mind recoils from a dualistic universe, and rationalized religion usually seeks to minimize even such remnants of dualism as the conception of a spirit of evil. It is easy for us now to regard the opponents of Galileo and Kepler as purblind fools. Base motives certainly prompted some of the opposition; but in essence the opposition expresses the reluctance of the human mind to adopt any teaching which disturbs its unitary conceptions. A reasoned view of the universe, physical and moral, had grown up during the Middle Ages. It would have been indeed a marvel if this had been relinquished without an embittered struggle, for faith is not necessarily accompanied by either wisdom or learning, or foresight.

Despite the failure of his first attempt, Kepler still pursued his life aim, the foundation of an astronomy in which demonstrable mathematical principles should replace arbitrary hypotheses. He examined the relation of the distances of the planets to their time of revolution round the central sun. It was clear that the time of revolution was not proportional to the distance. For that the outer planets were too slow. But why? There is, he suggested, 'one moving intelligence in the sun that forces all round, but most the nearest—languishing and weakening in the more distant by attenuation of its virtue by remoteness'. How different from the phraseology of modern astronomy which dates from Newton! In such phrases as 'moving intelligence', 'languishing of its virtue', etc., Kepler was employing the Aristotelian phraseology that had arisen during the Middle Ages. The conception was familiar to the medieval philosophers, Christian, Moslem, and Jewish. Aquinas (p. 182), Averroes (p. 140), and Maimonides (p. 146) all had a clear conception of intelligence moving the planets. They had derived this conception ultimately from Greek thinkers, and they had adapted it each to his own theology. The conception was quite familiar to all in the sixteenth and seventeenth centuries.

As the sixteenth century turned into the seventeenth, Kepler received a great incentive to work by joining Tycho Brahe (p. 183) as assistant. On the death of Tycho in 1601 Kepler became his literary legatee. The next nine years saw him largely occupied with the papers of Tycho and with work on optics, in the course of which he developed an approximation to the law of the refraction of light. In 1609 he issued his greatest work the *New Astronomy with Commentaries on the Motions of Mars*. It is full of important suggestions, notably that the earth attracts a stone just as the stone seeks the earth, and that two bodies near each other will always attract each other if adequately beyond influence of any third body. It also develops a theory of the tides in relation to attraction by the moon. But above and beyond all, the work sets forth the cardinal principles of modern astronomy, the so-called first two planetary laws of Kepler by which

(a) Planets move round the sun not in circles, but in ellipses, the sun being one of the foci.

(b) A planet moves not uniformly but in such a way that a line drawn from it to the sun sweeps out equal areas of the ellipse in equal times (Fig. 62).

Downfall of Aristotle. New Attempts at Synthesis

It was another nine years before Kepler enunciated in the *Epitome Astronomiae* (1618) his third law to the effect that

(c) The squares of the period of revolution round the sun are proportional to the cubes of their distance.

For one who accepted these principles of Kepler the Aristotelian cosmology lay derelict. Its foundations were undermined and

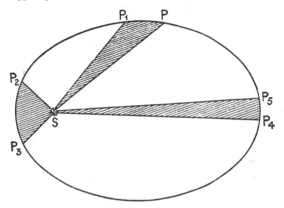

FIG. 62. Planets sweep out equal areas in equal times. PP_1, P_2P_3, P_4P_5 are distances along its orbit around the Sun S traversed by a planet in equal times. Areas SPP_1, SP_2P_3 and SP_4P_5 are equal.

their place taken by an intelligible mathematical relationship. The scholastic Aristotelianism was to become as much an embarrassment to official religion as the narratives of miracle became at a later date. It was, however, as hard for one section of the Church to rid itself of its scholastic heritage as it was for another at a later date to disembarrass itself of the dead-weight of miracle.

Certain further reflections on Kepler's work rise to the mind. It is a fundamental error to separate science from learning or, perhaps, it would be best to say from tradition. By the Greeks the study of conic sections had been prosecuted as an intellectual exercise. These figures, hyperbola, parabola, ellipse, existed, so far as they knew, in the mind and in the mind alone. They corresponded to nothing in the natural world. And then, after two thousand years, Kepler shows that these ancient concepts correspond to something that is also revealed by the use of the sense. Is not the mind then somehow attuned to nature? It has been well said by a great historian of science that 'if the Greeks had not cultivated conic sections, Kepler could not have superseded Ptolemy; if the Greeks had cultivated Dynamics, Kepler might have anticipated Newton' (Whewell).

Dynamics as we have seen, was in fact a creation of Kepler's contemporary, Galileo. In character and temper Kepler and Galileo form an extraordinary contrast. The German Protestant, mystic and dreamer rather than observer and experimenter, produced voluminous, numerous, and wholly unreadable volumes. He stands over against Galileo, the Italian Catholic, clear and cold of intellect, unrivalled in experimental skill, witty and happily endowed with artistic and literary prowess, who wrote never a work and hardly a line that was not significant. In sheer genius, however, the two men were not rivals but peers and comrades. On them, in equal measure, rest the foundations of the conception of a mathematical universe.

Galileo's astronomical activity began in 1604. In that year, in the constellation Serpentarius, there appeared a new luminous body. He demonstrated that it was without parallax, that is to say there was no difference in its apparent position in the heavens, from whatever point it was viewed. Now parallax decreases with distance. In Galileo's time the planets were known to have parallax, but the parallax of the fixed stars was so small, by reason of their vast distance, that it was unmeasurable by the instruments of the day. This new body was thus in the remote region of the fixed stars. Now that outer zone had been regarded by Aristotle and his followers as absolutely changeless (p. 47). New stars when previously noticed had been considered to belong to the lower and less perfect regions nearer to earth. To the same lower region were assigned such temporary and rapidly changing bodies as meteors and comets. Galileo had thus attacked the incorruptible and unchangeable heavens and had delivered a blow to the Aristotelian scheme, wellnigh as serious as the experiment on the tower at Pisa (p. 195).

In 1609 Galileo made accessible two instruments that had the profoundest influence on the subsequent development of science, the telescope and microscope. His earliest discoveries with the telescope were issued in a little pamphlet of 24 leaves, his *Messenger of the Heavens* ('Sidereus nuntius') in 1610. There are no 24 leaves in all scientific literature that contain more important revelations.

The first half of that famous booklet is occupied by observations on the moon. The surface of the moon, far from being smooth and polished, as it appeared to the naked eye, was now seen to be

FIG. 63. The Moon as seen by Galileo in 1609.

rough, with high mountains and deep depressions. The latter Galileo interpreted as rivers, lakes, and seas. From the appearance of illuminated mountain tops he could estimate the height of some of them. He found them to rise four or five miles above the general level (Fig. 63).

Galileo's lunar observations have an interesting relationship with English literature. In 1638 he was old and blind and nominally a prisoner of the Inquisition at Fiesole. He was visited there by Milton. The incident has inspired several artists and writers. In 1658, nine years after Galileo's death, Milton began his *Paradise Lost*, completing it in 1666. Its cosmology is deliberately Ptolemaic, not Copernican (p. 180). Nevertheless, *Paradise Lost* does recall the poet's induction into the new astronomy twenty-seven years previously. It describes Satan's shield of which the

broad circumference
Hung on his shoulders like the moon, whose orb

The Insurgent Century

Through optic glass the Tuscan artist views
At evening, from the top of Fesole,
Or in Valdarno, to descry new lands,
Rivers, or mountains, in her spotty globe.

(Paradise Lost, i. 286–91.)

The *Messenger of the Heavens* discusses the revelation by the telescope of an immense number of hitherto unobserved fixed stars. These were seen to be at least ten times as numerous as those that had been catalogued. The more conspicuous star clusters were found to contain many stars too faint for recognition by the naked eye. Parts of the Milky Way and some of the nebulous patches in Orion, the Pleiades and elsewhere were resolved into groups of stars of various magnitudes.

The remainder of the little book is devoted to an account of the satellites of Jupiter which Galileo discovered on one of the first occasions when he used his telescope. The existence of these bodies was of peculiar interest at the time, since the planet was seen to be itself a sort of little model of the solar system, with minor bodies centering round a great central body. The contemporary discussion as to the 'plurality of worlds' (pp. 186–7) was given a new turn by this discovery of a world modelled on the Copernican solar system.

There were other observations made by Galileo about this time that were later the subject of much discussion. Important were the observations on the inner planets and notably on Venus. It had been a real objection to the Copernican hypothesis that if the planets resemble the Earth in revolving round a central sun, they might be expected to be luminous only when exposed to the Sun's rays. In other words, they should exhibit phases like the Moon. Such phases were now actually observed in Venus by Galileo.

In the same year the outermost of the known planets, Saturn, was investigated. Peculiar appearances in him were noted, though their interpretation as rings was the work of Christian Huygens (1629–95) at a later date (p. 195).

Soon after this Galileo first observed dark spots on the surface of the Sun. These he saw narrowed continuously as they approached the edges of the Sun's disk. He rightly regarded the process as foreshortening and as indicating that they were on the surface of

208

the Sun's orb which was itself rotating. The date and circumstances of the announcement (1612) were unfortunate, since they
involved him in a controversy with a powerful Jesuit rival who
not only claimed priority of observation, but also put another
interpretation on the spots.

The controversy spread far beyond its original focus. An aspect
of the dispute was the question of the habitability of the Moon,
the planets, and even of the stars, for these, too, some thought to
be worlds. His critics believed this a natural corollary of Galileo's
development of the 'Copernican' view which he had now openly
espoused. The conception of the 'plurality of worlds' gave rise
to a very considerable literature. The doctrine, it was believed,
was contrary to Aristotelian and Christian teaching. It had been
enunciated by the heretical Bruno (p. 185).

Thus became united against Galileo a variety of interests. The
band of academic Aristotelians had long been fuming against him.
Jesuits who were actively engaged in teaching, as well as many
political churchmen, now joined them. Pious folk were outraged
by the conception of the plurality of worlds. To them were further
united many of that intellectually timid class that forms the mass
of every population in every age and is by no means rare in university circles. Deeper though less expressed was the great philosophic fear of the infinite universe that Bruno had suggested. The
matter came before the Inquisition early in 1616. Galileo was
admonished 'to abandon these opinions and to abstain altogether
from teaching or defending or even discussing them'. A few days
later a decree was issued ordering the work of Copernicus to be
'suspended till corrected'.

In 1624 Galileo published *Il Saggiatore* ('The Assayer'), a work
which contains a conception of great import for the subsequent
development of science. This conception, moreover, was destined
to colour deeply much of the philosophical thought of later ages.
He here distinguishes sharply between those qualities of an
object that are susceptible of exact numerical estimation and those
which cannot be treated in this way.

> 'No sooner', says Galileo, 'do I form a conception of a material or
> corporeal substance, than I feel the need of conceiving that it has
> boundaries and shape; that relative to others it is great or small;

that it is in this place or that; that it is moving or still; that it touches or does not touch another body; that it is unique, rare, or common; nor can I, by any effort of imagination, disjoin it from these (primary) qualities. On the other hand, I find no need to apprehend it as accompanied by such conditions as whiteness or redness, bitterness or sweetness, sonorousness or silence, well-smelling or ill-smelling. If the senses had not informed us of these (secondary) qualities, language and imagination alone could never have arrived at them. Wherefore I hold that tastes, colours, smells, and the like exist only in the being which feels, which being removed, these (secondary) qualities themselves do vanish. Having special names for them we would persuade ourselves that these (secondary qualities) have a real and veritable existence. But I hold that there exists nothing in external bodies for exciting (the secondary qualities) tastes, smells, and sounds, but (the primary qualities) size, shape, quantity, and motion. If, therefore, the organs of sense, ears, tongues, and noses were removed, I believe that (the primary qualities) shape, quantity, and motion would remain, but there would be no more of (the secondary qualities) smells, tastes, and sounds. Thus, apart from the (percipient) living creatures, I take these (secondary qualities) to be mere words.'

This distinction between *primary qualities* and *secondary qualities*, as they came afterwards to be called, has been made by men of science ever since. Galileo was the prime mover in that development which is summed up in the phrase *Science is Measurement*.

As to whether men of science have been right or wrong in their view that primary qualities have a reality lacking in secondary qualities, we need not for the moment consider. It is evident that ordinary experience is almost entirely made up of secondary qualities. The fact that men of science have dwelt chiefly on something else, something which ordinary men do not ordinarily consider, has separated them from their fellows. Since Galileo, men of science have formed a sort of priesthood which has been, not infrequently, opposed to another priesthood. Nor has the distinction which Galileo made remained with the working men of science. Through Thomas Hobbes (1588–1679) and John Locke (1632–1704) in England, and through Marin Mersenne (1588–1648) and René Descartes (1596–1650) in France, it has passed into general philosophy.

Downfall of Aristotle. New Attempts at Synthesis

By 1630, after many years' work, Galileo had at last completed his epoch-making *Dialogue on the Two Chief Systems of the World*, that is the Ptolemaic and the Copernican. Quite apart from the discussion of the relative position of Earth and Sun in the universe, the *Dialogue* is the consummation of the labours of Galileo in that it seeks to present the doctrine of uniformity in the working of the material universe.

The point of view expressed by the *doctrine of uniformity*, the view that corresponding causes are everywhere producing corresponding effects, is so familiar to us nowadays as to be a part of our manner of thinking. We are brought up to it from our earliest years. The only occasions on which it is ever questioned by educated men of our own time are (*a*) in the discussion of the nature or reality of miracles, and (*b*) in the discussion of the relation of mind and matter. But in the seventeenth century it was not so. The Aristotelian conception of the universe still ruled supreme. According to that view the events in the high supralunary spheres—'celestial physics' as we may call them—were of a very different order to our earthly happenings—'terrestrial physics'. A large part of medieval philosophy may indeed be regarded as a debate, prolonged through hundreds of years, of the relation of celestial to terrestrial physics. That there was a difference between the two had hardly yet been questioned, save by Bruno (p. 185). Even Galileo was in no strong position to discuss celestial physics. It is of interest, however, that he throws out a definite suggestion that it can be discussed on the terrestrial basis, thus foreshadowing the doctrine of universal gravitation.

'Since, as by a unanimous conspiracy of all the parts of Earth for the formation of its whole, those parts do congregate with equal inclination and, ever striving, as it were, at union, adapt themselves to the form of a sphere, so may we not also believe that Moon, Sun, and the other members of the solar system (*corpi mondani*) are likewise of spherical form by a concordant instinct and natural concourse of all their parts? And if any of their parts were violently separated from the whole, might we not reasonably suppose that they would revert spontaneously by natural instinct? May we not therefore conclude that as regards their proper motion, all members of the Solar System (*corpi mondani*) are alike?'

Permission to print this dialogue was obtained from the ecclesiastical authorities on the express condition that the subject was to be treated theoretically as a convenient hypothesis and not as representing the facts. It was issued in 1632.

The debate in this work is carried on by three persons, an open advocate of the Copernican doctrine, an obtuse and obstinate follower of Aristotle and Ptolemy, and an impartial participator open to conviction. The conditions of publication are only superficially complied with, and the tone leaves no doubt as to Galileo's real opinions. The work is full of prophecies of the development of cosmic theory.

The Aristotelian in the dialogue is represented as hopelessly stupid, and the Copernican has the best of the dispute. In fact, however, the Copernican passes far beyond Copernicus, notably in his total rejection of the idea of the stars as fixed in a crystal sphere. The stars, as in the works of Bruno, are held to be at inconceivable but differing distances from our Earth, and the absence of visible stellar parallax is considered as due to the vastness of this interval.[1]

The *Dialogue* brought matters to a head. Oddly enough, it was not the sweeping generalizations on which Galileo's opponents seized—maybe they did not realize their full significance. It was rather certain details opposed to the current view that were specially suspected. In August 1632 the sale of the book was prohibited and its contents submitted for examination to a special commission. They reported against Galileo. The end is well known.

5. *Implications of the Galilean Revolution.*

Galileo, more than any other man, had introduced the change in our manner of thinking that broke with ancient and led on to modern science. Contributions had also been made by Copernicus, by Vesalius, by Bruno, by Tycho, and by Kepler and others. The share of Galileo is, however, so overwhelming that it is not unfair to call it the 'Galilean Revolution'. The change was more than an addition to knowledge. It was more even than an alteration in the conception of the structure of the universe. It was

[1] The measurement of the parallax of a fixed star was not made until 1838, when it was achieved by Bessel (p. 269).

Downfall of Aristotle. New Attempts at Synthesis

rather a change in mood as to the kind of knowledge that was to be sought. It partook of the nature of a philosophical crisis. Its implications are so fundamental for science that we must attempt to review them. This we can most conveniently do under various headings which, it must be recognized, are incommensurable. They are not divisions of the subject, but themes which suggest themselves in connexion with Galileo's life work.

(a) The Mechanical World.
(b) Extension of the Senses.
(c) The Universe as Mathematical and Boundless.
(d) Religion and Science.

(a) *The Mechanical World.* The science of elementary mechanics exists to-day in substantially the state in which Galileo left it. Its formulation was his real life task. Among his earliest observations were those on the pendulum—made when he was eighteen years of age. In explaining its movements, in the draft of a work on mechanics prepared a few years after these observations, he invoked the action of gravity. Nevertheless Galileo conceived no exact idea of the action of gravity—of which the pendulum is a special case—until many years later. His conclusions on that topic are embodied in his *Discourses concerning two new sciences* (1638) published when he was seventy-four years old. The wide separation in time of these two events illustrates how wholly different is the order and manner of presentation of the thought of a scientific investigator from the order and manner in which he reaches his conclusions.

In this, his final work, the results of his investigation extending over more than half a century are placed in logical or rational order. Thus their historic sequence is concealed. The process of setting forth a scientific discovery involves of necessity the covering up of the true historic sequence. This is one of the reasons that make the history of science difficult to master.

Of all Galileo's contributions to mechanical conceptions perhaps the most fundamental was that the continuous application of a force produces either an increment or decrement of velocity *at every moment.* The conception of acceleration as a constantly changing velocity *accompanying the application of force* was in contradiction to the Aristotelian principle that terrestrial bodies

213

tend *of their own nature* to come to rest at a level which is natural for them. Acceleration, as we understand it, was one of Galileo's fundamental contributions. It involves the conception of the indefinite splitting up of time and thus of the application to time of the doctrine of limits as Archimedes had applied it to space. Through his mathematical teaching concerning moving bodies Galileo leads on to Newton.

Again, the philosophers of the Middle Ages and the mathematicians of the sixteenth century had found great difficulty in conceiving a body as the subject of several simultaneous movements. For them the type of 'perfect' motion was to be seen in the supposedly circular path of the heavenly bodies. Galileo by introducing the idea of acceleration, and especially of acceleration as natural to falling bodies, made familiar the idea of compounded motion. By his analysis of the path of projectiles (p. 199) he introduced that view into curvilinear as well as rectilinear motion. Thus he paved the way for Newton's synthesis of terrestrial and celestial mechanics.

Moreover Galileo's developments of the science of mechanics were applicable to all visible and tangible objects. His conception of a mechanical universe swiftly reacted even on the biological sciences. In the rebound of sentiment against Aristotle, biologists sought to explain the animal body as a machine. The first important biological works of the seventeenth century—for example those of Sanctorius (p. 236), of Harvey (p. 237), of Descartes (p. 191)—all sought thus to explain the body. Though Galileo in general eschewed the investigation of living things, in this matter he was himself a pioneer. He pointed out that a machine to be most efficient must be of a particular size. If one dimension is increased it is not enough to increase the others in proportion. The machine must be designed anew.

Arising out of this principle he shows it to be impossible for a swiftly moving land animal to increase its size, retaining its proportion of parts, and at the same time to maintain its agility. Increase in size increases weight as the cube of the length, but areas of cross-section of bones or muscles only as the square of the length. Thus if an animal's dimensions are doubled, its ability to overcome forces is quadrupled, but the forces to be overcome are increased eight times. But Galileo also saw that if an animal be immersed in water, then the weight is counterpoised to the extent of an equivalent volume of water. Under such circumstances the character of the physical barrier to increase in size is altered. The importance of this principle has only been appreciated in modern times. Thus each species has, of its physiological nature, a limit of growth, which is enormously higher for certain water animals, as whales among vertebrates and cephalopods among invertebrates, than for any land animals. The change in the proportions of the parts during growth had, in fact, already been made a subject of special study by the artist Albrecht Dürer a century earlier. But Dürer had not subjected the basic principle to mathematical analysis as did Galileo (p. 173)

In sum Galileo produced a conception of a world in which search might reasonably be made for mechanical principles alike in the movements of the heavens and the changes on the earth, in the circulation of a planet's satellites as in the structure of a minute insect. It is an increasingly mechanic world with which men of science have henceforth to deal. Astrology had laid sacrilegious hands on the heavens. The new determinism was to be a much more intimate thing which concerned the stars no less than men and men no more than mice. This was evident enough to the lofty genius of Spinoza, but these complications of the mechanical conception of the world were almost wholly missed by Galileo's leading opponents. They saw in him, as they had seen in the astrologers, merely another disturber of traditional religion. Had the real nature of the Galilean revolution been realized, it would have fared even worse than it did with its author and his followers.

We may here say a word concerning Galileo's opponents. They have been the objects of contumely because of his base and cruel treatment by the Inquisition and by those in high authority in the Church. We need not stop to defend these inane pomposities, nor need we pause to denounce the dishonesty and foolishness of other of his opponents. But not all those who were opposed to Galileo were fools or rogues. A great body of not unreasonable opinion hesitated to accept his physical philosophy. It is right to remember that a complete system of philosophy, weaving into one vast scheme the moral and physical, the terrestrial and celestial worlds, had been built up during the Middle Ages. This satisfied the need of the day. The fact that Galileo had made a breach in that scheme was no clear reason to abandon the whole. Would the fact that recognized scientific laws were shown to be inapplicable to some particular group of phenomena be a reason nowadays for abandoning the scientific method of exploring nature? Remember that Galileo had to offer his audience no complete system even of physical philosophy—that was reserved for successors of Newton. Even if the contemporary critic were a specialist in physics—and such were few in those days—a reasonable attitude would surely have been one of friendly and non-committal suspension of judgement not so much as to Galileo's findings, but as to their implications.

It is true that the older astronomical position had been shaken also by Kepler's demonstration that the movements of the planets are more easily understood if we suppose them to follow elliptical and not circular courses. But Galilean physics and Keplerian astronomy had not been linked together. That again was reserved for Newton's generation. Moreover many of the exemplars of Aristotle's science were taken from the world of life. Aristotle's biological system was still the best, and it was the Aristotelian physical system that Galileo was attacking. Further, as things then stood, abandonment of the Aristotelian scheme of the universe meant abandonment of much religious teaching. We are entitled to expect that judges should be both just and merciful. The judges of Galileo were perhaps neither. But the facts of human nature offer no warrant for the hope that all teachers will have the insight and understanding of a great master's immediate following. *Suspension* of judgement as to the validity of Galileo's arguments was thus a necessary consequence of the imperfect nature of man. To note this is not to justify either the ignorance, the duplicity, or the cruelty of some of his opponents.

Apart from professional theologians on the one hand and Spinozists on the other, most reasonable men in the seventeenth century were, in fact, content with a compromise. 'The heavens are the heavens of the Lord; But the earth hath he given to the

children of Men.' This was the atmosphere in which arose and
flourished the great scientific movement of the age.

(*b*) *Extension of the Senses.* Galileo is best remembered for his
wonderful astronomical observations. But at the back of these
observations lay his invention of the telescope and his successive
improvements in the construction of that instrument. And at the
back of that lies yet another movement, the introduction of the
skilled mechanic into the service of science. In this movement,
too, Galileo may be said to be an important figure.

Apart from the striking changes, artistic, literary, intellectual,
during the fifteenth and sixteenth centuries, there were other
changes, less dramatic but affecting even more closely and deeply
the lives of men. One of these was the refinement of craftsmanship
incident on the greater accessibility of good steel for tools. The
houses, the furniture, the apparatus of life of, say, the year 1600
represent great technical advances on the year 1450. A well-
known exhibition of that improvement was in the building of sea-
going ships which had made transoceanic exploration possible.
One reason for the forwardness of Germany and of Germans in the
art of printing was the excellence and reliability of German craftsmen. Regiomontanus (p. 171) left Hungary for Nürnberg (1470)
because he could there obtain good workers for his astronomical
instruments. But until the seventeenth century highly skilled
craftsmen were seldom invoked by the man of science. No
small part of Galileo's success as an experimenter was due to
his constant employment of specially trained mechanics. He thus
laid the foundation of the profession of scientific instrument
maker. The existence of that class became a main condition of the
advancement of science in the centuries which followed. Compound optical apparatus had been constructed by others before
Galileo. The results obtained were negative till the great discoverer perfected the method of manufacture.

With such instruments in his hand Galileo was in a position to
observe with an accuracy and a detail that had previously been
quite unknown. He is the effective inventor of the telescope and
the father of modern observational astronomy. There is, however, another aspect of his optical discoveries that is less often
recalled. He is the inventor also of the compound microscope, and,

The Insurgent Century

indeed, revelations of that instrument are mentioned in one of the earliest independent accounts of his work. The minute world revealed by this instrument was almost as wonderful as the new discoveries in the starry sphere. The heavens had always been recognized as vast almost beyond the power of thought. But the incomparable complexity of life and of matter close at hand was a wholly new conception. That beings, minute beyond the powers of our vision, could have structures as complete and complex as ourselves was a truly startling thought. If there was world beyond world in the heavens there was world beyond world within us.

It is interesting to see how these matters looked in the eyes of the first generation of professed microscopists. In England the pioneer of such studies was HENRY POWER (1623–68), disciple of Sir Thomas Browne, who writes in his *Experimental Philosophy* (1663):

> 'Dioptrical Glasses are but a Modern Invention neither do Records furnish us with anything that does antedate our late discoveries of the Telescope and Microscope. The want of which incomparable Artifice made the Ancients not onely erre in their fond Coelestial Hypothesis and Crystalline wheelwork of the Heavens but also in their nearer observations of the smallest sort of Creatures which have been perfunctorily described as the disregarded pieces and huslement of the Creation. . . . In these pretty Engines are lodged all the perfections of the largest animals: . . . and that which augments the miracle, all these in so narrow a room neither interfere nor impede one another in their operations. Ruder heads stand amazed at prodigious and Colossean pieces of Nature, but in these narrow Engines there is more curious Mathematicks.'

In the time of Galileo atomic views were coming again to the fore. There was as yet no experimental evidence for the existence and nature of atoms. It remained a philosophical doctrine. But it seemed to fit the revelations of the microscope. Were these tiny beings atoms? Were atoms alive? These questions gave rise to a considerable literature which, since it led nowhere, has been almost forgotten. Yet it fitted and stimulated current philosophical views, and the curiosity which it aroused had a very definite influence in directing the biological observation of the generations which followed.

Downfall of Aristotle. New Attempts at Synthesis

(c) *The Universe as Mathematical and Boundless.* With the advent of the Galilean physics and the Keplerian astronomy, it began at least to appear possible that all parts of the universe were mechanically interrelated. The astrological teaching of antiquity and of the Middle Ages had treated the inner spheres of the world scheme as dependent on the outer spheres. In this sense the extreme expression of astrological doctrine was determinist. But now Galileo, following Bruno and Gilbert, thought of the world as boundless. In such a universe no part could be said to be inner, none outer, none centre, none circumference. In such a universe the mechanics of one part are presumably the mechanics of another, though proof of this had to await Newton. Of such a boundless universe no beginning in time can be intelligibly predicated.

The implications of this view represent a series of enormous changes some of which we have already discussed. Especially it affected the conception of the task of the man of science. The physical world, in the thought of Galileo, was a separate and mathematical conception, a piece of machinery, the action of any part of which was calculable. It was thus quite separate from the moral world with which it had been united in the medieval scheme. The knowledge of the world as a whole—philosophy— was thus divided into two categories, *natural philosophy* and *moral philosophy*, a distinction which is still recalled in the naming of the departments of the university where Newton taught. In the main we may say that the division has held from Galileo's day to our own.

A further implication in the conception of a boundless physical universe and the separation of natural from moral philosophy is the movement known in modern times as 'scientific specialization.' Science, natural philosophy, proceeds on the information given by the senses. The line of its attack is thus limited and we cannot hope that anything but limited objectives can be reached. Science does not seek to solve universal problems. On the other hand it does seek to solve its limited problems with a known degree of accuracy and a known margin of error. The desire for exact expression and for the translation of observation into terms of measurement has penetrated every department of science from the time of Galileo onwards. Even the biological sciences have been affected. The physico-mathematical form in which the biological works of Sanctorius (1600), Harvey (1628), and Descartes (1637) are cast may be contrasted with the beautiful but not mathematically controlled works of the 'German fathers of botany' (1530-42, pp. 176-7) or of Vesalius (1543, p. 177). Since the work of Galileo there has always been a group of biologists that has sought to represent biology as a department of physics.

(d) *Religion and Science*. Medieval philosophy had presented a view of the world as a whole. Looking back on it, from our modern point of view, we can see two breaches of continuity. One between the celestial and the terrestrial, the other between the living and the not living. These two gaps were, however, well concealed from all but the most acute, until displayed in the seventeenth century by the work of such men as Galileo and Harvey. But thought could not rest content with the multiple system thus revealed. There is an insatiable demand for explanation of the world on a unitary basis. Law must reign, and if not divine law then physical law. This call for an explanation of all things in terms simpler than themselves was first met in modern times by the philosophy of Descartes (p. 221).

The conception of a mechanical and mathematical universe affected other philosophers whose world schemes have endured better than that of Descartes. The model suggested by the new science of mechanics involves the belief that any event in one part of the world must, of necessity, have its consequences in another part. Each event gives rise to its own chain, circle, sphere of events. Events are never without consequences which go on like waves caused by the dropping of a pebble in water producing ever widening if less apparent circles which are reflected and reflected again from the margins of the pool. This view of the world was essential to the thought of Spinoza (1632-77). In such a view we can think of the dissipation of neither matter nor energy. The belief in the conservation of both was implicit in all of Galileo's work though not expressed until he had been dead for two hundred years.

The whole question leads on to the philosophical problem of 'causality' where we cannot follow it. But science, true to its principle of limited attacks and limited objectives, has its own working rules of causality. It follows Galileo in agreeing to discuss only certain particular types of sequence and treating them as related, the relation being regarded as cause and effect. Thus the physicist will deal only with physical, the chemist only with chemical sequences, the biologist only with biological sequences. In the course of this process new relationships may be discerned or become more apparent, as for instance in the physical state of the heavenly bodies or of the relative constitution of parents and offspring. Thus will arise new sciences—astrophysics and genetics—which will limit their scope to the relations in their particular fields. All departments will agree, however, that only those sequences shall be considered that can be measured or at least estimated. From Galileo's day onward we see science as measurement.

But since science must limit its objectives, the world based on science, as Galileo the artist well knew, is not a complete world. The appearance of our world depends on how we look at our world —that is, on our 'mood'. We may be in a scientific, an artistic, an emotional, a social mood. The resultant of all the ways that we have of looking at our world—the resultant of our moods—is, in effect, our religion. Galileo founded a new conception of the world —he almost founded a mood in which to regard it. In doing so he certainly affected the religion of all men who are able to accept or partake of his mood. But to say that that mood was all of Galileo, to say that the universe as he looked at it was wholly mathematical and physical, is not only going beyond his teaching but also going beyond all that we can learn of the nature of the man. Reasons are doubtless at hand for the rejection of any established religious formula, but it would be perverting the historical record to ascribe the desire to do so to Galileo or to men of science in general and as a whole.

6. *Prophets of Science.*

DESCARTES (1596–1650), the 'first modern philosopher' and the most dominant thinker of the seventeenth century, made striking contributions both to scientific theory and practice.

(*a*) He set forth views as to how science should be prosecuted.

The Insurgent Century

(b) He was the first in modern times to propound a unitary theory of the universe that became widely current.

(c) He made important contributions to mathematical, physical, and physiological science.

These three activities of Descartes are not as essentially connected as he would have wished. In 1633 he was about to publish his cosmic view in a work which he termed *The World*, when he heard of the condemnation of Galileo. He promptly withdrew the book. In the event his first publication was the *Discourse on Method* (1637).

(a) Descartes on Scientific Method.

From an early date Descartes felt great dissatisfaction with the results of the usual studies of his time. It seemed to him that there was no clear distinction between facts, theories, and tradition. Want of clarity was abhorrent to him. He attempted to divest himself of every preconceived notion and then to build up his knowledge. With this end in view he tells us in his *Discourse* that he made certain resolutions:

(i) 'Never to accept anything for true which he did not clearly know to be such, avoiding precipitancy and prejudice, and comprising nothing more in his judgment than was absolutely clear and distinct in his mind.'

(ii) 'To divide each of the difficulties under examination into as many parts as possible.'

(iii) 'To proceed in his thoughts always from the simplest and easiest to the more complex, assigning in thought a certain order even to those objects which in their own nature do not stand in a relation of antecedence and sequence—i.e. to seek relation everywhere.'

(iv) 'To make enumerations so complete and reviews so general that he might be assured that nothing was omitted.'

He believed that such truth as is ascertainable is so only by the application of these principles. These, he thought, are the true principles of science, and only by their application can science advance. They apply, he held, as much in the sphere of religion as in mathematical or physical matters. In essence, therefore, revealed religion in the ordinary sense is superfluous. For him the fundamental test of a truth is the clearness with which we apprehend it. *I think, therefore I am*, is the most clearly apprehended of all truths, and, therefore, personality cannot be an illusion. Similarly, to him, the conception of the soul as separate from the body was clear and even obvious; therefore, he maintained, it must be true. Moreover, he considered that the mind could not create something greater than itself. Therefore, the conception of infinite perfection transcending humanity must have been put into our minds by infinite perfection itself; that is, by God.

It is noteworthy that in reaching his scientific results he did not employ the method that he advocates. It is doubtful if any one actively and successfully prosecuting scientific discovery has ever or could ever proceed on the lines that he lays down. It may, indeed, be doubted whether scientific discovery ever follows any prearranged system. The spirit bloweth where it listeth and discovery is a thing of the spirit. There is no one method of discovery but as many methods as there are discoverers. There is no human faculty or power that has not at times been pressed into the service of scientific discovery. There *is* a method of scientific *demonstration*, but that is a very different thing from a method of discovery. The setting forth of the one must almost necessarily conceal the nature of the other. We, therefore, consider Descartes separately as a scientific discoverer and as a prophet and critic of science.

Of the achievement of Galileo, Descartes formed no high estimate. Galileo was eliciting mechanical laws. Descartes belittled this effort since it included no analysis of the basic conceptions with which Galileo was dealing, force, motion, matter, space, time, number, extension, and the like. The obvious retort is that had Galileo done these things, philosophy might have been richer, but science would certainly have been poorer in being deprived of the most successful experimenter and the most acute exponent of natural law that had yet arisen.

(b) Descartes' Cosmology.

We may now turn to the conception of the material universe as formed by Descartes. Here, too, we may honour him as a pioneer, while we regret that he is less critical of himself than of others. The form of the world, according to him, is inevitable, in the sense that, had God created more worlds, 'provided only that He had established certain laws of nature and had lent them His concurrence to act as is their wont, the physical features of these worlds would inevitably form as they have done on ours.' Descartes accepts the probability of creation of matter as a momentary act, but holds that this act of creation was the same as that by which creation is now sustained.

Descartes regards the universe as infinite and devoid of any empty space. The primary quality of matter is extension, but there are also the secondary and derived qualities of divisibility and mobility, which are created by God. We may connect the assertion of Descartes that divisibility and mobility are derived qualities with the formulation of the law that matter, in so far as it is unaffected by extraneous forces, remains in motion or at rest.

He regarded matter as uniform—i.e. made of the same basic stuff—though divided and figured in endless variety. Matter is closely packed, without any vacuum. Therefore, the movement of any part of matter produces the movement of all matter. It thus follows that throughout the universe there are circular vortices of material particles that vary in size and in velocity. If one considers any limited part of the universe, the particles in it, as they whirl around their vortices, get their corners rubbed off. These being rubbed finer and finer become a minutely divided dust which tends to centripetal action. This fine dust is 'first matter' and forms the sun and the stars. Ultimately these spherical globules acquire a contrary or centrifugal action. They then form 'second matter', which constitutes the atmosphere enveloping first matter. The centrifugal tendency of the second matter produces rays of light which come in waves from the sun or the stars to our eyes. In the process of vortex formation particles are liable to get detained on their way to the centre. These settle round the edge of the sun or star, like froth or foam. This 'third matter' can be recognized as the sun-spots (p. 208) and certain other celestial phenomena. Major vortices are responsible for planetary movements, minor vortices for terrestrial phenomena. The action of gravity is identified with centripetal action of a vortex.

The theory of vortices failed to explain a multitude of known phenomena, including Kepler's laws of planetary motion (p. 204). It became, however, very fashionable. It was elaborated and a whole system of physics and cosmology erected on it. It survived in France until near the middle of the eighteenth century though it had less influence in other countries. From the first it was subject to destructive criticism, and it was made untenable by the work of Newton.

(c) Descartes on the Nature of Man.

For the completeness of his system it was necessary for Descartes to include the phenomena presented by living things. Here, too, his work was of a pioneer character though he invented a number of structures and functions that had no existence outside his mind. The analogies that he draws, however, are sometimes both striking and valuable.

'I remained satisfied that God first formed the body of man wholly like to ours, as well in outward shape as in inward conformation, and of the same matter; that at first He placed in it no rational soul, nor any other principle, beyond kindling in the heart a flameless fire similar, as I think, to the heat generated in damp hay, or to that which causes fermentation in must.'

Descartes is here trying to co-ordinate combustion, metabolism, respiration, and fermentation.

'For, when I examined the kind of functions which might, as consequences of this supposition, exist in this body, I found precisely all those which may exist in us independently of all power of thinking, and consequently without being in any measure owing to the soul; in other words, to that part of us which is distinct from the body, and from that of which it has been said above that the nature distinctly consists in thinking—functions in which the animals void of reason may be said wholly to resemble us; but among which I could not discover any of those that, as dependent on thought alone, belong to us as man, while, on the other hand, I did afterwards discover these as soon as I supposed God to have created a rational soul, and to have annexed it to this body.'

He thus considered that man once existed without a rational soul and that animals are still automata. He knew, for instance, William Harvey's account of the circulation of the blood, and he based upon it a most elaborate and carefully worked-out theory of the action of the animal body. Man, however, at least in his present state, Descartes considered to differ from animals, in the possession of a soul. This he believed to be especially associated with a particular part of the body, the pineal gland, a structure within the brain which, in his erroneous opinion, was not found in animals. In the pineal gland two clear and distinct ideas produce an absolute mystery. It is there that the mystery of creation is concentrated.

The Cartesian philosophy was the first complete and coherent system of modern times. It rapidly found adherents, spread in every country, and was popular for several generations. In Descartes' native land it won its way even among churchmen. Gradually, however, the numerous physical errors which it involved were exposed. Towards the end of the century the theory of vortices became quite untenable. It was then shown to be inconsistent with astronomical observation, and to harmonize neither with the cosmical system of Newton nor with the revived atomic theory. As an explanation of cosmic phenomena it could no longer be held. Important scientific works that professed to be based on the Cartesian system appeared, however, as late as the middle of the eighteenth century.

Further, the advance of physiological knowledge exposed basic errors of Descartes in the interpretation of the workings of the animal body. Descartes, however, had laid the foundations of modern philosophy, and from his time on there has been a continuous chain of thinkers who have claimed to interpret the world by the unaided powers of their own minds.

(d) Francis Bacon as Prophet of Science.

Less adapted than Descartes by his powers, his temper, and his outlook, to make a great philosophical synthesis was FRANCIS BACON Lord Verulam (1561–1626). The Englishman was, moreover, less efficient in the actual handling of scientific material and incomparably below Descartes in scientific achievement. Despite the fact that Bacon was the older of the two, his influence made itself felt somewhat later than that of Descartes. Bacon's scientific ineffectiveness prevented his works and their author from gaining an entry into circles occupied in the actual advancement of science. 'He writes philosophy (i.e. science)', said William Harvey, 'like a Lord Chancellor.' While no one ever worked on the scientific principles laid down by Descartes, we must, nevertheless, remember that there were three Descartes, the cosmic philosopher, the prophet and critic of science, and the investigator. There was, however, only one Bacon, the author of the *Proficiencie and Advancement of Learning* (1605) and the *Instauratio magna* or *Novum Organum* (1620).

Let us consider Bacon's attitude towards the investigation of Nature as set forth in these works. What was this new scientific process which he practised worse than he preached? Bacon was for conducting his investigations by collecting all the facts. This done, he thought, the facts might be passed through a sort of automatic logical mill. The results would then emerge. But this method cannot be applied in practice, since facts, phenomena, are infinite in number. Therefore, we must somehow choose from among them, though Bacon thought otherwise. How then shall we choose our facts? Experience shows that they only choose profitably who have a knowledge of how their predecessors have succeeded or failed in their choosing. In other words, the process of choosing facts is an *act of judgement* on the part of the *learned chooser*, the man of science. So it is also with the process of choosing words on the part of the word-chooser whom we call a poet. The choice of the man of science, as of the poet, is controlled by knowledge of his art—of 'his subject' as we are wont to call it at the universities or in the laboratories. The man of science, like the poet, exercises his judgement to select those things which bear a certain relation to each other. And yet no skill in reasoning, however deft, no knowledge of the nature of scientific method, however profound, no acquaintance with his science, however complete, will make a scientific discoverer. Nor, for that matter, will any learning in the lore of metre or in the nature and history of poetry make a poet. Men of science, like poets, can be shaped, but they cannot be made. They must be born with that incommunicable power of judgement.

The scientific man in the prosecution of his art of discovery has to practise three distinguishable mental processes. These may be distinguished as firstly, the choosing of his facts; secondly, the formation of an hypothesis that links them together; and thirdly, the testing of the truth or falsehood of the hypothesis. When this hypothesis answers numerous and repeated tests, he has made

227

what is usually called a 'scientific discovery'. It is doubtless true that the three processes of choosing facts, drawing a hypothesis or conclusion, and testing the conclusion, are often confused in his own thinking by the man of science. Often, too, his demonstration of his discovery, that is the testing of his hypothesis, helps him, more or less unconsciously, to new acts of judgement, these to a new selection of facts, and so on in endless complexity. But essentially the three processes are distinct, and one might be largely developed while the others were in a state of relative arrest.

In this matter scientific articles, and especially scientific text-books, habitually give a false impression. These scientific works are composed to demonstrate the truth of certain views. In doing so they must needs obscure the process by which the investigator reached those views. That process consists, in effect, of a series of improvised judgements or 'working hypotheses', interspersed with imperfect and merely provisional demonstrations. Many hypotheses and many demonstrations have had to be discarded when submitted to a further process of testing. Thus a scientific article or book, which tells nothing of these side issues, blind alleys, and false starts, tends, in some sort, to conceal the tracks of the investigator. For this reason, among others, science can never be learned from books, but only by contact with phenomena.

The distinction between the process of discovery and the *demonstration* of discovery was constantly missed during the Middle Ages. On this point, in which our thought is separated from that of the men of those times, Bacon remained in darkness. He succeeded, indeed, in emphasizing the importance of the operation of collection of facts. He failed to perceive how deeply the act of judgement must be involved in the effective collection of facts.

As an insurance against bias in the collection and error in the consideration of facts, Bacon warned men against his four famous *Idols*, four false notions, or erroneous ways of looking at Nature. There were the *Idols of the Tribe*, fallacies inherent in humankind in general, and notably man's proneness to suppose in nature greater order than is actually there. There were the *Idols of the Cave*, errors inherent in our individual constitution, our private and particular prejudices, as we may term them. There were the *Idols of the Market-place*, errors arising from received systems of thought. There were the *Idols of the Theatre*, errors arising from the influence of mere words over our minds (*Novum Organum*, 1620).

But did not Bacon himself fail to discern a fifth set of idols? These we may term the *Idols of the Academy*. Their worship involves the fallacy of supposing that a blind though learned rule can take the place of judgement. It was this that prevented Bacon from entering into the promised land, of which but a Pisgah view was granted him.

Yet despite Bacon's failure in the practical application of his method, the world owes to him some conceptions of high importance for the development of science.

(*a*) He set forth the widening intellectual breach which separated his day from the Middle Ages. He perceived the vices of the scholastic method. In the clarity and vigour with which he denounced these vices, he stands above those of his contemporaries who were striving toward a new form of intellectual activity.

(*b*) He perceived, better than any of his day, the extreme difficulty of ascertaining the facts of nature. He forecast the critical discussion that characterizes modern science. He missed, however, the important point that the delicate process of observation is so closely interlocked with discussion that both must almost necessarily be performed by the same worker.

(*c*) English writers of the later seventeenth century concur in ascribing to the impetus of Bacon's writings the foundation of the Royal Society. Thomas Sprat (1635–1713), Bishop of Rochester, the first historian of the Society, assures us of this (1677), as do Oldenburg and Wilkins, its first secretaries. The opinion is fully confirmed by Robert Boyle (1627–91), the most effective of its founders, and by John Locke (1632–1704), the greatest of English philosophers.

(*d*) It is, perhaps, in the department of psychological speculation that the influence of Bacon has been most direct. The basic principle of the philosophy of John Locke is that all our ideas are ultimately the product of sensation (*Essay concerning Human Understanding*, 1690). This conception is implicit in Bacon's great work, his *Novum Organum* (1620). Through the 'practical' tendency of his philosophy and especially through Locke, Bacon was the father of certain characteristically English schools of thought in psychology and ethics. These have affected deeply the subsequent course of scientific development.

Whatever his scientific failures, we may thus accord to Bacon his own claim that 'he rang the bell which called the wits together'.

7. Character and Conduct of Matter.

Our word *matter* is derived from the Latin *materia*, which in its turn is connected with *mater*, 'mother'. Originally *materia* was a general term for the stuff of which things are composed and especially things employed in buildings. So in the medieval nomenclature and in that of the alchemists *materia prima* was the stuff of which all things were built, the 'primal matter' that lay at the back of all four elements. Both alchemists and the medieval philosophers were prepared to believe that matter of one type, by a mere rearrangement of its four elements, could be transformed into matter of another type. Nor were they convinced that, in some circumstances, matter might not appear 'out of the air' or out of nothing. They did not in general regard air as possessing weight, and some of them would have claimed that it had 'negative weight' since like fire it tended to rise. Nevertheless, it is not exactly true to say that the medieval writers had no idea at all of what we call the 'conservation of matter' Had that been so no trade that used weights would have been possible. Had there been no constancy in weight such stories as that of Hiero's crown (p. 64) would have been meaningless. We would rather say that in the Middle Ages the idea of conservation of matter was indefinite, inexact, unexpressed, and implicit, whereas now it has become definite, exact, formulated, and explicit. Three centuries of application of experimental methods has made this difference.

There was one particular aspect of matter that had special bearings on the early development of modern ideas on the subject. The question as to the nature of the air that we breathe and whether or not it has weight had been debated since antiquity. One of the most popular of the pagan systems of physical philosophy

—to which Galen adhered—held that the 'pneuma' of the world soul is inhaled during the act of breathing, which on that account is necessary for life. On the cessation of breathing the individual soul joined again the world soul (p. 91). Such a view was contrary to the medieval Christian attitude. Medieval Christian thought generally ignored the objective existence of air as either a material or spiritual entity. Nevertheless, Peter of Abano (p. 163) in the fourteenth century on theoretical grounds, and Cardinal Nicholas of Cusa (p. 171) in the fifteenth century on experimental grounds, had held that something material was in fact drawn from the air. The problem was given a new aspect by van Helmont.

The Belgian, JAN BAPTIST VAN HELMONT (1577–1644), was a pious mystic who devoted his life to the investigation of chemical processes basing himself largely on the views of Paracelsus (p. 174). He published little. Soon after he died his son, who occupied himself with similar pursuits, collected his father's writings and issued them as *The Fount of Medicine (Ortus medicinae* 1648). These writings are in extremely obscure language. Moreover, the alchemical school, to which van Helmont belonged, was justly despised by the clear thinkers, such as Galileo and Descartes, who were attacking Aristotelianism and contributing to the upbuilding of the new physical philosophy. Thus van Helmont exerted little influence on scientific writings until his works were translated and interpreted in the 'sixties of the seventeenth century.

Van Helmont concluded from a repetition of the experiments of Nicolas of Cusa (p. 171) that plants draw their whole substance from water. (He did not, of course, know the part played by the atmosphere, especially by carbon dioxide, in the growth of plants.) Further, he showed that vapours, though similar in appearance, may be very different in character and conduct. In other words, there are many kinds of 'gas' The idea is so familiar to us that it is hard to realize it as an innovation. Yet the very word *gas* was invented by van Helmont. Etymologically it is *chaos* phonetically transmuted in his native Flemish speech.

Galileo also was well aware that the atmosphere has weight. Nevertheless, he failed to invoke it to explain the failure of a suction pump to lift water higher than 35 feet. The explanation was

adduced by Galileo's pupil and secretary, EVANGELISTA TORRI-
CELLI (1608–47). He reasoned that as mercury is about 14 times
as heavy as water the atmosphere should support $\frac{35}{14}$ —i.e. about
2½ feet of mercury. He selected a glass tube of ¼-inch calibre and
4 feet long and closed at one end. This he filled with mercury,

applied his finger to the open end and inverted it in a basin of mercury. The mercury
sank at once to 2½ feet above the basin,
leaving 1½ feet apparently empty (1643). This
was the *Torricellian vacuum* as it came to
be called.

Torricelli had in fact constructed a barometer (Greek 'weight measurer'). He observed that at times his barometer stood
higher than at other times. He inferred that
when the barometer stood high the air was
heavier, when low, lighter. Descartes predicted that at greater altitudes the mercury
column would stand lower since there was
less atmosphere to support it. Experiments
suggested by Pascal confirmed this. The
matter was further investigated by Huygens,
Halley, Leibniz, and others. The barometer
has since been greatly improved, but in
essence it is still that suggested by Torricelli.

FIG. 64. Galileo's
thermometer.

The thermometer has had a somewhat
different history. An air thermometer was
invented by Galileo. It consisted of a glass
bulb containing air connected to a glass tube dipping into a liquid
(Fig. 64). It was very sensitive to temperature changes, but was
very inexact as it was also subject to barometric changes. About
1612 Galileo invented the modern type of sealed tube with glass
bulb filled with liquid. Technical difficulties in construction,
however, prevented a delicate and accurate instrument from being
made until the eighteenth century.

Very great advances in our knowledge of physical and chemical states are due to the air-pump. This instrument was invented in 1656 by OTTO VON GUERICKE (1602–86), burgomaster of Magdeburg in Prussia. With it he gave a direct and convincing demonstration that air has weight. Guericke is remembered by the 'Magdeburg hemispheres' which, though easily separable under normal conditions, could not be separated by two teams of sixteen horses each when he had drawn out the air with his air-pump. Guericke also invented the first electrical machine. It consisted of a globe of sulphur which was made to rotate. Pressure of the hands upon the rotating globe charged it electrically. He also showed that bodies charged with the same kind of electricity repel each other.

The air-pump of Guericke was considerably improved (1658–9) by ROBERT HOOKE (1635–1703) working at Oxford for his employer ROBERT BOYLE (1627–91). Hooke was one of the most skilful and ingenious of physical experimenters, Boyle one of the ablest and most suggestive of scientific investigators. A large part of the foundations of the modern sciences of chemistry and physics in their various departments was laid down by these two men.

By means of the air-pump Boyle and Hooke examined the elasticity, compressibility, and weight of the air (1660). The necessity of air for respiration and combustion was later demonstrated by means of the same instrument (1662). Finally, Boyle showed that a part only of the air was used in the process of respiration or combustion. The matter was well expressed by Hooke in his great work *Micrographia* (1665):

'The dissolution of sulphureous bodies is made by a substance inherent and mixt with the Air, that is like, if not the very same, with that which is fixt in Salt-peter. . . . That shining body which we call flame is nothing else but a mixture of Air and volatile parts of combustible sulphureous bodies which are acting upon each other whilst they ascend.

This substance 'inherent and mixed with the air' is oxygen, of which Hooke and Boyle may be regarded as the discoverers.

Boyle's name is familiarly recalled in 'Boyle's law' which states that the volume of a gas varies inversely as the pressure upon it, provided temperature be constant. Boyle took a U-shaped tube with a shorter closed and a longer open limb. By pouring mercury into it he cut off air in the short limb and, by shaking, the mercury was brought to the same level in both limbs. The air in the short limb was now under atmospheric pressure. Adding mercury to the long limb he could increase the pressure continuously, thereby reducing the bulk of contained air. Thus when the barometric pressure stood at 30 inches, by adding mercury in the long limb till it stood 30 inches above the level in the short limb, the pressure on the imprisoned air was doubled. The bulk of that air was then found to be reduced to one half. Under three times the atmospheric pressure it was reduced to a third and so on. Moreover, he could reverse the process.

Boyle's more purely chemical investigations and speculations were of high importance. His most famous work, the *Sceptical Chymist* (1661), opens the modern period of chemistry, and marks the end of the doctrine of the four elements of the Aristotelians.

'To prevent mistakes,' he says, 'I must advertize to you, that I now mean by Elements . . . certain Primitive and simple . . . bodies ; which not being made of any other bodies, or of one another, are the Ingredients of which all those call'd perfectly mixt Bodies are immediately compounded and into which they are ultimately resolved.'

This, in effect, is the modern definition of an element. There can be little doubt that he derived his view of chemical elements in part from the modest German teacher, JOACHIM JUNG (1587–1657) of Hamburg. Jung had enunciated similar views as early as 1634 and published them in 1642. Boyle had received a draft of Jung's physical philosophy in a letter received by him in 1654.

Among other important contributions of Boyle must be included the suggestion of chemical 'indicators' for testing the acidity or alkalinity of liquids, and his isolation of elemental phosphorus. He was extremely active in the scientific life of the later seventeenth century. Almost every aspect of contemporary science is discussed in the course of his numerous and diffuse works.

There is one doctrine popularized by Boyle to which we must pay especial attention. In his *Origin of Forms and Qualities* (1666) he definitely 'espoused the atomical philosophy, corrected and purged from the wild fancies and extravagancies of the first

234

inventors of it'. He assumes the existence of a universal matter, common to all bodies, extended, divisible, impenetrable. This matter consists of innumerable particles, each solid, imperceptible and of its own determinate shape. 'These particles are the true *prima naturalia*.' There are also multitudes of corpuscles built up from several such particles and substantially indivisible or at least very rarely split up into their *prima naturalia*. Such secondary 'clusters' have each their own particular shape. 'Clusters' and 'prima naturalia' may adhere to form characteristic and similar groups which are not without analogy to molecules and atoms in the modern acceptance of these terms. Nevertheless, the analogy of Boyle's atomism to either modern or ancient atomism is far from close.

Boyle had certainly derived his atomic views from the French philosopher, PIERRE GASSENDI (1592–1655), 'the reviver of Epicureanism'. Gassendi adapted that system of thought to the exigencies of the philosophy of his time. Boyle's nomenclature is taken direct from Gassendi who devoted at least twenty years to his great work on atomic philosophy (1649).

Some form of corpuscular philosophy was widely accepted by Boyle's contemporaries, especially in England, where it was espoused by the philosopher, JOHN LOCKE (1632–1704). The corpuscular philosophy, however, though much discussed, was not developed on the experimental side for more than a century. Chemical observations were collected in plenty and science became overwhelmed by a vast number of disconnected chemical facts and records, inadequately linked by generalizations.

An idea of the estimate which seventeenth-century thought placed upon a corpuscular (or atomic) hypothesis can be gathered from John Locke's *Essay concerning Human Understanding* (1690). Whenever he deals with the ultimate physical cause of secondary qualities and of powers of material substances, it is to 'the corpuscularian hypothesis' that he appeals. 'These insensible corpuscles', 'the active parts of matter and the great instruments of nature', are for him the source of all secondary qualities. He maintains that if the figure, size, texture, and motion of the minute constituent parts of any two bodies could be known, then the mutual operations of bodies could be foretold.

The Insurgent Century

Thus 'the dissolving of silver in *aqua-fortis* and gold in *aqua-regia* and not vice versa, would then, perhaps, be no more difficult to know than it is to a smith to understand why the turning of one key will open a lock and not the turning of another'.

8. Mechanization of Physiology.

(a) First Application of Physics to Physiology.

Biological science, it is often said, always lags behind physical science and is always in a more elementary stage. The statement is hardly borne out by history. It depends for any truth that it may possess upon a particular conception of the nature of science. In antiquity, in the hands of Aristotle, biological science was far ahead of physical. Again the earliest modern scientific work of a monographic character, the great book of Vesalius (p. 177), is exclusively biological. The treatise of Copernicus, published in the same year, is medieval by comparison, and contains very few original observations (p. 179). To justify the doctrine of the relative backwardness of biological science it is necessary to postulate that the aim of biology is to represent biological phenomena in physical terms. Thus expressed the statement becomes a self-evident proposition for, if the postulate be granted, biology can never advance beyond its physical data. A large school of biological thinkers does not accept this postulate. Nevertheless, it is true that the most significant biological advances of the insurgent century were, in fact, attempts to express biological findings in physical terms.

The first to apply the new physical philosophy to biological matters was SANTORIO SANTORIO (1561–1636), a professor of medicine at Padua, in his little tract *De medicina statica* (1614). Inspired by the methods of Galileo who had been his colleague at Padua, he sought to compare the weight of the human body at different times and in different circumstances. He found that the body loses weight by mere exposure, a process which he assigned to 'insensible perspiration'. His experiments laid the foundation of the modern study of 'metabolism'. Santorio also adapted Galileo's thermometer to clinical purposes. It marks the medieval character of much of the thought of the day that his account

Downfall of Aristotle. New Attempts at Synthesis

of this (1626) is concealed in a commentary on a work of Avicenna (p. 134).

The Englishman, WILLIAM HARVEY (1578–1657), is also to be regarded as a disciple of Galileo though he himself was, perhaps, little aware of it. Harvey studied at Padua (1598–1601) while Galileo was active there. By 1615 he had attained to a conception of the circulation of the blood. He published his demonstration in 1628. The story of that discovery is very accessible. We would emphasize that the essential part of its demonstration is the result not of mere observation but of the application of Galileo's principle of measurement. Having shown that the blood can only leave the ventricle of the heart in one direction, he turns to measure the capacity of the heart. He finds it to be two ounces. The heart beats 72 times a minute so that in the hour it throws into the system $2 \times 72 \times 60$ ounces $= 8,640$ ounces $= 540$ pounds, that is to say about three times the body weight! Where can all this blood come from? Where can it all go to? The answer to that is that the blood is a stage army which goes off only to come on again. It is the same blood that is always returning (Fig. 65).

The knowledge that the blood circulates has formed the foundation on which has since been built a mass of physical interpretation of the activities of living things. This aggregate forms the science of *physiology*. The blood is a carrier, ever going its rounds over the same route to return whence it came. What does it carry? And why? How and where does it take up its loads? How, where, and why does it part with them? The answering of these questions has formed the main task of physiology since Harvey's time. As each generation has obtained a more complete and a more rational answer for one organ or another, so it has been possible to form a clearer picture of some part of the animal body as a working mechanical model.

Yet despite the triumphs of physical methods in physiology, we cannot suppose, with Descartes, that the clearest image—which is certainly at first sight the most satisfying—is of necessity also the truest, for the animal body can be shown on various grounds to be no mechanical model. A machine is made up of the sum of its parts. An animal body, as Aristotle perceived, is no more the sum of its parts than is a work of art. The Aristotelian

world-system was falling. The Aristotelian biology held and still holds.

(b) Physiological System of Descartes.

Nevertheless, the physical discoveries of Galileo and the demonstrations of Santorio (p. 236) and of Harvey (p. 237) gave a great impetus to the attempt to explain vital workings on mechanical grounds. A number of seventeenth-century investigators devoted themselves to this task. The most impressive exponent of physiological theory along these lines was Descartes himself. His account of the subject appeared posthumously (1662 and 1664). It is important as the first modern book devoted to the subject of physiology.

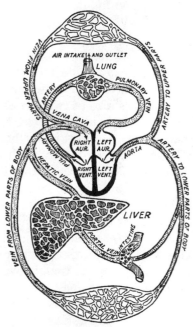

FIG. 65. Diagram to illustrate Harvey's Theory of the Circulation of the Blood.

Descartes had not himself any extensive practical knowledge of physiology. On theoretical grounds he set forth a very complicated apparatus which he believed to be a model of animal structure. Subsequent investigation failed to confirm many of his findings. For a time, however, his ingenious scheme attracted many. A strong point in his physiological teaching was the stress laid on the nervous system, and on its power of coordinating the different bodily activities. Thus expressed, his view may sound modern, but it is, in fact, grotesquely wrong in detail.

An important part of Descartes' theory is the position accorded to man. He regarded man as unique in his possession of a soul. Now in the view of Descartes the special prerogative of the soul is to originate action. Animals, he thought, are machines, automata. Therefore, given that we know enough of the works of the machine, we can tell how it will act under any given circumstances. But the human soul he regarded as obeying no such laws, nor any laws but its own. Its nature he believed to be a complete mystery for ever sealed to us. Descartes conceived that the soul governs the body through the action of the nervous system, though how it does so he again leaves as a mystery. The two insoluble mysteries come, he believed, into relationship to each other in a structure or organ in the brain, known to modern physiology as the 'pineal body'. This organ he wrongly believed absent in animals other than man. All their actions and movements, even those which seem to express pain or fear, are purely automatic. It is the modern 'behaviourism' with man expressly excluded.

The word 'mystery' is not popular among modern men of science. It is, therefore, right to point out that the processes by which a sensory impression passes into sensation, by which sensation educes thought, and by which thoughts are followed by acts, have been in no way elucidated by physiological science. In these matters we are in no better case than Descartes. If we have abandoned his terminology we are no nearer a solution of his leading problems. The basic defects of Descartes system were errors in matters of fact. It was on account of these that he ceased to have a physiological following with the first generation after the publication of his essay on man.

(c) Iatrophysicists.

One of the ablest critics of the physiological system of Descartes was the Dane, NIELS STENO (1648–86), whose scientific work was done mostly in Italy and France. Steno, like Descartes, was a mechanist, but unlike Descartes applied himself to the exploration of bodily structure. He found a pineal gland like that of man in other animals, and he could not persuade himself that it had the connexions, material or spiritual, described by Descartes. His criticism of Descartes in detail was very damaging (p. 278).

More constructive was the achievement of GIOVANNI ALFONSO BORELLI (1608–79), an eminent Italian mathematician, astronomer, and polymath, a friend of Galileo and Malpighi. Borelli's work *On motion of animals* (1680) is the classic of what is variously called the 'iatrophysical' or 'iatromathematical' school. It stands as the greatest early triumph in the application of the science of mechanics to the working of the living organism. Stirred by the success of Galileo in giving a mathematical expression to mechanical events, Borelli attempted to do the like with the

FIG. 66. Modified from Borelli to illustrate bodily action as mechanism.

animal body. In this undertaking he was, in fact, very successful. That department of physiology which treats of muscular movement on mechanical principles was effectively founded and largely developed by him. Here his mathematical and physical training was specially useful. He endeavoured, with some success, to extend mechanical principles to such activities as the flight of birds and the swimming of fish. His mechanical analyses of the movements of the heart, or of the intestines, were less successful, and he naturally failed altogether in his attempt to introduce mechanical ideas in explanation of what we now know to be chemical processes, such as digestion.

(d) Iatrochemists.

Just as Descartes and Borelli sought to explain all animal activity on a mechanical basis, so others resorted to chemical interpretation. Forerunners of this point of view were Paracelsus (p. 174) and van Helmont (p. 231). A more coherent attempt was made by FRANCISCUS SYLVIUS (1614–72), professor of medicine at

Downfall of Aristotle. New Attempts at Synthesis

Leyden. That university had become in the second half of the seventeenth century the most progressive scientific centre north of the Alps. It was the seat of the first university laboratory, which was built at his instance.

Sylvius devoted much attention to the study of salts, which he recognized as the result of the union of acids and bases. Thus he attained to the idea of chemical affinity—an important advance. With a good knowledge of anatomy and accepting the main mechanistic advances, such as the doctrine of the circulation of the blood and the mechanics of muscular motion, Sylvius sought to give a chemical interpretation to other vital activities, expressing them in terms of 'acid and alkali' and of 'fermentation'. In this attempt he made no clear distinction between changes induced by 'unorganized' ferments, as gastric juice or rennet, and changes induced by micro-organisms, as alcoholic fermentation or leavening by yeast. Nevertheless, he and his school added considerably to our knowledge of physiological processes, notably by their examination of the body fluids, especially the digestive fluids such as the saliva and the secretions of the stomach and of the pancreas.

The views of yet another group of biological theorists were best expressed by another expert chemist, GEORG ERNST STAHL (1660–1734). He is remembered in connexion with *phlogiston* (p. 288) and also stands as protagonist of his age of that view of the nature of the organism which goes under the term *vitalism* (p. 42). Though expressed in obscure and mystical language, Stahl's vitalism is in effect a return to the Aristotelian position and a denial of the views of Descartes, Borelli, and Sylvius. To Descartes the animal body was a machine, to Sylvius a laboratory. But for Stahl the phenomena characteristic of the living body are governed neither by physical nor chemical laws, but by laws of a wholly different kind. These are the laws of the *sensitive soul*. This sensitive soul in its ultimate analysis is not dissimilar from the *psyche* of Aristotle (p. 41). Stahl held that the immediate instruments, the natural slaves of this sensitive soul, were chemical processes, and his physiology thus develops along lines of which Aristotle could know nothing. This does not, however, alter the fact of his hypothesis being essentially of Aristotelian origin (p. 347).

The Insurgent Century

(e) *Plant Physiology.*

Most of the physiological discussion of the seventeenth century turned on the vital process of animals and especially those of man. The plant physiology of the age was of a more elementary character.

Van Helmont had shown that plants draw something of nutritive value from water (p. 231). This was contrary to the Aristotelian teaching that plants draw their food, ready elaborated, from the earth. The generation following van Helmont sought to erect a positive scheme of plant physiology without, however, very much success. MARCELLO MALPIGHI (1628–94), the great Bolognese microscopist (p. 243), held wrongly that the sap is brought to the leaves by the fibrous parts of the wood. The leaves, he thought, form from the sap the material required for growth. This, he knew, is distributed from the leaves to the various parts of the plant. He conceived a wholly imaginary 'circulation of sap' comparable to the circulation of the blood in animals. The respiration of plants, he falsely believed, is carried on through the 'spiral vessels' which bear a superficial resemblance to the breathing tubes or tracheae of insects with which he was very familiar.

The earliest experimental work on the physiology of plants was that of the French ecclesiastic, EDMÉ MARIOTTE (died 1684). This able physicist observed the high pressure with which sap rises. This he compared to blood pressure. To explain the existence of sap pressure he inferred that there must be something in plants which permits the entrance but prevents the exit of liquids. He held that it is sap pressure which expands the organs of plants and so contributes to their growth (1676).

Mariotte was definitely opposed to the Aristotelian conception of a vegetative soul (p. 41). He considered that this conception fails to explain the fact that every species of plant, and even the parts of a plant, exactly reproduce their own properties in their offspring, as with 'cuttings'. He was, so far as plants are concerned, a complete 'mechanist', and, therefore, anti-Aristotelian. All the 'vital' processes of plants were for him the result of the interplay of physical forces. He believed, as a corollary to this view, that organisms can be spontaneously generated (p. 245).

Downfall of Aristotle. New Attempts at Synthesis

(f) The Classical Microscopists.

The interpretation of vital activity in chemical and physical terms has had a continuous history to our own time. It is far other with the very striking microscopical researches with which the second half of the seventeenth century is crowded. Five investigators of the front rank, MARCELLO MALPIGHI (1628–94) at Bologna, ROBERT HOOKE (1635–1703) and NEHEMIAH GREW (1641–1712) in London, JAN SWAMMERDAM (1637–80) at Amsterdam, and ANTONY VAN LEEUWENHOEK (1632–1723) at Delft, all busied themselves with microscopic investigations of the structure and behaviour of living things. Their results impressed their contemporaries as deeply as they have modern historians. Nevertheless, their labours gave rise at the time to surprisingly few general ideas. Moreover, none of these microscopists inspired a school. Thus the following century hardly extended their observations, and we have to turn to the nineteenth century for their true continuators. On this account the 'classical microscopists' must be accorded a less prominent place in a general history of science than the great interest of their biological observations might suggest. We may briefly consider the general ideas that they initiated.

(i) The infinite complexity of living things in the microscopic world was nearly as philosophically disturbing as the unexpected complexity and ordered majesty of the astronomical world which Galileo and Kepler had unveiled to the astonished gaze of a previous generation. Notably the vast variety of minute life gave at once new point and added new difficulty to the conception of 'Creation'.

(ii) In a few notable respects the microscopic analysis of the tissues of animals aided the conception of the living body as a mechanism. Thus Harvey had shown that the blood in its circulation passed from arteries to veins. The channels of passage were unknown to him. They were revealed as 'capillary vessels' by Malpighi and Leeuwenhoek. These observers also discovered the corpuscles of the blood, the secretory functions of 'glands', and the fibrillary character of muscles, thus helping to complete details of the 'animal machine'.

(iii) The nature of sexual generation had been a subject of secular dispute. The discovery (1679) in the male element of 'animalcules' —'spermatozoa', as we now call them—aroused new speculations. The sperm then was organized. How was it organized? The eye of faith, lit within by its own light, looking through an imperfect microscope, lit without by a flickering candle, saw many a 'homunculus' in many a spermatozoon and even the piercing eye of a Malpighi or a Leeuwenhoek saw that which was not (Fig. 67).

FIG. 67. Spermatozoa as seen in the seventeenth century: *a, b, c,* by Leeuwenhoek (1679), *d,* by Hartsoeker (1694), in man, *e, f, g,* by Plantades (1699), in man.

The faith of others demanded that the homunculus should be carried by the female element, by the germ rather than by the sperm. That, too, was seen by the eye of faith. The more sober and conservative Harvey insisted that the production of the complex embryo in the simple substance of the egg was a 'new appearance', a recurring miracle, induced or excited by that magic imponderable, the 'generative force'.

(iv) Microscopic analysis revealed some similarity between the structures of plants and animals. False analogies were drawn and carried at times to fantastic lengths. For some such fantasies, justification at least appeared. The 'loves of the plants', on which poets had dwelt, were not wholly fables. It began to be realized that flowers contained the sexual elements, and a real parallel was perceived between their reproductive processes and those of animals.

(v) Lastly, there is an aspect of minute life that came to the fore

Downfall of Aristotle. *New Attempts at Synthesis*
in the later seventeenth century that requires some special discussion. It is the theme of *spontaneous generation* of living things,
that is, the generation of living things from non-living matter.

(g) *Spontaneous Generation.*

Neither ancient nor medieval nor renaissance scientific writers
doubted that spontaneous generation took place on occasion. The
subject has a considerable literature. In familiar language corpses
were said to 'breed' worms, dirt to 'breed' vermin, sour wine to
'breed' vinegar eels, and so forth. The doctrine of spontaneous
generation is often fathered on Aristotle and is certainly encountered in his writings, but in truth it was not so much a doctrine as
a universal assumption. It so fell out that when the reality of
spontaneous generation was first questioned, the authority of
Aristotle—or rather the contemporary misunderstanding of him—
was a very real obstacle to scientific advance. It is also true that
Aristotle gave spontaneous generation a place in his biological
scheme. But his error was shared by every naturalist until the
seventeenth century, and indeed it is hard to see how these men,
with the knowledge at their disposal, could take any other view.

With the advent of effective microscopes in the second half of
the seventeenth century, new tendencies set in. On the one hand,
exploration of minute life showed many cases of alleged spontaneous
generation to have been falsely interpreted. Thus plant galls had
been regarded as spontaneously generated, but Malpighi showed
that these curious growths are related to the action of insect larvae.
On the other hand, the microscope revealed minute organisms
which seemed to appear out of nothing. Thus Leeuwenhoek saw
excessively small creatures in infusions of hay and other substance.
Such infusions, perfectly clear when first prepared, become in a
few days or even hours cloudy with actively moving microscopic
forms. These seemed to be spontaneously generated.

The first scientific treatment of the question was made by
FRANCESCO REDI (1621–97), a physician of Florence. He tells us
(1668) that he

    'began to believe that all worms found in meat were derived from
    flies, and not from putrefaction. I was confirmed by observing

245

that, before the meat became wormy, there hovered over it flies of that very kind that later bred in it. Belief unconfirmed by experiment is vain. Therefore, I put a (dead) snake, some fish, and a slice of veal in four large, wide-mouthed flasks. These I closed and sealed. Then I filled the same number of flasks in the same way leaving them open. Flies were seen constantly entering and leaving the open flasks. The meat and the fish in them became wormy. In the closed flasks were no worms, though the contents were now putrid and stinking. Outside, on the cover of the closed flasks, a few maggots eagerly sought some crevice of entry.

'Thus the flesh of dead animals cannot engender worms unless the eggs of the living be deposited therein.

'Since air had been excluded from the closed flasks I made a new experiment to exclude all doubt. I put meat and fish in a vase covered with gauze. For further protection against flies, I placed it in a gauze-covered frame. I never saw any worms in the meat, though there were many on the frame, and flies, ever and anon, lit on the outer gauze and deposited their worms there.' [Abbreviated.]

It is odd that, despite these admirable experiments, Redi continued to believe that gall insects were spontaneously generated. This subject was taken up by another eminent Italian physician, ANTONIO VALLISNIERI (1661–1730), who again demonstrated that the larvae in galls originate in eggs deposited in the plants (1700). Vallisnieri compared the process of gall formation, as well as infection of plants by aphides, to the transmission of disease. Other investigators showed that fleas and lice—to this day popularly thought to be 'bred by dirt'—are, in fact, bred only by parents like themselves.

Thus the matter closed in the seventeenth century with the general balance of opinion against spontaneous generation. The possibility had been disproved—so far as a universal negative can be disproved—for visible organisms. The question was still open for the minute organisms encountered in infusions, the miscellaneous biological group classed in the language of the day as *Infusoria*.

In summary we may say that for Biology the Insurgent Century closed with a strong mechanistic bias. The microscopic world, however, remained an enigma, a land of wonders where all laws seemed at times to be broken. *De minimis non curat lex* ('The law does not concern itself with the most minute things') was not infrequently quoted, but the *lex* of the lawyer was a very different thing from the *lex naturae.*

VIII. THE MECHANICAL WORLD
Enthronement of Determinism
1. The Newtonian Key to the Mathematics of the Heavens.

ST. AUGUSTINE, about A.D. 427.

'This glorious doctor, as he went by the sea-side studying on the Trinity, found a little child which had made a little pit in the sand, and in his hand a spoon. And with the spoon he took water and poured it into the pit. And St. Augustine demanded what he did. And he answered: "I will lade out all the sea into this pit." "What?" said St. Augustine, "How may it be done, sith the sea is so great, and thy pit and spoon so little?" "Yea", said he, "I shall lightlier draw all the water of the sea and bring it into this pit than thou shalt bring the mystery of the Trinity into thy understanding, for it is greater to the comparison of thy wit than is this great sea unto this little pit." And therewith the child vanished.'—Abbreviated from 'The Golden Legend', as englished by William Caxton in 1483.

ISAAC NEWTON, A.D. 1727, shortly before his death.

'I do not know what I may appear to the world, but to myself I seem to have been only like a boy playing on the sea shore, and diverting myself in now and then finding a smoother pebble or a prettier shell than ordinary, while the great ocean of truth lay all undiscovered before me.'—From the *Anecdotes* of Joseph Spence (1699–1768).

Nothing emerges more clearly from a survey of the history of science than the lasting and essential sameness of the human spirit. The same aspiration for a coherent and comprehensive plan of his universe has characterized the mind of man from his very dawn and has survived a thousand defeats. It is therefore by no means strange that two men widely separated in time, genius, mood should take refuge in the same image to express their thought of infinity.

St. Augustine (354–430; p. 124) marks the effective beginning of a great epoch—a space of thirteen centuries—of which the effective end is marked by the arrival of ISAAC NEWTON (1642–1727). In his *Confessions* Augustine says that the sole fundamental truth lacking to the 'Platonists'—by which he means his Neoplatonic teachers (p. 124)—was the doctrine of the Incarnation. It was Augustine who determined that Christian thought should

248

be cast in a Neoplatonic mould, the impress of which it has borne
to our own day. It was his specifically Christian contribution to
award to man a unique dignity that was denied by certain pagan
philosophers. The Augustinian Neoplatonist is still working in
John Dryden (1631–1700). He is still straining his ears to hear
the 'music of the spheres' in the very year in which Newton's
greatest work appeared:

> From harmony, from heavenly harmony,
> This universal frame began.
> From harmony to harmony
> Through all the compass of the notes it ran,
> The diapason closing full in Man.
>
> (*A Song for St. Cecilia's Day*, 1687.)

In the Neoplatonic Christian world there was a hierarchy of
existences from purely spiritual to purely physical, the whole
linked together in God's heavenly harmony. The centuries rolled
on, and still that music of the spheres lulled man's mind to sleep
while his spirit waked. At last 'Aristotle'—a strangely changed
Aristotle—was recovered by the Latins from his Arabian custo-
dians (p. 162), and Scholasticism was born. Thus the ancient
cosmic scheme was enlarged by a Neoplatonic Aristotelianism
and the 'Dark Ages' of Faith gave place to the 'Middle Ages' of
Reason. Yet the spell of Plato and of his mouthpiece Augustine
still remained unbroken. The spiritual realm of the medieval
Christian stretched to the infinite, aspiring to the timeless God.
But the Christian's material world, the world of Augustine, of the
Neoplatonists, of the Stoics, and of 'Aristotle' remained limited by
those flaming ramparts beyond which even thought could hardly
penetrate.

The change came with the sixteenth century. Copernicus put
Earth from her ancient seat (p. 179) in a new form of an old con-
vention. But it was Bruno who proclaimed a universe of world
beyond world, without centre or circumference, in which all place
and all motion were relative. For him the stars were no longer
fixed and the frontiers of the universe were an idle dream. Next
Kepler reduced the movements of the heavenly bodies to intelligible mathematical rules. Galileo developed the system of
earthly mechanics with which, he hinted, the heavenly bodies

must somehow show accord. The conduct of matter was explored by Boyle and the new experimental school in a new and exact spirit, without the older presuppositions. While Harvey, Descartes, Borelli, expounded the living body as a mechanical system, Malpighi, Hooke, Grew, Leeuwenhoek, Swammerdam revealed, with their microscopes, vast and unsuspected regions and forms of life and the endlessly complex structure of even the minutest living things whose very existence had not been conceived.

In the third quarter of the seventeenth century learned societies in France, England, and Italy became centres for the exchange of scientific ideas. Perhaps the greatest achievement of these societies was the development and perfection of the manner of presenting inquiries. Thus the form of scientific communications became standardized and the demand for rigorous demonstration insistent. To quote authority was useless. *Nullius in verba* ('On the word of no man') stands on the crest of the Royal Society, whose publications began in 1664. The demand for evidence, for tangible data, for experience that can be repeated at will, had created science as we know it.

A fruitful source of misunderstanding of the aims and methods of the new science has been the unfortunate necessity that its technique of presentation must conceal the investigator himself. With the advent of the 'scientific journal' it becomes increasingly difficult to reach behind the text to the mind of the author. The new method of scientific publication does not allow us to see the trial attempts and tentative views of the men who wrote these books and papers. The point comes out admirably in the career of Newton himself.

The demonstrations of Galileo and Kepler, while they banished the earth-centred universe, did not at once destroy the conception of a sun-centred universe. No one had proved that the fixed stars were at various distances from our planetary system, and that view was not generally expressed. Nevertheless, such an opinion was certainly widely held in scientific circles. The varying size of the stars, the occasional appearance of new stars and many other phenomena, suggested that the stars were of the same order as our sun, or earth, and the planets of our system. The leaven of Bruno had worked. In 1686, the year before the publication of

Newtonian Key to the Mathematics of the Heavens

Newton's *Principia*, appeared the very famous work *On the Plurality of Worlds* by the French writer Le Bovier de Fontenelle (1657–1757). There were many who were thinking the same thought. 'I am of like opinion with all the great philosophers of our age', wrote Huygens, 'that the sun is of the same nature as the fixed stars. And may not every one of the stars or suns have as great a retinue of planets with moons to wait upon them as has our own sun?' (1698). The earth, then, being but a moving particle in space, space itself must be infinite, as Bruno had claimed. The Cosmos, not Man, must be the prime reality. In that new-found Cosmos the philosophers vied with one another in tracing laws, and the music of the spheres grew more distant and, at times, even discordant.

The change was at first one of degree rather than of kind. Law had been traced in the heavens from of old. The rules of planetary and stellar motion had been gradually developed from the astronomical theories of antiquity. Even in the Middle Ages a few new mathematical relationships of the heavenly bodies had been discerned. In the sixteenth century astronomy under Tycho (p.183) put her house in order for the Great Instauration (p. 227) of the coming age. And then Galileo startled the world with his proof of change in the uttermost heavens (p. 206) in the very region held by the Aristotelian and Platonic schemes to be utterly changeless.

By 1618 Kepler had enunciated his 'three laws of planetary motion', bringing these movements into an intelligible relation with each other (pp. 204–5). Then Galileo determined the rule of action of gravitation and came near to the 'three laws of motion' which we call Newton's (pp. 199–200). Others, Hooke and Wallis among them, were feeling their way in the same direction. But it was Newton who first affirmed these laws and succeeded in linking them with Kepler's laws of planetary movement. Before Newton, no man had shown, or clearly and demonstrably perceived, how the complex movements of the heavenly bodies were in relation to the natural succession of earthly phenomena. Reason no less than Faith would have been against such a view. Newton's unique achievement was to prove that this relationship amounted to identity. It was Newton who moved men's minds to see that the force that causes a stone to fall is that which keeps the planets

251

in their path. It was Newton who first enunciated a law the writ of which ran no less in the heavens than on the earth. With Newton the Universe acquired an independent rationality quite unrelated to the spiritual order or to anything outside itself. The Cosmology of Plato, of Aristotle, of Augustine, of the theologians was doomed.

Newton knew that if a stone be let drop, its weight—which is another name for Earth's attraction—will cause it to fall a certain measurable distance in the first second of its fall. He came early to suspect that the force which kept the moon in her orbit was none other than this terrestrial attraction. The period of the moon's revolution round the earth, and the dimensions of her orbit, were alike susceptible of estimation, so that her velocity could be calculated. Now the moon, like any body pursuing a curved course, is moving at any particular moment in a direction tangential to her orbit. But the moon, as we know, does not continue to move along the tangent, but is constrained to follow her elliptic path round the earth. At the end of the second, she, like the stone, has 'fallen' a certain distance toward the earth (Fig. 68). The earth has drawn her to herself. Now, from Kepler's laws, Newton had reason to suspect that the attractive power of the earth on any body decreases as the square of the distance from the centre of the earth. If the conjecture were correct, he had the equation:

$$\frac{\text{Distance fallen by Moon}}{\text{Distance fallen by stone}} = \frac{(\text{Distance of stone from Earth's centre})^2}{(\text{Distance of moon from Earth's centre})^2}.$$

When Newton first approached this problem (1666) he found that the moon's 'fall' was but seven-eighths of what he expected. But he had seized on the conception of universal gravitation, that is, that every particle of matter attracts every other, and he suspected that the attraction varied directly as the product of the attracting masses, and inversely as the square of the distance between them. It was still years before he was armed with the knowledge and means to show that the 'fall of the Moon' had the value required by his theory. By that time (1671) he had developed the wonderful mathematical method of dealing with curves which has since, with another nomenclature, become familiar under the name of 'Calculus'.

252

Newtonian Key to the Mathematics of the Heavens

The action of gravity on the earth and in the heavens was now seen to be the same, at least for a particular case. Newton's grand hypothesis was launched, though not yet worked out in detail. We owe it to the astronomer EDMOND HALLEY (1656–1742)—whose name is recalled periodically by his comet (p. 260)—that Newton undertook to attack the whole problem of gravitation. He had years of labour before he could show that the attraction of a spherical body on an external point was as if the spherical body

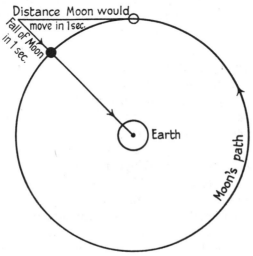

FIG. 68. Illustrating the orbit of the moon as compounded of tangential and centripetal movements.

were concentrated at its centre (1685). He had no expectation of so beautiful a result till it emerged from his mathematical investigations. With this theorem in his hands, all the mechanism of the universe lay spread before him. The vision was set forth in the *Philosophiae Naturalis Principia Mathematica* of 1687. Halley bore all the stress, set aside his own researches, sacrificed himself to forward what is regarded as the greatest of all scientific works. The *Principia*—as the work is usually called—established a view of the structure and workings of the universe which survived to our own generation.

The full extent and revolutionary character of the change that Newton was working in men's minds was not at first recognized even by himself, but it became apparent in the course of the eighteenth century. The essential revolutionary element was that Newton had conceived a working universe wholly independent of the spiritual order. This was the profoundest break that had yet been made with all for which the Middle Ages stood. With Newton there set in an age of scientific determinism.

But if the nature of the Newtonian revolution was not at first apparent, the scientific importance of the *Principia*, as of Newton's other contributions, was recognized immediately on publication. Newton wrote for mathematicians, and his full significance was beyond the comprehension of any others. He needed interpreters. Of these the ablest and most effective was VOLTAIRE (1694–1778), who spent the years 1726–9 in England. To him we owe the well-known story of Newton and the falling apple. Voltaire was aided in the preparation of his version of the Newtonian philosophy by his mistress, Émilie de Breteuil, Marquise du Chastelet (1706–49), who was a competent mathematician and herself translated the *Principia* into French (published posthumously 1759). Voltaire's delightful and lucid exposition (1737) marks the real victory of the Newtonian philosophy and the final submergence of Aristotelianism.

The changes in method and outlook introduced by Newton were so great that their general conformity as members of an historical series is sometimes lost to view. The issue is further obscured by the use or misuse of certain well-worn phrases. Newton's phrase 'I invent no hypotheses' is often quoted. The prestige of his name led to the assertion that 'whereas his predecessors *described* the motions of the heavenly bodies, Newton was the first to *explain* them'. Scrutiny of these statements throws light on the nature of scientific process.

Newton's famous phrase *Hypotheses non fingo* occurs at the end of the *Principia*. 'I have not yet been able to deduce from the phenomena the reason of these properties of gravitation and *I invent no hypotheses*. For whatever cannot be deduced from the phenomena should be called an *hypothesis*.'

Now Newton is here giving to the word hypothesis its exact original meaning. In the works of Plato[1] as well as in yet earlier

---

[1] e.g. *Phaedo*, 101 D, E.

works bearing the name of Hippocrates (p. 30) the word 'hypothesis' is used for a postulated scheme or plan which must be accepted if discussion is to take place. It is literally a 'foundation' (Greek *hypo thesis*, 'a thing placed under'). We have such hypotheses constantly before us in law. Some are mere legal fictions, as that 'the King can do no wrong'; others are convenient presentations of a remote possibility, as 'the lease that runs for 999 years'; others refer to procedure, as that 'a man is innocent (i.e. *treated* as innocent) until proved guilty'. All these are hypotheses in the Platonic, Hippocratic, and Newtonian sense. None are deduced from the phenomena. None are verifiable. All are parts of a working scheme into which certain phenomena can be conveniently and tidily fitted. In this use of the word Newton was certainly right when he said 'I invent no hypotheses'.

But if *hypothesis* be taken to mean what we usually understand by a scientific hypothesis, that is a generalization drawn from a series of observations which, it may reasonably be hoped, will be confirmed by yet further observations, then we must say that Newton was constantly both inventing and employing hypotheses. His application to the movements of the moon of the doctrine of gravity as he knew it on earth (p. 252) was an obvious example. Once he had such an 'hypothesis' that would fit the moon, he could and did apply it to other members of the planetary system. Its verification from the planets strengthened his conviction of the value of his first inference. The whole of his scientific activity was remarkable for invention of hypotheses. The successful invention of hypotheses is indeed the mark of his scientific eminence.

As regards the distinction between description and explanation, the position is somewhat the same. Newton knew that a property which we call *gravity* is associated with all matter of which we have direct experience. Having reached an exact conception of this property, he proceeds to examine the motions of the planetary bodies and finds that they may be re-expressed in terms of gravity. To do this is to give a description, not an explanation. It may reasonably be claimed that 'description is the true aim of science'. Let us apply the claim to some of Newton's predecessors.

Ptolemy represented the apparent movements of the heavenly bodies in terms of epicycles. This was his method of description.

If he were asked 'Why were the epicycles thus disposed?' he could have given no answer. He described; he did not explain.

Copernicus displaced the geocentric scheme. He expounded the appearances more simply and fully by ascribing them to the motion of the earth round a sun that was at rest. If asked 'Why does the earth move so?' he could have given no answer. He described; he did not explain.

Kepler represented the appearances more simply and fully by a system of ellipses. If asked 'Why should this form have been chosen?' he could have given no answer. He described; he did not explain.

Newton's completer scheme was based on the mutual attraction of bodies. If asked 'Why do they mutually attract each other?' he could have given no answer.[1] If, therefore, his account of the planetary system may be called an explanation, then such an explanation is indistinguishable from a description. The distinction between description and explanation cannot be ultimately maintained. It is the function of science to describe in terms that are as simple as possible. Ultimately the description must be in terms that defy further analysis, if such terms there be.

There is a significant change in nomenclature that expresses epigrammatically the change that came into men's minds with the acceptance of a mechanical world. For fourteen centuries, between St. Augustine and Newton, the Christian philosophic synthesis had reigned supreme; undisputedly at first, a little uneasily at last. But during the succeeding two centuries the results of the investigation of Nature appeared to fit less and less neatly with the accepted philosophic scheme. Changes in the meanings of words are sometimes straws that tell how the winds of thought are blowing. It is no accident that, precisely during these two centuries, certain kinds of 'philosophical enquiries'—as Newton and his contemporaries always described their labours—came gradually to be known as 'scientific researches'. Science, the knowledge of nature, was separated from philosophy, the search for the key to the universe. The change represents a fragmentation of interests that has lasted

---

[1] Newton did attempt to give an answer. He sought to 'explain' gravitation in terms of ether. Even had his attempt been successful, which it was not, it would have been of the nature of a re-description.

beyond the period that we are considering. For this reason, among others, it is peculiarly difficult to present the history of modern science as a coherent whole. From now on, our narrative, to become intelligible, needs a minuter subdivision. Science does not describe the world as a whole, but only a little bit of it at a time, each science choosing its own bit. This departmentalism now becomes self-conscious.

2. Morphology of the Universe.

Investigations on the general structure of the cosmos associated with Newton's conceptions fall naturally under three heads:

(i) *Observational astronomy*, that is, the direct investigation of the heavenly bodies by means of the telescope.

(ii) *Dynamical astronomy*, that is, the reduction to mathematical form of the movements of the heavenly bodies and the prediction, on a gravitational basis, of the movements of those bodies based on the mathematical expressions thus reached.

(iii) *Astrophysics*, that is, the investigation of the physical and chemical constitution and state of the heavenly bodies.

(i) Observational Astronomy.

At the command of Louis XIV, the great scientific architect CLAUDE PERRAULT (1613–88) built an observatory at Paris. This was the first State observatory of modern times. It was expressly intended to provide there facilities for men of science, whatever their country of origin. Soon after its completion the Frenchman Jean Picard, the Hollander Christian Huygens, the Dane Olaus Roemer, and the Italian G. D. Cassini were all at work there.

JEAN PICARD (1620–82) was an exact and careful observer, remembered for his measurements of the dimensions of the earth (1671, p. 271). These formed the basis of Newton's calculations. He recognized the astronomical value of the pendulum clock invented by Huygens, and he was the first to introduce the systematic use of telescopic 'sights'.

CHRISTIAN HUYGENS (1629–95, pp. 193–4), before coming to the new observatory, had already completed much important scientific work. Thus, he had improved the telescope, and had proved that

257

The Mechanical World

the changes in the appearance of Saturn—its 'horns' as Galileo called them—were due to a ring inclined at 28 degrees to the ecliptic (1653–6). The *micrometer*, a telescopic device for measuring small angular distances, was effectively introduced by him (1658).[1] His astronomical experiences raised in him a desire for an exact mode of measuring time. With this in view he attached a pendulum to a clock driven by weights, so that the clock kept the pendulum going but the pendulum regulated the rate of movement of the clock. The device was made public in his *Horologium* (1658), a work universally regarded as the foundation of the modern clock-maker's art.

Huygens began work at the royal observatory at Paris in 1671, and in 1673 published his famous *Horologium oscillatorium*,[2] a work of the highest genius which has influenced every science through its mastery of the principles of dynamics. It is second in scientific importance perhaps only to the *Principia*, which is in some respects based on it. It is primarily a mathematical analysis of the principles of the pendulum clock. It devotes attention to the composition of forces in circular motion. A memorable sentence in the work is the formulation of what has since become known as Newton's 'first law of motion' (p. 199) Huygens writes: 'If gravity did not exist nor the atmosphere obstruct the motions of bodies, a body would maintain forever, with equable velocity in a straight line, the motion once impressed upon it.' The work presents the modern view of the nature of inertia with great clearness.[3]

Huygens measured the acceleration due to gravity by experiments with a seconds pendulum, that is to say, a pendulum the oscillations of which occupy exactly one second. It is possible to calculate this acceleration at any spot of the earth's surface from the accurate measurement at that spot of the distance between

[1] The micrometer had been invented about 1640 by the Englishman William Gascoigne (1612–44). Huygens's device was improved about 1666 by the Frenchman Adrien Auzout (d. 1691).

[2] Not to be confused with the *Horologium* of 1658.

[3] The ideas of mass and of inertia were implied by Huygens in his statement of the laws governing the collision of elastic bodies as presented to the Royal Society in 1669. In this matter he had been preceded to some extent (1668) by Wallis (p. 193) and Christopher Wren (1632–1723).

258

the point of suspension and the centre of gravity of a seconds pendulum. Huygens's own result was 32·16 feet per second.

In 1681 Huygens returned to Holland and devoted himself once more to optical investigations and devices. He introduced a principle of optical construction which obviated much of the difficulty of chromatic aberration by employing lenses of enormous focal distance for his very long 'aerial telescopes'. The 'Huygenian eyepiece' invented by him is still in use.

OLAUS ROEMER (1644–1701) was the first to show that light has a definite velocity (1675). His conclusion was based on his observation that the intervals between the eclipses of Jupiter's moons were less when Jupiter and Earth were approaching each other than when they were receding. His discovery was of the highest importance, but it was rejected by the conservative Cassini, the astronomical dictator of the age.

G. D. CASSINI (1625–1712) began life as an engineer in the papal service. He established an astronomical reputation by his writing on comets (1652) and by his observations of the rotation periods of Jupiter, Mars, and Venus (1665-7). He was called to Paris by Louis XIV in 1669 and became the most influential figure in the observatory. Under his auspices it was shown that the earth was flattened towards the poles, a discovery that had important astronomical implications (p. 272). Under him, too, the parallax of Mars was measured. This led to an estimate of the distance of Mars from the sun (1673). His estimate of the distance of the sun from the earth, though by far the best up to its date, was some 7 per cent. in error.

Cassini was a man of conventional piety and—remarkable at that date—was an anti-Copernican. He was succeeded at the Paris observatory by three generations of descendants. The Cassini régime at the observatory lasted for a century and a quarter (1671–1794) and their lives extended over more than two centuries (1625–1845). Their conservative bias gradually weakened as the dynasty came to an end, but it was very injurious to French science.

In England, interests were increasingly maritime, and a scheme for finding longitude at sea was propounded in 1675. JOHN FLAMSTEED (1646–1719), already recognized as a promising

The Mechanical World

astronomer, showed this to be impracticable without a more accurate knowledge of the positions of the fixed stars than was then available. Charles II, hearing of this, declared that 'he must have them anew observed, examined and corrected for his seamen'. An observatory was erected for Flamsteed at Greenwich. His industry there was enormous, and between 1676 and 1689 he determined the positions of some twenty thousand fixed stars. His best observations were made with a mural arc, which he erected in 1689. This marked a great instrumental advance, and made possible far more accurate determinations than had before been attempted. His star catalogue forms the basis of modern astronomy.

Flamsteed was succeeded at Greenwich (1720) by EDMOND HALLEY (1656–1742). This remarkable man had detected discrepancies between the observed and the theoretical paths of Jupiter and Saturn before he was twenty. Perceiving that observations in the southern hemisphere were needed for the adjustment of these differences, he embarked for St. Helena (1676), where he observed for eighteen months. During this period he improved the seconds pendulum (p. 258) and determined the position of 341 stars of which no accurate record then existed. At the same time he made many other contributions to science and, notably, made a series of meteorological observations. These led to his publication of the first map of the winds of the globe (1686) and an attempt at their explanation (p. 275). He also made the first complete observation of a transit of Mercury.

In 1680 Halley began the study of the orbits of comets. In 1682 a comet appeared, the course of which was watched by several observers. Newton had suggested that comets might move in very elongated ellipses, indistinguishable from parabolas—as such ellipses must be—when near the sun (Fig. 69). Halley calculated the form, position, and measurements of the path of the comet of 1682, and noted their likeness to those of similar comets of 1531 and 1607. He inferred that his comet was a return of these. Other returns were traced. In 1705 he expressed the view that his comet returns every seventy-five and a half years, following an immensely long elliptical orbit extending far beyond the orbits of the planets (Fig. 70). Halley's comet is now known to have reappeared at about that interval from 12 B.C. to A.D. 1910—

twenty-six appearances in all. A famous appearance of this comet was that of 1066, which undermined Harold's morale, being interpreted as indicating his defeat by William the Conqueror. It is represented in the Bayeux tapestry.

Halley was succeeded at Greenwich by JAMES BRADLEY (1693–1762), who contributed to observational astronomy two important conceptions, *aberration of light* (1729) and *nutation of the earth's axis* (1748).

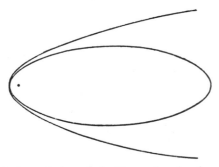

FIG. 69. Parabola and elongated ellipse, showing how they become indistinguishable from each other as they approach their common focus.

The aberration of light is most simply explained by the very illustration which suggested the idea to Bradley himself. Imagine travelling in a boat in a wind and with a flag at the mast-head. If the course be changed, the flag alters its apparent direction. Replace, in imagination, the wind by light coming from a star, and the boat by the earth moving round the sun and ever changing its direction. The result must be a cyclic change in the apparent position of a star. This Bradley was the first to observe and to explain.

The nutation (Latin 'nodding') of the earth's axis is an undulatory movement grafted on to that simple movement of the axis which corresponds to the precession of the equinoxes (p. 77). Thus the movement of the axis is not in a circle, as it would be if the precessional movement were uncomplicated, but in a figure of crenated outline (Fig. 71). Since Bradley's time many astronomers have studied the conduct of the earth's axis. It has transpired that nutation is only one of a whole series of complications of its motion.

261

The Mechanical World

The most impressive figure among eighteenth-century observational astronomers was FREDERICK WILLIAM HERSCHEL (1738–1822). Born in Hanover—then a possession of the British Crown—he came to England (1757), turned early to astronomy, and acquired great technical skill in constructing instruments. He conducted four complete reviews of the heavens, with telescopes of increasingly greater power. The second review revealed Uranus (1781), the first new planet to be discovered in historic time.

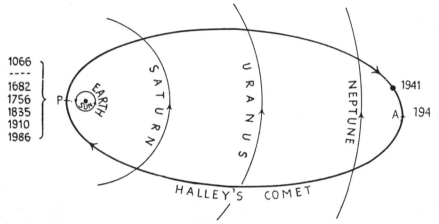

FIG. 70. Path of Halley's Comet. The position at various dates, with reference to the Perihelion, P, and Aphelion, A, is indicated.

Further improvements in his instruments led to his discovery of the satellites of Uranus (1787) and of Saturn (1789).

Herschel's industry and accuracy as an observer were unrivalled and his skill as an instrument maker was of the highest order. His most striking investigations were directed to the distribution of the stars. He concluded that the entire sidereal system is of lens shape, the edge being formed by the Milky Way.[1] The diameter of the lens is about five times the thickness. Our sun is not far from the centre of this lens (Fig. 72).

Closely linked with Herschel's conception of the form of the Universe was his immense series of observations on nebulae, of which he discovered many hundreds. He found, as had Galileo

[1] A similar conclusion had been reached in 1750 by Thomas Wright (1711–86) and in 1755 by the philosopher Immanuel Kant (1724–1804).

before him, that some of the nebulous appearances could be resolved into star clusters by instruments of sufficiently high power. At first he considered that all nebulae were of this nature and that they represented 'island universes' outside our own. Later, however, he concluded that some nebulae, at least, were composed of 'a shining fluid, of a nature totally unknown to us' (1791). He finally came to the conclusion that such shining fluid might gradually condense, the points of condensation forming stars and the

FIG. 71. Precession and Nutation. The axis of the earth moves, in the course of centuries, in such a way that a point on it, the North Pole for instance, describes a circle (dotted line). This produces the phenomenon known as 'precession of the equinoxes'. Added to this motion, as Bradley showed, was another, that of 'nutation', producing waves in the circle, in fact a 'gently undulating ring'. In the figure the undulations are enormously exaggerated.

whole forming a star cluster which might pass into a single star or star group (1814).

Linked also with his conception of the general form of the sidereal system was his view as to the movement within it of the solar system. It had been known since the time of Halley that certain stars move relatively to each other. Basing his opinion on the nature of their apparent movement, Herschel concluded that the entire solar system is itself progressing towards a point in the constellation Hercules (1805).

Herschel always emphasized the fact that stars are not merely scattered at random. In considering their distribution he noted that many were in closely contiguous pairs, 'double stars'. On an average the less bright would be the more distant. Owing to the orbital displacement of the earth, such pairs can be viewed, at intervals of six months, from two points 180 million miles apart. The perspective relations thus involved make it theoretically pos-

Fig. 72. Section of the Universe according to Herschel's Lens-theory.

sible to estimate the relative distances of the two members of a pair. Herschel pursued this idea with extraordinary tenacity over a period of many years, mapping out the places and aspects of numerous double stars. At last (1802) he was able to show that some of these stars circulate round each other. In their manner of doing this they follow the mathematical formulae of the laws of gravitation. Those laws, enunciated by Galileo for bodies on our earth and shown by Newton to rule the solar system, were now to be demonstrated among the distant stars.

(ii) *Dynamical Astronomy.*

In the eighteenth century, in the absence of any knowledge of the exact distances and movements of the stars, mathematical analysis could be applied only to the solar system. The distances from each other of the members of this system as well as their proportional sizes became fairly known. The demonstration of Newton for certain of them had left a presumption that all attracted each other according to the law of gravitation. The problem was to fit the exact consequence of that law to the movements which were revealed by progressively more exact observation.

Morphology of the Universe

This was the main task of the mathematicians of the age. Among them a foremost place must be accorded to the German philosopher and statesman G. W. LEIBNIZ (1646–1716), a man of very varied talents. His mathematical and scientific activity began after a visit to Huygens in Paris (1672) and to Boyle and others in London (1673). During three years' subsequent residence in Paris he devoted himself to mathematical study under Huygens. From

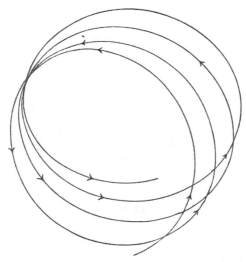

FIG. 73. Illustrating the path of a point moving in a varying ellipse.

this there resulted the conception of the 'differential calculus' on which the work of subsequent mathematicians was based.

The first formal publication of the method (1684) was preceded and followed by many years of controversy in the learned world on the question as to whether the priority rested with Newton (p. 252) or Leibniz. In fact, however, the nomenclature adopted by subsequent investigators was that of Leibniz.

LEONHARD EULER of Basel (1707–83), who early became blind, showed that certain irregularities in the earth's movement between the time of Ptolemy (p. 83) and his own were best explained by supposing that our planet is moving in a path which is a 'varying ellipse' and not a fixed one (1756, Fig. 73). This variation had pursued such a course that the axis of the earth's orbit had altered about five degrees since the time of Ptolemy.

The Mechanical World

J. L. LAGRANGE (1736–1813), of Turin and Paris, one of the greatest mathematicians of all time, made an important contribution concerning certain irregularities in the moon's motion. It had been known since Galileo that while the moon always turns the same face to us, yet there are parts near her edge that are alternately visible and invisible to us. Lagrange showed that this was best explained on the assumption that neither earth nor moon is truly spherical. Neither could therefore be treated as though the force of gravity acted at its centre (1764), as Newton originally thought (p. 253).

Lagrange distinguished two types of disturbance of members of the solar system: (a) *periodic*, which complete a cycle of changes in a single revolution or a few revolutions of the disturbing body, and (b) *secular*, in which a continuous disturbance acts always in the same direction and presents no evidence of a cyclic factor. The disturbance of one member of the solar system by another depends both on the relative position of the two bodies and also on their orbital sizes, shapes, planes of movements, &c., the quantities that are known mathematically as the *elements* of the orbit. The relative position of the planets is constantly changing. Thus they produce changing disturbances one upon the other, the effects going through periodic cycles. But apart from these, there are disturbing forces based on the orbital elements themselves which give rise to changes in the orbital elements of other bodies. These secular changes in the orbital elements are in general very small, but they accumulate continually.

In the discussion of the periodic and secular movements of the members of the solar system there was a constant interdigitation of the work of Lagrange and that of P. S. LAPLACE (1749–1827). That remarkable man spent his life at Paris pouring out a stream of books on astronomical and mathematical subjects. He did not permit his activities to be greatly interrupted either by the Revolution or by later successive governmental changes. His first major contribution was to show that an observed, very slow increase in the moon's rate of motion round the earth is explicable as due to a corresponding slow decrease of the eccentricity of the earth's orbit. This change in its turn is being produced by the gravitational action of the planets (1787). The order of change is

Morphology of the Universe

such that the length of the month decreases by about $\frac{1}{30}$ second per century.

As long ago as 1650 irregularities in the motion of Jupiter and Saturn had been suspected. Halley had noted them (1676). They were thought to be of a secular nature. Laplace, working on suggestions of Lagrange, showed that the inequalities corresponded to a period of about 900 years. This was the starting-point of a series of most remarkable investigations by Lagrange and Laplace on secular inequalities (1773–84). The final result was the following general law:

Take for each planet the product

$$\text{mass} \times \sqrt[2]{(\text{axis of orbit})} \times (\text{eccentricity})^2.$$

Add together these products for all the planets.

The resulting sum is then invariable, except for periodic inequalities.[1]

This law establishes the existence of a constant stock or fund of eccentricity for the solar system. The total of this fund cannot be altered. If the eccentricity of one planet be increased, that of another must be diminished. (In fact nearly the whole fund is absorbed by Jupiter and Saturn.) The law forms a sort of guarantee of the stability of the solar system.

The work of the eighteenth-century astronomers was summed up by Laplace in his great *Celestial Mechanics* (1799–1825). Its object he declares to be 'to solve the great mechanical problems of the solar system and to bring theory to coincide so closely with observation that empirical equations should no longer be needed'. It is the most comprehensive attempt of its kind ever made. With its completion the Newtonian problem seemed solved. The movements of the known members of the solar system were deducible from the law of gravitation. The discrepancies were so small, compared to those which had already been removed, that the impression was created that they too would be removed by more careful observation or by some correction of calculation.

Laplace's name is indissociably linked with his 'nebular

[1] 'Eccentricity' is the technical term for the ratio, in an ellipse, of the distance between the foci to the whole length of the major axis. For ellipses approaching a circle it is very small and it approximates to unity as the ellipse lengthens (see Fig. 26).

hypothesis' which appeared in his popular but nevertheless scientifically valuable *Essay on the System of the World* (1796). He pointed out that the motions of all the members of the solar system—some thirty to forty motions—were in the same direction.[1] All the motions were in planes but slightly inclined to each other, and the orbits of none were very far from circular. Attention was at the time being drawn to the nebulae by Herschel (p. 262). Laplace suggested that the whole solar system had condensed out of a vast rotating atmospheric mass, a huge gaseous nebula, that filled the bounds of the present solar system. The conception struck the imagination of the age and has remained an integral part of general thought concerning the cosmos.

The death of Laplace took place just a century after that of Newton. The two events provide convenient landmarks in the history of science.

Two most remarkable observations, the direct result of theoretical considerations, were made in the first half of the nineteenth century.

The first of these was made on the basis of the numerical sequence known as 'Bode's law' (J. E. BODE, 1747–1826) which had been set forth as early as 1772. If to each member of the simple sequence 0, 3, 6, 12, 24, 48, 96 (each figure being double the previous) the number 4 be added, producing 4, 7, 10, 16, 28, 52, 100, we obtain approximately the proportionate distances from the sun of Mercury, Venus, Earth, Mars, Jupiter, Saturn with blank for the number 28. Unsuccessful search was long made for this missing planet. In 1801 GIUSEPPE PIAZZE (1746–1826) of Palermo found a very small planet, which he named *Ceres*, about a quarter the size of the moon, at the required distance. This directed the general attention of astronomers to the possibility of finding more such small bodies. Since that time over a thousand of these 'minor planets' or asteroids have been found, most of them in very similar orbits to that of Ceres and nearly all circling between the orbits of Mars and Jupiter. It is suggested that they represent an exploded larger planet of which meteors may also have been parts.

---

[1] The motion of the satellites of Uranus is, in fact, in the opposite direction, but this had not emerged very clearly at the time Laplace was writing.

Morphology of the Universe

The second and more famous of these discoveries anticipated on theoretical grounds was that of a major planet. The existence of this body was betrayed by irregularities in the movement of the planet Uranus. In 1846 JOHN COUCH ADAMS (1819–92) of Cambridge and U. J. J. LE VERRIER (1811–77) of Paris, working quite independently, indicated the part of the heavens where the perturbing body was to be found. Telescopic search revealed it as foretold and it was given the name Neptune.

A constant desideratum of astronomy has been a determination of the distance of stars. This can be done by measuring the angle that the earth's orbit subtends to a star. The angle is so excessively small that its observation presents great experimental difficulties. These were first overcome in 1832 by THOMAS HENDERSON (1798–1844). His result was not published till 1893, while that of F. W. BESSEL (1784–1846) appeared in 1838.

(iii) *Astrophysics.*

By the first quarter of the nineteenth century there had developed clear ideas of the general structure of the universe and mathematical conceptions of the forms, dimensions, and relations of its constituent members. There was, however, little positive knowledge of their physical and none of their chemical constitution.

The possibilities of a science of astrophysics may be said to have opened with the nineteenth century. W. H. WOLLASTON (1766–1828), examining the solar spectrum in 1802, observed dark streaks crossing the coloured band, which he took to be boundaries of the natural colours. Some twelve years later a self-educated Bavarian instrument-maker, JOSEPH FRAUNHOFER (1787–1826) attached a telescope to the prism and examined the spectrum much more closely. He found that the resulting spectrum exhibited numerous black transverse lines of constant position (1814). Similar lines were visible in all forms of sunlight, whether direct, as from the sun itself, or reflected as from the clouds, moon, or planets. In the spectra from the stars, on the other hand, the distribution of lines was different.

In 1859 the two Heidelberg professors, GUSTAV ROBERT KIRCHHOFF (1824–87) and R. W. BUNSEN (1811–99), succeeded in showing that there was an invariable connexion between certain rays of the spectrum and certain kinds of matter. The assurance of their conclusion was certified by their discovery, through the spectra alone, of two new elements (*Caesium* and *Rubidium*). Kirchhoff went on to demonstrate certain essential characteristics of spectra and so was able to determine the existence in the sun of a large number of elements.

With the advent of the spectroscope and its application to the heavens, all departments of astronomy became intimately linked. It must suffice to attempt a mere enumeration of some of the results of this modern phase which opened with William Herschel.

The subject of double stars, to which Herschel drew attention, was particularly developed by F. G. W. STRUVE (1793–1864) and his successors at St. Petersburg, working at first with telescopes constructed by Fraunhofer. A great many multiple stars have been made known. Their numbers render it certain that the forces that have given rise to our universe have a special tendency to the production of these multiple bodies.

No general picture of the universe can be formed unless the laws of the motions of the stars are known. The proper motions of a few stars were known to Herschel. In 1837 F. W. A. ARGELANDER (1799–1875) knew about 400. The number now known is many thousands. In recent years great stress has been laid on the prevalence among brighter stars of opposite stream-flows towards two regions in the Milky Way. This is presumably due to the motion of the solar system as a whole, which can thus be estimated.

Spectroscopic research from Kirchhoff's time has been persistently directed towards the sun. The majority of elements have been identified in the sun. During an eclipse of 1869 the solar spectrum was found to include a gas to which the name 'helium' was given. Twenty-seven years later the gas was obtained on our earth.

The conception of the physical conditions of the sun has undergone a very great change in the century since Herschel. Much attention has been paid to the sun-spots which were shown, as early as 1843, to have a definite period, a definite distribution and order of appearance, and a rate of rotation which is different in different solar latitudes. The relation of sun-spots to terrestrial magnetic storms is remarkably constant.

Morphology of the Universe

The solar prominences observable by the eye only during eclipses can be examined by means of the spectroscope during full daylight (1868). Investigations have shown that the prominences increase and decrease in harmony with the sun-spots. The prominences originate in a shallow gaseous layer, the *chromosphere*, which is distinguished from the brilliantly incandescent inner layer the *photosphere*. Between the two is a narrow 'reversing layer' detectable only during eclipses and exhibiting special spectroscopic properties.

A very important principle associated with the name of CHRISTIAN DOPPLER (1803–53) was introduced in 1842. According to 'Doppler's principle' the movement of a spectrum-yielding body or part of a body can be measured by the shifting of lines in its spectrum. This has rendered possible the estimation of the sun's rotation rate and also of the rate of approach and recession towards or away from us of various stars.

3. The Terrestrial Globe.

(i) Measurement of the Earth.

The size of the earth was the subject of discussion from an early date. That it was an exact sphere was assumed at least from Aristotelian times (p. 47). An exacter mode of measuring angular elevation became possible with the invention of the telescope. With its aid an estimation of the length of a degree was undertaken (1669–71) for the Académie des Sciences by JEAN PICARD (1620–82, p. 257). The figure reached was 69·1 miles, which was a large variant from that of 60 miles which had been the estimate generally accepted. The method adopted by Picard was in principle that of Eratosthenes (p. 70), a star being used instead of the sun. Picard's result was issued in a somewhat inaccessible form (1671). Thus it was at first missed by Newton, who, in ignorance of it, abandoned for some years his calculations, based on earlier measurements, seeking to identify gravity as the force that kept the moon and planets in their orbits (p. 252).

Soon after Picard's determination the Académie organized an astronomical expedition (1671–4) to Cayenne in French Guiana, then occupied by a French commercial company. Cayenne is in latitude 5°. It was found that, to keep time there, the pendula of the clocks set for Paris in latitude 49° had to be shortened. The explanation of this, as we now know, is the bulging of the earth in the region of the equator. Gravitation decreases as we pass southward, since we are also getting farther from the earth's centre, and the pendulum therefore swings slower and has to be shortened if it is to keep time. The results of the Cayenne expedition were published in 1684. In 1673 Huygens in his *Horologium oscillatorium* (p. 258) had set forth the relation between the length of a pendulum and time of oscillation. This principle, together with the measurement of Picard, was utilized by Newton for the investigation of the figure of the earth in the *Principia* (1687).

Between 1684 and 1714 long series of pendulum measurements were undertaken in France by G. D. Cassini (p. 259) and his son Jacques (1677–1756). The results of these suggested that the form of the earth is that produced by the rotation of an ellipse round its major axis (a *prolate spheroid*).

This conclusion was in discord with that of Huygens and Newton. Thus the form of the earth became a main subject of scientific discussion, and several expeditions went forth to make measurements and to take pendulum observations. Of these, the most important left Paris in 1735 for South America under c. m. DE LA CONDAMINE (1701–74) to determine the length of a degree of longitude in the neighbourhood of the equator. The expedition laid down a famous and well-measured base, still spoken of as the 'Peru line'. In 1738 it was proved by P. L. M. de MAUPERTUIS (1698–1759), who had been a member of a similar expedition to northern Sweden, that the form of the earth was that derived from the rotation of an ellipse round its minor axis (an *oblate spheroid*). These results came to be finally accepted about the middle of the century, when the era of exact geodetic survey begins.

If the French excelled during this period in the exactness of their observations, the English made such observations possible by the skill and ingenuity of their instrument-makers. Thus GEORGE GRAHAM (1673–1751) invented the so-called 'dead beat escapement' of clocks and also the mercurial pendulum which remains always of the same effective length, since any expansion

The Terrestrial Globe

by heat of the rod is compensated by expansion of mercury in a suspended jar. He constructed astronomical instruments for Halley and Bradley and geodetic instruments for Maupertuis. JOHN HARRISON (1692–1776)—'Longitude Harrison'—devised the self-compensating gridiron pendulum (1726) and also a maintaining mechanism by which a clock continues to go during the process of winding. He is especially remembered for his chronometer which made possible, for the first time, the exact determination of longitude at sea. The instruments of JESSE RAMSDEN (1732–1800) were no less renowned. Best known of them was his instrument known as an 'Equatorial' (1774), which can be adjusted so as to cause a telescope to follow by clockwork the apparent motion of any point in the heavens to which it was directed. Modifications of it are in use in every modern observatory. Of comparable value was his engine for dividing mathematical instruments. He also completely transformed the surveying instrument for measuring angles, known from Elizabethan days as the 'theodolite'.

(ii) *Cartography.*

It was a period of great exploratory activity. Exacter determinations of the position of geographical points were constantly being recorded, and a more scientific cartography came into being. The numerous longitudes observed by Picard and his associates were utilized in 1679 for a map of France drawn up for the Académie by G. D. Cassini (p. 259), who also issued a good map of the world in 1694. The interest thus aroused produced a number of firms of map-makers, and several States appointed cartographers. At Venice was founded the earliest geographical society, the Accademia Cosmographica dei Argonauti. The French excelled in cartography for most of the eighteenth century. Especially prominent was J. B. BOURGUIGNON D'ANVILLE (1697–1783), many of whose admirable maps were in current use until a century ago. He was merciless to legend, preferring to leave the interior of Africa blank to filling it fancifully, and rejecting the conception of an Antarctic continent covering half the southern hemisphere. He portrayed China (1718) according to surveys conducted by Jesuit missionaries under the Emperor Kanghi (reigned 1661–1721). D'Anville devoted much attention to the history of his

The Mechanical World

science. For long the best topographical work was the *Carte géométrique de la France*, based on surveys carried out (1744–83) by c. f. cassini (1714–84) and his son jacques dominique (1748–1845), and issued in 1793.

In the second half of the eighteenth century a number of factors contributed to the furtherance of maritime exploration. The accurate determination of longitude at sea was made possible by the chronometers of 'Longitude Harrison'. The conditions of seamen were ameliorated by the use, on the recommendation of the British naval surgeon james lind (1736–1812), of orange and lemon juice as a preventive of scurvy, then the main obstacle to long sea voyages. The three voyages of Captain james cook (1728–79) which occupied the last twelve years of his life will always be memorable. It has been said that Cook's monument is the map of the Pacific. In cartographical achievement he is, however, rivalled by the two French officers j. f. de galaup, comte de la pérouse (1741–89), and j. a. bruni d'entrecasteaux (1739–93), who began the exact record of geographical points in Chinese and Japanese waters and in the Eastern Archipelago.

The labours of explorers of this type mark the opening of the exact scientific stage of topographic development. In 1787, working with a theodolite provided by Jesse Ramsden (p. 273), General william roy (1726–90) measured a base line for the triangulation of the British Isles that was to lead up to the Ordnance Survey. The primary triangulation was not completed till 1858, but the detailed survey was begun in 1791, the first inch-to-the-mile sheet was issued in 1801, and the first six-inch-to-the-mile sheet (that is 1 in 10,560) in 1846.

Other countries have followed along somewhat similar lines but at later dates. Proposals in France to replace the Cassini map were held up by war, and no steps were taken till 1817. The map was brought to final completion only in 1880. Among continental surveys, of special interest as presenting peculiar difficulties is the beautiful map of Switzerland published in 1842–65 and based on a triangulation completed in 1833. The scale, however, as with all continental maps, is less than that of the Ordnance Survey.

The Terrestrial Globe

(iii) *Wind and Water.*

Along with the exploration of the globe there developed a desire to reach some generalized conception of its phenomena, its magnetism, the watery atmospheric envelope, the tides, the currents, the winds, and the climates. 'Geophysics', the body of knowledge thus collected, is a quite modern term (1888), but the kind of inquiry that it represents came into prominence in the eighteenth century.

The knowledge of the prevalent winds was brought into relation with the study of the earth as a whole by Halley (p. 260), who published in 1686 his account of the trade winds and monsoons. The map which accompanies it shows a clear line of demarcation between the variable winds of the temperate zones on the one hand and the more reliable tropic winds on the other, along a line which runs at about 30 degrees both north and south of the equator. Halley was the first to connect the general circulation of the atmosphere with the distribution of the sun's heat over the earth's surface. In a later version of this map (1700) he added observations of the deviations of the magnetic compass, indicating the lines of equal variation (see p. 277).

GEORGE HADLEY (1685–1768) enunciated in 1735 the still current theory of trade winds as the resultant of the rotation of the earth and the displacement of air by tropical heat. Later the same view was taken by Dalton (1793). The first general work on winds was produced in 1742 by the French mathematician JEAN LE ROND D'ALEMBERT (1717–83). Of the meteorological advances during the century, following the appearance of this work, the most significant were perhaps the investigations on the watery content of the atmosphere (1783) by H. B. DE SAUSSURE (1740–99) of Geneva, the balloon ascents to ascertain the properties of air at high altitudes, notably by Gay-Lussac (1804), the introduction of the 'wind scale' (1805) by Admiral Beaufort (1774–1857), and the theory of dew set out (1814) by the American CHARLES WELLS (1757–1817).

A new outlook on geophysics was introduced by the American naval officer MATTHEW FONTAINE MAURY (1806–73). From 1839 onward he occupied himself in extracting from logbooks great numbers of observations of winds, currents, temperature, and so forth. By collating these he was able to draw up marine charts which led to such shortening of passages that an international conference was called in 1853 to consider further organization of such observations. Maury's *Physical Geography of the Sea* (1855) is the foundation work of modern knowledge on the subject. Largely as the result of his work, meteorological offices were established by several governments, and the international meteorological services initiated. In England the first director of the Meteorological Office, Admiral ROBERT FITZROY (1805–65), was appointed in 1855. Darwin had sailed with him twenty years previously in the *Beagle*, and he is still remembered by the 'Fitzroy barometer'.

Of all aspects of geophysics, the theme of the tides has perhaps attracted the greatest amount of scientific ability. Kepler and Galileo devoted attention to the subject. Newton, in the *Principia* (1687), placed the theory of the tides on a gravitational basis.

An adequate exposition of the tides is a very difficult task, nor is the tidal theory of Newton applicable to the prediction of the times or the height of tide at any required place. Newton, however, did give a satisfactory explanation of many of the characteristics of tides. The Newtonian view was expounded by Halley for the benefit of King James II, and this exposition has since become traditional in text-books. It is illustrated by a diagram to be found on the first plate of nearly every school atlas. The diagram is misleading since the problem is represented as one of statics when it is, in fact, one of dynamics. An easy presentation of the problem of tides is one of the desiderata of the art of scientific exposition.

(iv) *Terrestrial Magnetism.*

The subject of terrestrial magnetism has been especially studied because of its importance to navigation. An immense mass of data was collected, though there were few general ideas to connect them until long after our period. That the magnetic compass does not normally point to the true north is said to have been discovered by Columbus during his first voyage to America in 1492. The degree by which it departs from this line is known as the *declina-*

The Terrestrial Globe

tion or *variation*. That the compass suspended about a horizontal needle in the magnetic meridian will also dip was discovered at the end of the sixteenth century. The degree of dip is known as the *inclination*. Gilbert (1600, p. 188) knew that both declination and inclination were different in different places. In the early years of the seventeenth century it was found that the declination varied in the course of years in the same place. George Graham (p. 272) showed in 1724 that there was also a diurnal change in the declination. Much work was done by Halley on the difference in the degree of declination in different parts of the world. In 1700 he drew up an interesting chart in which the distribution of equal degrees of declination in the earth's surface is represented by lines, *isogonic lines* as we now call them. The method, here used for the first time, has since been adopted for innumerable other terrestrial variations such as isoclinals (lines of equal magnetic dip), isomagnetics (lines of equal magnetic force), isobars (lines of equal barometric pressure), isotherms (lines of equal temperature), and the like.

Between 1756 and 1759 a number of observations by John Canton showed that on certain days the movements of the compass were conspicuously irregular and that the Aurora borealis was then often visible. These phenomena, it was soon realized, were related to the occurrence of sun-spots.

Another landmark in the history of terrestrial magnetism was the discovery, towards the end of the eighteenth century, that the intensity of the magnetic force varies at different parts of the earth. The first published observations on this subject were those made in equatorial America (1798–1803) by Humboldt. In 1827 Arago showed that this intensity also exhibits diurnal variation. In 1834 the mathematician K. F. GAUSS (1777–1855) instituted at Göttingen the first special observatory for terrestrial magnetism. He greatly improved the type of instrument for magnetic observations. In 1840 a number of magnetic laboratories were established in various parts of the British Empire under the general superintendence of EDWARD SABINE (1788–1883), who had long been occupied on the subject. His numerous publications on terrestrial magnetism issued between 1823 and 1871 are still currently referred to.

The Mechanical World

(v) *Early Views of Earth History.*

That something of the history of the earth might be learned by a study of its crust was believed from of old. Much positive mineralogical knowledge accumulated from the mining industry. Among the most puzzling phenomena presented by the crust of the earth was that of fossils. The Dane NIELS STENO (1648–86), who spent some years in Italy, discussed the formation, displacement, and destruction of the stratified rocks in Tuscany (1669) and recognized the organic origin of fossils. A number of Italian, English, and French writers concurred with Steno, and during the first three-quarters of the eighteenth century there was an extensive accumulation of geological data and many theories were proposed to explain them (p. 239).

The first comprehensive general account of the history of the earth, which included a consideration of the nature of fossils, was put forward by GEORGES LOUIS LECLERC, COMTE DE BUFFON (1707–88) in his *Époques de la Nature* (1778). Buffon in forming his theory laid special stress upon certain data not all of which can now be interpreted as he would have had them. He held in mind primarily (*a*) the oblate spheroid form of the earth; (*b*) the contrast between the small amount of heat received from the sun and the large supply possessed by the earth; (*c*) the effect of the earth's internal heat in altering the rocks; and (*d*) the presence of fossils in all sorts of situations, even mountain tops. In association with the last he noted that limestone in north Europe, Asia and America often consists largely of the remains of marine organisms; and that the remains of large terrestrial animals, more or less similar to living forms, often occur near the surface, showing that they were recently living, whereas the deeper-lying remains of marine creatures in the same region belong to extinct forms or to forms related only to the inhabitants of far distant seas. He conceived that the earth (and other planets) arose from the collision of a comet with the sun. Thus arose a molten spheroid, the history of which can be divided into seven epochs, thus:

> 1st epoch. Incandescent to molten. 3,000 years.
> 2nd epoch. Gradual consolidation. Rents in crust allow influx of molten metallic ores. 35,000 years.

278

The Terrestrial Globe

3rd epoch.   Atmospheric vapours precipitated as the primitive universal ocean. Continents appear. Life begins in waters and marine sediment accumulates. 15–20,000 years.

4th epoch.   Access of internal heat. Period of violent volcanic activity. 5,000 years.

5th epoch.   Calm restored. Equatorial regions still too hot for habitation. Life over polar areas where dwell huge terrestrial animals, elephants, mastodons,. rhinoceroses, &c., which now came into existence. Fauna and flora gradually migrate southward.

6th epoch.   Land mass broken up. Man appears.

7th epoch.   Man asserts his supremacy. This epoch will continue till the earth cools and life becomes extinct.

The scheme is historically important both as the first effective attempt to explain observed and collected facts bearing on the history of the earth, and also as an estimate of many geological formations as of very slow growth and of great antiquity. It provided a basis for inquiry. In common with most early schemes it laid great stress on volcanic activity, earthquakes, explosions, and other dramatic events.

Despite the remarkable insight of the accomplished Buffon, and the attractiveness and popularity of his literary style, the geological dictator of the age was ABRAHAM GOTTLOB WERNER (1750–1817), a teacher at the school of mines at Freiburg, who wrote hardly anything at all, did not travel, and whose teaching was vitiated by his belief that the sequence of rock masses which he recognized in his native Saxony was of universal application.

Werner was an unusually successful teacher, and through his pupils the physical features of rocks all over the world became more widely known. His main doctrine was that of the aqueous origin of rocks, and his followers, known as Wernerians or 'Neptunists', were opposed by those who stressed the influence of subterranean heat, the 'Vulcanists'. The influence of Werner continued long after his death and reached the youthful Charles Darwin.

Very important in the history of geology is the influence of the French naturalist Cuvier (p. 329). He realized that the evidence of the rocks reveals a succession of animal populations. He perceived that vast numbers of species, many no longer existing, appeared upon the earth at different periods. Following Linnaeus, he was a firm believer in the fixity and unalterability of species, though his contemporary Lamarck (p. 379) was engaged in putting forward the opposite view. Cuvier had, however, to account for the extinction of some forms of life, and for what seemed the creation, or at least the appearance, of new forms. His explanation of these remarkable facts was that the earth has been the scene of a series of great *catastrophes*. He believed that of the last of these catastrophes we have an historic record. It is the flood recorded in the Book of Genesis. He expressly denied the existence of fossil man of great antiquity.

(vi) *Stratigraphy*.

The work of JAMES HUTTON (1726–97) initiates a more modern attitude. He travelled widely in order to study rocks, and satisfied himself that it is mostly in stratifications that fossils occur. He saw clearly that the imposition of successive horizontal layers is inexplicable as a result of a single great flood but suggests rather a quiet orderly deposit over a long period. In his *Theory of the Earth* (1795) he interpreted the strata as having once been the beds of seas, lakes, marshes, &c.

It was soon recognized that rocks often contain fragments from lower layers, nor could the fact be missed that stratified series are often tilted, bent, or broken. Many, encouraged by Cuvier's doctrine of 'catastrophes', ascribed these irregularities to violent upheavals. In this connexion it is interesting to observe that the *Essai sur la géographie minéralogique des environs de Paris* (1811) of ALEXANDRE BRONGNIART (1770–1847), though written in collaboration with Cuvier, inclines more to the views of Hutton.

WILLIAM SMITH (1769–1839), a civil engineer, obtained an insight into the nature of strata while cutting canals. He published the first coloured geological map (1815). His *Stratigraphical System of Organised Fossils* (1817) showed that certain layers have each their characteristic series of fossils. Some members of a series are wont to occur also in the layer below, others in the layer above, others in all three. Therefore changes in the flora

The Terrestrial Globe

and fauna which these fossils represent could not have been sudden. He saw, too, that the farther back we go, the less like are the fossils to forms still living.

A third British geologist, CHARLES LYELL (1797–1875), finally exorcized the catastrophic demon. He took to the study of geology while at Oxford, travelled considerably, and was influenced both by William Smith and by Lamarck. He saw that the relative ages of the later deposits could be determined by the proportion they yielded of living and of extinct molluscan shells. In his great *Principles of Geology* (1830–3) he showed that rocks are now being laid down by seas and rivers and are still being broken up by glaciers, rain, sandstorms, and the like: that, in fact, geologically ancient conditions were in essence similar to those of our time. Few books have exercised more influence on the course of biological thought. Darwin's early observations were made in the light of Lyell's great work.

We are struck by the overwhelming share of British investigators in the early development of geology as a science. The very names of the formations suffice to establish the British share in the development of the science. Lyell is responsible for *Pliocene* (Greek, 'more recent'), *Miocene* ('less recent'), and *Eocene* ('dawn of recent'); Sedgwick, the Cambridge geologist with whom Darwin went on geological excursions, invented *Devonian* (from its predominance in Devonshire), *Cambrian* (Cambria = Wales), *Palaeozoic* (Greek, 'ancient life'), and *Cainozoic* ('new life'). Between the last two formations John Phillips of Oxford (1800–74) interpolated *Mesozoic* ('intermediate life'). Other British contemporaries are responsible for *Carboniferous* (or 'coal-bearing'), *Ordovician* and *Silurian* (the Ordovices and Silures are British tribes mentioned by Caesar), *Permian* (from the province of Perm in east Russia), and *Cretaceous* (Latin 'chalky'). On the other hand, *Triassic* (Latin *Trias*, the number 'three') and *Jurassic* (from the Jura mountains) were titles given by German geologists at the beginning of the nineteenth century. The term *Tertiary* is older and was used by eighteenth-century Italian writers. The tertiary formations were held to be the third of a series of which the *Secondary* correspond roughly to the *Mesozoic* and *Palaeozoic*, and the *Primary* to the non-fossil-bearing rocks. The word

281

The Mechanical World

Geology itself was effectively introduced (1779) by H. B. de Saussure (1740–99) of Geneva, founder of modern mountaineering.

Of all writers on geophysics none has treated the subject so comprehensively and philosophically as ALEXANDER VON HUMBOLDT (1769–1859). His life was largely spent in travel and exploration of the most varied kind, and his occupation as diplomatic agent in Paris brought him contact with nearly all the leading scientific men of his day. Among his positive additions to science is the introduction of isothermal lines (1817), and he was the first to make a general study of temperature and pressure over the globe which has been essential to the modern science of meteorology. He was the first to investigate the rate of decrease of mean temperature with increased altitude. He made many studies of volcanoes and showed that they occur in linear groups, presumably corresponding to subterranean fissures. He showed .that many rocks thought to be of aqueous were really of igneous origin. He discovered that the magnetic force of the earth decreases from the poles to the equator (1804). He made the preliminary steps to a real geography of plants, studying them in relation to the physical conditions in which they grow. None of his services, however. is greater than the magnificent production in which he summarizes the work of his life, his *Kosmos*, of which the publication was begun in 1845 and completed posthumously in 1862. This book has been said to combine the large and vague ideas, typical of eighteenth-century thought, with the exact and positive science of the nineteenth. It is a truly transitional work, but still forms an excellent introduction to the study of geophysics.

During the first half of the nineteenth century, as geology grew into an independent science, the structure of the earth was studied from the point of view of the distribution and arrangement of its rocks (stratigraphy), from the point of view of the structure and composition of its rocks (petrography), and from the point of view of the nature and affinities of its fossils (palaeontology). Perhaps no country in the world presents so much geological variety within so small an area as does England. It is thus not inexplicable that geology became an especially English science. 'The Geological Survey of England and Wales' was begun by Sir .THOMAS DE LA BECHE (1796–1855) in 1832. It was far earlier in inception and execution than any comparable work produced in any other country.

A series of other English investigators gave to geology its rational framework for the detailed research of the next century. Thirty years of work by G. POULETT SCROPE (1787–1876), beginning with his *Considerations on Volcanos* (1825), marked the end of the Wernerian view. He laid the foundations of the current theory of volcanic origin and drew attention to their very peculiar distribution. RODERICK MURCHISON (1792–1871) in his great *Silurian System* (1839) expounded the chronological correspondence of rocks, introduced much of the nomenclature now in use, and explained the nature and incidence of many scenic details. His views were shown to be applicable over a wide area by his geological exploration of Russia (1841–5). Behind the band of British geologists stood ADAM SEDGWICK (1785–1873), who worked with them all and among them with his pupil Charles Darwin (pp. 379 ff.).

4. Transformations of Matter.

(i) Rise of Quantitative Method.

A belief in the indestructibility and uncreatability of matter is, in some degree, implicit in many operations outside the scientific sphere (p. 230). In the seventeenth century the belief sometimes became explicit. Thus Francis Bacon wrote: 'It is sufficiently clear that all things are changed, and nothing really perishes, and that the sum of matter remains absolutely the same' (*Cogitationes de natura rerum*, published posthumously, 1653), and there are comparable passages in the writings of Boyle (p. 233). The doctrine was given express form by Newton.

The law on which gravity acts, that of inverse squares, implies that the *weight* of a body is not constant, but varies according to its relation with other bodies. But Newton's second law of motion, that 'change of momentum[1] is proportional to the impressed force', implies that quantities of matter, that is to say *masses*, are equal if they suffer equal changes of motion under the action of equal forces, and that, conversely, forces are equal if they produce the

[1] Newton's word is *motion*, not momentum, but he means what we mean by the latter word.

same changes of motion in the same body. Thus Newton distinguished clearly between *mass* and *weight*. The *mass* of a body is proportional to the force that produces a given acceleration in the body. This force, in the case of a freely falling body, is the weight. Since all bodies fall at the same place with the same acceleration, their masses are proportional to their weights at the same place.

This exact and express doctrine of the *constancy of weight at the same place* (provided that other attracting bodies are unmoved) was a condition for the development of conceptions concerning the nature of physical changes. Without that doctrine the belief in any sudden inexplicable or magical appearance is possible. With it all changes in the state of matter can, in theory, be expressed in terms of number, weight, and measure. The changes that are specially investigated on the basis of weight are those known as 'chemical'. Thus the Newtonian conception gave a special impetus to the rationalization of chemistry and provided, in effect, the doctrine of the indestructibility and uncreatability of matter.

The investigation of chemical processes in the seventeenth century had yielded, by the dawn of the eighteenth, a vast accumulation of data for which no satisfactory system of classification had been suggested. Antitheses, as *acid and alkali*, were emphasized and tests for them were devised. Categories were invented and defined, such as *salts* (that is soluble, sapid, and crystalline substances), *earths* (that is friable, fire-resisting, and tasteless substances), and *calces* (that is powdery products of heated minerals). JAN BAPTIST VAN HELMONT (1577–1644) had indicated the existence of various aeriform substances, for which he devised the name *gas* (1644),[1] which could be condensed, as he supposed, into solid bodies and released therefrom by chemical change. He had, however, no method of collecting gas. Chemical theory, though it had emerged from the alchemical stage, was a confused mass of doctrine and tradition.

The Rev. STEPHEN HALES (1677–1761) devised an apparatus for collecting gases by leading them, from the retorts in which

---

[1] Helmont introduced the word 'gas' as a representative of the Greek word *chaos*.

they were produced by heating, through a pipe to a vessel filled with and inverted over water, in the so-called 'pneumatic trough'. He was able to measure the volumes of gases produced from weighed amounts of solids. He made, however, no further chemical examination of these gases because he supposed the product to be, in all cases, 'air' which had functioned as a kind of cement binding together the particles of the solids that he had heated.

Chemical technique was, in other respects, advanced and refined. This process was aided, from about 1670 onwards, by apparatus made from the newly introduced transparent flint glass in place of the older opaque vessels. The knowledge of the age was admirably summarized by the distinguished Dutch physician HERMANN BOERHAAVE (1668–1738). His *Elements of Chemistry* (1732) is among the very few great works written expressly as a students' text-book. Though exhibiting few new departures, it is firmly based on personal experience and is exceptionally lucid. Boerhaave held that all chemical events are ultimately reducible to relatively few and simple categories, and he believed vital processes to be expressible in chemical terms. Boerhaave's attitude gave to experimental chemistry a hopeful outlook which supported it for more than a generation, despite the paucity of important general laws.

The most notable chemical development of the earlier eighteenth century was the idea of 'affinity'. In 1718 the French physician ÉTIENNE FRANÇOIS GEOFFROY (1672–1731), influenced by Boerhaave, drew up tables in which acids were arranged in the order of their affinity for certain bases, and metals were arranged in the order of their affinity for sulphur. The relative degrees of affinity were estimated by ascertaining whether one base turned out another base or one metal another metal from a given compound. This idea of Geoffroy was further pursued by Black and others and notably by Bergman and Berthollet (p. 291).

(ii) *Intensive Study of Chemical Reaction.*

The Scottish investigator JOSEPH BLACK (1728–99) published in 1756 his *Experiments upon Magnesia alba, Quicklime, and some other Alcaline Substances*. Perhaps no brief chemical essay has

The Mechanical World

ever been so weighted with significant novelty. Black was a cautious investigator and his success was due to the accuracy of his measurements. He knew that chalk, by being heated and thus turned into quicklime (equation 1), ceases to effervesce with acids but gains the power of absorbing water (equation 2). As we would now formulate it:

(1) $CaCO_3 = CaO + CO_2$ (Chalk into quicklime. Gas evolved).

(2) $CaO + H_2O = Ca (OH)_2$ (Slaking of quicklime. Water absorbed).

Moreover, Black showed that, in the process of heating, the chalk loses weight, a loss which, by applying the methods of Hales (p. 284), he attributed to the removal of air in the process. And it had long been known that if the slaked lime be treated with a mild alkali, e.g. carbonate of soda, it is changed back into the state in which it was before heating, in fact, into chalk, while the mild alkali is converted into a caustic alkali. The process would now be represented thus:

(3) $Ca(OH)_2 + Na_2CO_3 = CaCO_3 + 2NaOH$.

Moreover, he showed that a definite amount of chalk, whether heated into quicklime or not, neutralizes an equal weight of acid, the only difference being that the neutralization takes place with effervescence and loss of weight if the chalk is unheated, and without effervescence or loss of weight if the chalk is first heated into quicklime. Thus:

(4) *Unheated* $CaCO_3$ (chalk) $+ 2HCl = CaCl_2 + H_2O + CO_2$.

(5) *Heated* CaO (quicklime) $+ 2HCl = CaCl_2 + H_2O$.

The gas given off by chalk in (1), transferred from one alkali to the other in (3) and given off in the effervescence produced by the reaction (4), he named 'fixed air', thus differentiating it from the ordinary air of the atmosphere more clearly than van Helmont (p. 284) had been able to do in his tentative and chemically incomplete work. We now call it 'carbon dioxide'. The conversion of quicklime into ordinary chalk by exposure to air:

(6) $CaO + CO_2 = CaCO_3$,

proved that carbon dioxide is a normal constituent of the air.

Black's work is of very great importance as the first intensive and detailed study of a chemical reaction. His especial triumph consisted in showing that the chemical changes occurring in this series of reactions could, without isolating the 'fixed air', be detected by subjecting them at every stage to the arbitrament of the balance. Black had thus discovered a gas different from air, which could exist in either the free or combined state, could be transferred from combination with one substance to another, and had many properties peculiar to itself. It had not hitherto been generally and clearly realized that there were any kinds of gases distinct from air. Attention was now drawn to this fact. The development of a technique for the isolation and study of gases and the discovery of the characters and laws of combination of gases was the main task of chemical endeavour of the later eighteenth and early nineteenth centuries.

(iii) *Gases.*

In 1766 the eccentric philosopher HENRY CAVENDISH (1731–1810), as exact an experimenter as Black, sent his first paper to the Royal Society. It bore the title *On Factitious Airs*, by which he meant gases produced artificially in the laboratory as distinct from 'natural' air. He discovered that a definite, peculiar, and highly inflammable gas, which he called 'inflammable air'—'hydrogen', as we now call it—is produced by the action of acids on certain metals. Continuing his investigations on an exact quantitative basis he published his *Experiments on Air* (1784). These demonstrated that the only product of the combustion of 'inflammable air' (hydrogen) and 'dephlogisticated air'—that is óxygen—is water. His figures give an approximately correct estimate of the proportions of the two in water.

Cavendish ascertained the amount of hydrogen evolved by action of acids on different metals. Adjusting his figures according to modern findings, we may say that he found that one part by weight of hydrogen was displaced by twenty-four parts of iron, twenty-eight of zinc, or fifty of tin. These numbers correspond to the 'equivalents' of these elements, and were used by him in 1766 to describe the different weights of different bases that neutralized a fixed amount of a given acid. He was the first to determine the weights of equal volumes of gases, a very fruitful line of research.

The chemical activities of the Unitarian divine, JOSEPH PRIEST-

The Mechanical World

LEY (1733–1804), were contemporary with those of Cavendish. He greatly developed and improved the technique of the preparation, manipulation, and study of gases. A series of important observations was made by him in the seventies and eighties. He showed that green plants would make respired air again respirable, and that they gave off a respirable gas. He prepared and studied a number of gases (ammonia, hydrogen chloride, sulphur dioxide, nitric and nitrous oxides, nitrogen peroxide), he investigated nitrogen and silicon tetrafluoride, and he isolated oxygen (1774–5) by heating certain oxides. He was hampered by his obstinate adherence to the old phlogiston theory. Phlogiston was a hypothetical substance supposed to exist in all combustible bodies and to be disengaged during combustion. It was 'the matter of fire' The word phlogiston had been introduced by Stahl in 1702 (p. 241).

Contemporary also with Cavendish and Priestley was CARL WILHELM SCHEELE (1742–86), a Swedish apothecary and one of the greatest of chemical experimenters and discoverers. His *Treatise on Air and Fire* (1777) proved that air consisted of two different gases now known as oxygen and nitrogen. Most of this research had been carried out before 1773. Thus Scheele's recognition and isolation of oxygen really preceded Priestley's. Scheele's numerous chemical discoveries include, not only oxygen, but also chlorine, manganese, baryta, silicon tetrafluoride, hydrofluoric acid, various inorganic acids, and the first extensive range of organic acids, glycerol, arseniuretted hydrogen, copper arsenite (still known as 'Scheele's green'), and many other substances. His *Treatise* and his many memoirs mark him as a rigorous experimenter and a concise writer.

(iv) *The Elements.*

The work of Black, Cavendish, Priestley, and Scheele assumed that matter was completely 'conserved', that is to say, neither came into being nor passed out of being in the course of their experiments. Further, they assumed weight to be the measure of the amount of matter.

In their day the old view of the four elements, earth, water, air, and fire, had not quite gone out of currency. It was, in fact,

288

Transformations of Matter

widely mooted that prolonged boiling converted water into earth. This question was taken up and finally resolved by the great French chemist ANTOINE LAURENT LAVOISIER (1743–94). He began the investigation with a simple but extremely carefully conducted series of experiments. By exact weighing he showed (1770) that if ordinary water be boiled in a suitably designed vessel in such a way that the steam produced is condensed and it and the residue weighed, then the weight of the solid particles that remain behind corresponds to the weight lost by the water. Thus nothing is lost and nothing gained.

Lavoisier next investigated the phenomena of calcination of metals. This process, it had long been known, results in the increase in weight of the calcined metal, an increase which Lavoisier was able to show as due to something taken from the air (1774–8). This was a serious blow to the phlogiston theory (p. 297). He proceeded to an extensive and quantitative investigation of the changes occurring during breathing, burning, and other forms of combustion (1772–83). In the course of these he discovered the true nature of respired air, and showed how both carbon dioxide and water are products of the normal act of breathing.

If clear grasp of its implication be accepted as the test of a discovery, Lavoisier was the discoverer of oxygen. We owe the word *oxygen* to him. He proved that in all cases of combustion there is a combination of oxygen with the substance burned. He repeated the experiments of Cavendish on exploding 'inflammable air' (hydrogen) and 'dephlogisticated air' (oxygen), and thence concluded that water was a compound of these two gases (1784). These experiments mark the end of the phlogiston theory. Men of science had now in their hands a technique by which the laws of chemical combination could be investigated.

Among Lavoisier's major contributions to science was his establishment, once and for all, of the conception of chemical 'elements' in the modern sense—'simple radicles' was the title attached to them by one of his French contemporaries. These 'simple radicles', following Boyle, he defines as substances which cannot be further decomposed. He divides them into four groups: (*a*) The gases oxygen, nitrogen, and hydrogen, and the

289

'imponderables' light and caloric; (*b*) Elements such as sulphur, phosphorus, and carbon which, on oxidation, yield acids; (*c*) Metals, of which he distinguished seventeen; (*d*) The 'earths', lime, magnesia, baryta, alumina, and silica. These last had not yet been decomposed. The same might be said of the 'alkalis', potash and soda, but Lavoisier was so certain that the alkalis were compound substances, produced by the union of oxygen with other 'simple radicles', as yet undiscovered, that he refused to include them among the 'simple radicles'.

Lavoisier was able to recognize correctly twenty-three elements in the modern sense, though his actual list was considerably longer. Together with de Morveau and Berthollet, in their joint work, *A New Chemical Nomenclature* (1787), he introduced a new system of naming substances according to their chemical composition, a reform that contributed greatly to the progress of chemistry by its rejection of the fanciful and often ridiculous alchemical names and the substitution of many now in use.

Lavoisier is generally regarded as the founder of the modern phase of chemistry, which he set forth in his classic *Elementary Treatise on Chemistry* (1789). His writings were widely studied. His experiments were models of painstaking ingenuity. Perhaps his numerous and varied achievements may be summed up in the statement that he gave coherence and clarity to the conception of the conservation of matter. All his work was based on the explicit assertion of the principle that, within experimental limits, the same weight of simple bodies can be drawn from compound bodies as had been put into them, no more and no less, and that compound bodies represent the combined weight of the simple bodies of which they are composed. This view became, with Lavoisier, explicit and axiomatic.

(v) *Atomism.*

As the eighteenth century turned into the nineteenth, the question of the innate constitution of matter was again raised. In the seventeenth century 'Epicureanism' based on atomic views had become a philosophic vogue. It was opposed to the current Cartesianism which it survived. Among early exponents of the atomic view were Gassendi (p. 235), whose main work appeared in 1649, and Boyle (p. 233), who treated the subject at intervals between 1661 and his death in 1691. Huygens also supported the atomic view. Newton, in his calculations of the motions of the planets, found it necessary to assume interstellar space to be a vacuum. He extended this conception to terrestrial matter and thus the conception of atoms naturally arose (*Principia*, 1687). It is, however, difficult to find any definite formulation regarding the exact nature of his 'corpuscles' or 'particles' in his works. But from his time onward, despite the opposition of Leibniz (p. 265), the constitution of matter was generally considered as atomic by physical investigators. The view was popularized and widely disseminated by Voltaire (p. 254).

The older investigators had great difficulty in obtaining their substances in a pure state. Indeed, chemical purity is an idea of very gradual growth, and is perhaps hardly consistent with the older doctrine of the four elements. The work of Black, Cavendish, and Lavoisier, however, drew general attention to the high degree of exactness possible in chemical operations. This conception was pressed by Lavoisier's fellow countryman JOSEPH LOUIS PROUST (1755–1826), who was the first to emphasize the constant composition of chemical compounds. With the improved methods available for the preparation of pure substances he was able to show that a definite compound, however formed, whether in Nature or by the hand of man in the laboratory, always contains the same 'simple bodies' (i.e. elements) combined in the same proportions by weight. This fact is expressed as the so-called 'Law of Definite Proportions'. Working on this law were several chemists. Notable among them was E. G. FISCHER (1754–1831), who prepared a table of equivalents (1802) from the figures of J. B. RICHTER (1762–1807) to correspond to the law of equivalent proportions.

Proust's conclusions were disputed by CLAUDE LOUIS BERTHOLLET (1748–1822), who in his *Essay on Chemical Statics* (1803) had set forth his views on chemical affinity and had criticized the development of Geoffroy's affinity table (p. 285) by TOBERN OLAF BERGMAN (1735–84). Bergman, recognizing (1773) that affinity tables should be double, one table showing the affinities for reactions in solution (the 'wet way') and the other showing the affinities when the substances were heated together (the 'dry way'), had drawn up large duplicate tables in his *Elective Attractions* (1775–83).

Bergman, moreover, recognized that in some reactions the chemical change could be carried to completion only if the amount of the reacting substance added exceeded that demanded by the amount of the substance acted upon. In more familiar phraseology he showed that it was necessary to add more than the amount 'chemically equivalent'. Berthollet clearly demonstrated that the relative amounts of the substances concerned in a chemical reaction, together with such factors as insolubility and volatility, affected the completeness of the reaction; that increasing proportions of one reactant caused the reaction to proceed still farther in one direction; and that chemical reactions in general were incomplete, the substance upon which two other substances acted with opposing forces being divided between them in proportion both to their affinities for that substance and to the quantities of those substances present.

From these theoretically sound principles, unfortunately neglected for many years, but later to become the basis of modern chemical dynamics, Berthollet erroneously concluded against Proust that chemical compounds were produced in analogous ways, and that their constituents were therefore combined, not in fixed and constant proportions, but in proportions that varied with the conditions under which the compounds were formed. Proust's conclusions were, however, accepted by chemists, and his law presently received a new and wider interpretation as a result of the atomic speculations of Dalton.

JOHN DALTON (1766–1844), a Quaker teacher of Manchester, had greater philosophic insight than Proust. Dalton's first important contribution to science was his rule that all gases expand equally with equal increments of temperature (1801). This law was about that time more explicitly formulated by the French chemist JOSEPH LOUIS GAY-LUSSAC (1778–1850), and his name is not unjustly associated with it.[1] His own 'law of partial pressure' (1801) Dalton decided might be explained on the atomic basis, 'a conclusion', he assures us, 'which seems universally adopted'.

[1] Gay-Lussac himself indicated that J. A. C. Charles (1746–1823) had preceded him in this discovery but had published no results.

Transformations of Matter

Dalton's line of thought on the constitution of matter had come to him first through his interest in meteorology. His analyses of air showed that it was always composed of the same proportion of oxygen and nitrogen, with small quantities of water vapour and carbon dioxide. He knew that these gases are not in combination and have different densities. Why then does the heaviest not sink to the bottom and the lightest rise to the top? These facts might be explained if they were all composed of minute particles of different sizes in the manner suggested by philosophers of antiquity such as Lucretius (p. 95). Adding to the ancient atomic conception the new view that matter was composed of a large number of elementary, homogeneous, and distinct substances, themselves composed of indivisible, indestructible, uncreatable atoms, it must also be assumed that all the atoms of any particular element are like each other but different from the atoms of other elements.

This view fitted well to Proust's recently formulated 'Law of Definite Proportions' (p. 291). In applying his theory to the facts of chemistry, Dalton started with the assumption that chemical combination takes place in the simplest possible way, one atom of one element combining with one atom of another, water being composed of H and O in a 1 : 1 ratio, and ammonia of N and H also in a 1 : 1 ratio. He assumed also that when two elements form more than one compound, higher ratios are possible, as for instance with the oxides of carbon (CO and CO_2).

Dalton had been working on his theory since the beginning of the century and gave it formal enunciation in 1808. The first number of his *New System of Chemical Philosophy* (1808–27) which appeared in that year has gained general acceptance as a classic. In it he pointed out that, though atoms must be far too small to measure or weigh directly, yet nevertheless it should be possible to determine the relative weights of atoms of different elements. For this we need only know the relative number of atoms combining to form a compound, and the relative weights in which the constituent elements combined to form that compound.

Dalton had very little real experimental guidance as to the number of atoms that form compounds. Thus he wrongly assumed that, in water, hydrogen and oxygen are combined in the ratio of

1 atom to 1 atom, instead of in the ratio 2 to 1. He then introduced experimental error in estimating the relative weight of the hydrogen and oxygen in water as 1 to 7 (instead of 1 to 8). Thus he ascribed to oxygen the relative atomic weight of 7 instead of 16.

(vi) *Molecular Theory.*

The publication of the atomic theory attracted much attention in France. GAY-LUSSAC (p. 292) was already working on similar lines. He was interested in the combination of gases and showed that, when gases combine, their relative volumes bear a very simple numerical relation to each other and to the volume of their product, if gaseous (1808). Thus one volume of oxygen combines with two volumes of hydrogen to form two volumes of water vapour; one volume of nitrogen combines with three volumes of hydrogen to form two volumes of ammonia gas, and so on.

The atomic theory and the findings of Gay-Lussac were clearly linked together in the exposition of the Italian AMEDEO AVOGADRO (1776–1856). Avogadro pointed out (1811) that if there is a simple numerical relation between combining volumes of gases and if they combine into uniform atomic groups, then there must be some simple connexion between the actual numbers of these atomic groups in equal volumes of combining gases. The simplest relation—and that which has been shown to be the real one—is that equal volumes of all gases contain in similar conditions the same number of atomic groups. Avogadro assumed that the atomic groups, as conceived by Dalton, were not indivisible but in the simplest case consisted of two parts, separable during chemical reaction. The divisible groups he named *molecules* (Latin = 'little masses'). Avogadro also assumed that these molecules, and not the individual atoms, were equally distributed throughout space in the case of all gases (1811). Both assumptions, he observed, fitted Gay-Lussac's law.

Avogadro's hypothesis, that 'equal volumes of all gases under the same conditions of temperature and pressure contain the same number of molecules', was, to the confusion of their subject, unfortunately not received by chemists, owing, firstly, to the small number of cases to which it could then be applied, and, secondly, to the fact that several of those cases gave anomalous results not understood until much later. It was not until 1858, after Avogadro's death, that authoritative attention was called to it by another Italian chemist, STANISLAO CANNIZZARO (1826–1910). This long eclipse of an important law rendered the results of physical chemistry far less profitable than they might have been for nearly half a century.

During this period there was enunciated a hypothesis that has had a somewhat similar history. In an anonymous paper published in 1815, the English physician WILLIAM PROUT (1785–1850) called attention to the closeness with which the atomic weights of the elements, expressed in terms of relation to hydrogen, approximated to whole numbers. Hydrogen, therefore, he regarded as the universal substance. In more modern times there was a general movement towards Prout's hypothesis of a *materia prima*, and his conception of atomic weights approximating to whole numbers has assumed a new significance.

Much of the chemical activity of the first half of the nineteenth century naturally went to the exact determination of atomic and molecular weights. Notably the Swede JÖNS JAKOB BERZELIUS (1779–1848) devoted himself to this task from 1811 onwards, ascertaining the molecular weights of thousands of substances. He also did important work as the founder of electrochemical theory. He developed the conception that a group of atoms or *radicle* can form an unchanging constituent through a series of compounds, behaving as though it were an element. He rendered a great service in establishing chemical nomenclature and developed the convenient mode of formulating elements by the capital initial letters of their Latin names, adding numerals to indicate the numbers of the various atoms present in a compound.

Many of the most fruitful lines of Lavoisier's work were continued by SIR HUMPHRY DAVY (1778–1829). Notably he succeeded by means of the electric current (p. 307) in resolving the alkalis, potash, and soda, and the alkaline earths, baryta, strontia, lime, and magnesia, into their elements. Those elements were oxygen on the one hand, and a series of metals which he called potassium, sodium, barium, strontium, calcium, and magnesium, deriving these names from the old terms for the substances in which the respective elements were contained (1807–8). He also showed that the gas chlorine, prepared by the Swedish chemist Scheele in 1774 and thought to contain oxygen, was of elementary character (1810).

Davy was especially fortunate in the practical application of much of his work. His 'safety-lamp' still bears his name, and deservedly so, for his detailed and important researches on flame and explosions made it practicable, though the principle on which it is based was discovered by George Stephenson, the engineer. He performed a great service to agriculture by codifying, for the first time, the mass of chemical knowledge applicable to it. His *Elements of Agricultural Chemistry* (1813) contains the first use in the English language of the word *Element* defined in the modern chemical sense:

> 'All the varieties of material substances may be resolved into a comparatively small number of bodies, which, as they are not capable of being decomposed, are considered in the present state of chemical knowledge as *elements*.'

At that date Davy recognized forty-seven of these elements.

An impressive figure in the scientific world of the thirties and forties was JUSTUS VON LIEBIG (1803–73), professor of chemistry first in Giessen and then in Munich. He applied to organic substances the exact methods that had been developed in the previous decades. Over his laboratory was inscribed 'God has ordained all things by measure, number and weight'. His great achievement was his application of exact chemical knowledge to the processes and products of vital activity. (For Liebig's physiological work see p. 352.)

With the work of LOTHAR MEYER (1830–95) and DMITRI MENDELEEFF (1834–1907) the study of chemistry passed into an entirely different phase. Their work demonstrated (1869–70) that there is a connexion between the atomic weights of the elements and their properties. The periodic table which is known by Mendeleeff's name enabled him and other workers to prophesy the existence and properties of elements, then undiscovered, but subsequently isolated. The table in an elaborated and modified form is still the basis of modern systematic chemical exposition.

5. *Transformations of Forces.*

(i) *The Imponderables.*

The seventeenth century—the age of Galileo—and the eighteenth —the age of Newton—established a view of a universe maintained by a balance of *forces* acting on *bodies.* There was still much vagueness as to the limits of the two. Thus, 'phlogiston', which was supposed to go forth from a body on combustion, and 'ether' which was at once agent and medium of light, no less than the electric and magnetic 'fluids', remained ambiguous conceptions to the very end of the eighteenth century and even into the nineteenth. This group of imagined entities, phlogiston, ether, the electric and magnetic fluids, were regarded as weightless substances: 'imponderables'. The confusion of language created by the 'imponderables' persisted in its crudest form. 'It is the imponderables—heat, electricity, love—that rule the world', wrote Oliver Wendell Holmes—himself a man of science—as late as 1858 (*The Autocrat of the Breakfast Table*).

Among the imponderables a place of special importance was occupied by the supposed substance of heat: 'caloric'. During the earlier eighteenth century two views of the nature of heat were current. On the one hand, it was generally conceived as a fluid held in greater or less quantity within the pores of all bodies. Thus when a metal grows hot on being hammered, the heat becomes more perceptible because the caloric, it was thought, was squeezed out by the pressure. The material and fluid nature of heat was a generally accepted idea which was not greatly disturbed by the victorious advance of the Newtonian philosophy.

On the other hand, there were adherents to the suggestion made by Boyle (1664), Hooke (1665), and Huygens (1690) that all basic physical phenomena—heat, light, chemical action, electricity, magnetism—were susceptible of mechanical explanation. It was believed that all were due to the movements on the part of small particles of the affected bodies, varying in form, velocity, order of arrangement, attractive power, and the like.

Certain relations between forces of different kinds were, of course, evident to every observer. This was the case, for example, with the general interconnexions of light with heat and again with electricity, and especially of heat with work. The production of fire by friction was a device of the highest antiquity; frictional electricity was well known; the steam-pump was becoming familiar; the production of heat, light, and sound in a variety of chemical and physical operations was also naturally very familiar. Nevertheless no exact relation between these various phenomena was yet recognized.

(ii) *Temperature Measurement.*

Methods of estimating temperature were greatly improved even before the elucidation of what may seem now the obvious distinction between heat and temperature. An air thermometer or rather thermoscope had been invented by Galileo about 1592, and an open-ended water thermoscope had been described by Jean Rey in 1632. A distinct advance, making the passage from the thermoscope to thermometer, was the sealed alcohol indicator, invented about 1641, probably by Ferdinand II, Grand Duke of Tuscany. It was used for the experiments of the Italian Accademia del Cimento during its brief life (1657–67). All these instruments were provided with arbitrary scales.

At the very beginning of the eighteenth century (1701) Newton suggested an oil thermometer with a rational thermometric scale, in which the temperature of freezing water was taken as 0° and that of the human body in health as 12°. By assuming that the rate of cooling of a hot body is proportional to the 'whole heat' [= temperature] of that body, he was able to estimate higher temperatures, such as 'red-heat', by observing the times taken by hot bodies to cool down to temperatures measurable on his thermometer. The proportionality here assumed has since become known as 'Newton's Law of Cooling'. This, more exactly, is that, for small ranges of temperature, the rate of cooling of a hot body is proportional to the difference in temperature between that body and the medium by which it is surrounded.

The mercury thermometer was introduced and thermometric standards fixed about 1715 by D. G. FAHRENHEIT (1686–1736) and described in a communication to the Royal Society in 1724. A maximum and minimum thermometer was constructed in 1757 by CHARLES CAVENDISH (1703–83), whose son Henry (p. 287 f.) ex-

Transformations of Forces

plored the thermometric conduct of mercury in 1783. This instrument was improved in the later years of the eighteenth century into the form we know. The maximum and minimum thermometer assumed its modern form in the hands of DANIEL RUTHERFORD (1749–1819) in 1794.

The invention of a satisfactory instrument for the measurement of temperature, with fixed points giving concordant readings in all circumstances to investigators in different places, had, as its most immediate important result, the foundation of the quantitative science of heat by JOSEPH BLACK (1728–99). About 1760 Black introduced the method of measuring quantities of heat by the number of degrees of temperature imparted to a definite quantity of matter, a method destined to have far-reaching effects. At the same time Black set forth clearly the distinction between *heat* and *temperature*, or *quantity of heat* and *intensity of heat*. Rejecting the older view that the quantities of heat necessary to produce equal increments of temperature in different bodies were proportional to the quantities of matter in these bodies, he showed that every kind of substance had its own characteristic 'capacity for heat', which appeared to bear no relation to the quantity of matter in the body investigated. Black's term 'capacity for heat' has since been replaced by the term *specific heat*.

(iii) *Heat a Mode of Motion.*

In 1761–4 Black showed that definite quantities of heat disappear during certain changes of physical states, such as melting and evaporation. He also demonstrated that the same quantities of heat reappear during the reverse changes, freezing and condensation. Black called this disappearing and reappearing factor the 'latent heat'.

Black's discovery of latent heat was shortly afterwards applied by the engineer, JAMES WATT (1736–1819), then occupied in improving the steam-engine (p. 302). Watt found that water, on conversion into steam at boiling-point, expanded at atmospheric pressure to about 1,800 times its liquid volume. He also found that steam at boiling-point, when passed into ice-cold water, could raise about six times its weight of that water to

299

boiling-point (1764). This puzzling result Black explained to him in accordance with his discoveries of 1761–4 on latent heat; and Watt (1765) applied Black's discovery in his contrivance of the separate condenser, the greatest of all his many improvements of the steam-engine (Fig. 74). This simple principle is still in use and has made possible many subsequent developments.

The conception of the nature of heat, from being a subject of

FIG. 74. Diagram of Watt's model illustrating condensing principle for steam-engine, 1765. In older engines the cylinder itself had been cooled at each stroke, after entry of steam. Watt attached a condenser and 'air-pump' to empty the cylinder, which could then be kept permanently at steam heat while the vacuum produced by condensation did its share of the work and thus added to the efficiency of the engine.

speculation, was now on what seemed an exact basis, susceptible of practical application. Heat was held to be an elastic, uncreatable, indestructible, measurable fluid. To emphasize this new outlook, Lavoisier and the French Academicians introduced for it the name 'calorique' (1787).

The theory of caloric was, however, already being undermined by the adventurous American, BENJAMIN THOMPSON, COUNT RUMFORD (1753–1814). Employing a balance, sensitive to one part in 1,000,000, he showed (1799) that there was no measurable alteration of weight in a mass of water on conversion into ice or on the reconversion of the ice into water, despite a heat-change of an order that would raise 9¾ oz. of gold from freezing-point to a red heat. Heat, therefore, if a fluid, 'must be something so infinitely rare, even in its most condensed state, as to baffle all our attempts to discover its gravity'. Therefore to Rumford it did not appear likely that heat was a substance distinct from, and accumulated in, the heated body. If, however, heat were 'nothing more than an intestine vibratory motion of the constituent parts of heated bodies', then no change of the weights was to be expected on heating, since only the internal motions, not their mass, would be affected.

In 1798 Rumford had published his *Inquiry concerning the Source of the Heat which is excited by Friction*. In boring cannon he estimated the heat produced by measuring the rise in temperature of a mass of water contained in a box suitably arranged around the boring-point. The heat generated by the friction of the borer and the cannon appeared to be inexhaustible, and he reasoned 'that any thing which any *insulated* body, or system of bodies, can continue to furnish *without limitation*, cannot possibly be a *material substance*'. Heat was therefore, he concluded, 'a kind of motion'.

Soon after these experiments there appeared the first publication (1799) of Humphry Davy (1778–1829, p. 295), describing work that he had carried out at the age of nineteen. It contains the often misquoted account of an attempt to melt two pieces of ice by the heat developed on rubbing them together in a vacuum. The arrangement of the experiment was very imperfect and, since Davy's recorded results were thermodynamically impossible, there can be little doubt that the proper technique was lacking. Perhaps the experiment is even now beyond the powers of any experimenter. The results were assumed, and throughout his brilliant career Davy held fast to his youthful and correct conclusion—unjustified or at least unconfirmed by his premises—that heat was a vibratory motion of the corpuscles of bodies. It is a remarkable case of that feeling or instinct for the correct solution that is the special gift of some talented investigators.

Count Rumford had come very near to a more demonstrable treatment of the transformation and conservation of energy, for he was not far from revealing the nature of the relation between heat and mechanical effort. He observed in his experiments on the boring of cannon that two horses, working steadily against frictional resistance, produced heat at a steady rate. He even compared the heat thus produced with the heat that would result from the combustion of the food consumed by the horses. Yet since he had no exact and transferable conception of *work* as a measure of mechanical action, he could not develop a complete doctrine of the transformation of one form of energy into another.

The development of the steam-engine by Watt, and its use in the pumping of Cornish mines was, about this time, much in men's minds. When the firm of Boulton and Watt first began to manufacture their engines, the terms of sale devised by Watt involved the annual payment by the buyer, over a period of years, of one-third of the value of the savings in fuel effected by the new engine where it replaced an older type. But since the new engines were often for use in new mines, or were to do more work than those they replaced, or were required to pump from greater depths, a method of comparing engines was needed. Thus the determination of the *duty* of an engine was introduced (1778) as a quantitative relation between output of work and consumption of fuel. The 'duty' was the number of pounds of water raised by the engine through a vertical height of one foot per bushel of coal consumed. From this could be calculated the *power* of an engine, i.e. its *rate of doing work*. A standard of power was introduced by Watt in 1782–3 from calculations of the rate of working of a mill horse, and the term *horse-power* was applied to define a rate of doing work equivalent to the raising of 33,000 pounds one foot per minute. It was not, however, until the middle of the nineteenth century that the general convertibility of heat into work was finally recognized.

(iv) *Static Electricity.*

In the field of electricity, until the end of the eighteenth century, only the static form was recognized. The process of electrical conduction was demonstrated in 1731 and it was shown that, while some bodies would conduct electricity, others would

Transformations of Forces

not. Thus 'insulation' became possible. It was shown also that all bodies are capable of electrification.

Early attention was drawn to electrical attraction and repulsion. To explain them a theory of *two fluids* was introduced (1730) by the French experimenter C. F. DU FAY (1698–1739). These fluids were supposed to be separated by friction and to neutralize each other when in combination.

The striking way in which an electric charge may be fixed by two conductors separated by a non-conductor, as in the familiar 'Leyden jar', was discovered at that town in 1746 by two Dutch experimenters. About this time BENJAMIN FRANKLIN (1706–90) began to take an interest in electricity and soon observed that electric charges could be drawn off with peculiar facility by metal points. He supposed that 'electric fire is a common element' existing in all bodies. If a body had more than its normal share it was called *plus*, if less *minus* (1747). This was the 'one fluid theory' which held the field until the time of Faraday (p. 310). Franklin explained lightning as of electrical origin, suggested lightning conductors (1749), and put the idea to a practical test (1752). Through the survey by Priestley of the general state of electrical knowledge in his *History and Present State of Electricity* (1767) such phenomena became generally recognized. A number of types of frictional electrical machines were introduced and the subject attracted much attention. Electrical investigation had hitherto been almost entirely qualitative. In 1767, from the observation that there was no charge on the inner surface of a hollow electrified metal body, Priestley had suggested that the law of electrical attraction was the same as that of gravitational attraction, namely, the law of the inverse square of the distance. Cavendish gave an experimental proof of this in 1771. Unfortunately, however, he did not publish his experimental verification, and it remained unknown till 1879.

The first method of measurement applicable to electricity was the action of an electrified object on light suspended bodies such as threads, metal foil, or pith-balls. An early attempt at quantitative expression was made in 1786 with a gold-leaf electroscope by measuring the angular divergence of the leaves when charged. But the first effective verification of the law of attraction was made by the French engineer CHARLES AUGUSTUS COULOMB (1736–1806), who adapted to electricity Hooke's principle 'ut tensio sic vis'. Using hairs and wires he constructed a 'torsion balance' (1785). The principle was to measure the amount of torsion required to bring a charged pith-ball within various distances of another pith-ball, equally charged with electricity of the same sign and therefore repelling it (Fig. 75). This method

FIG. 75. Coulomb's Torsion Balance. Within a closed chamber two charged balls are insulated. One is fixed to the framework, the other attached to a wire that can be turned by a milled head. The degree of torsion needed to bring them together is a measure of the force of their mutual repulsion.

was peculiarly adapted for the investigation of the distribution of electricity on surfaces and of the laws of electrical and magnetic action. Coulomb was the founder of the mathematical theory of these subjects, and by the use of his 'balance' was able to prove that Newton's law of inverse squares (p. 252) holds good for electric and magnetic attraction and repulsion.

In the later eighteenth century there was considerable interest in the shock-producing fishes, the skate-like Torpedo, and the electric eel or Gymnotus. Accounts of them were given by John Hunter (1773–5), Ingenhousz (1773), and Cavendish (1776), and it was realized that their shocks were of an electrical nature. The

attention thus drawn to electricity in the animal body led LUIGI
GALVANI (1737–98) of Bologna to investigate the susceptibility of
nerves to irritation. H⌐ s⊦ . that muscular contraction could
be produced by el⌐ and conversely that electric phenomena coul⌐ the muscular contraction. (1791.
Fig. 76.)

(v` ricity.

nimal electricity' was of its own

s on effects of metallic contacts on nerves
). To left a metal rod establishes electric
ies. In one lies the end of the nerves with the
er the feet. In the middle there is contact
amp mats, on one of which lies the spinal
and feet. To the right there is a similar
act which can be completed by bringing the

dubbed 'galvanism'. ALESSANDRO
ia, working on the results of Galvani,
ge through a nerve or sense organ not
ontraction but also sensation. If one
ibs of different metals were held in the
t was immediately produced when the
th the eye. A silver and a gold coin held
saltish taste when the coins were con-
ential thing was the contact of different

against
nected by ⌐
metals. Volta iat a muscle can be thrown into continuous contractio. epeated electrical stimulation, but he
was also able to demo.. rate (1800) that the animal relationship
of 'galvanism' is in no way essential, as had previously been
thought. Volta's device of the 'voltaic pile', in which the electric

discharge of coins of the original experiment was replaced by a whole series of pairs of coins or disks between cards soaked in brine, soon developed into his famous 'couronne de tasses' (1800. Fig. 77) and was the foundation of electrochemistry. The invention drew immediate and widespread attention. It was the first instrument for producing an electric current.

FIG. 77. Volta's Pile below and his Crown of Cups above. The pile is a series of paired disks of silver and zinc, sandwiched between paper strips soaked in salt water. They are supported by glass rods *m m*. From the lowest disk a metal strip goes to vessel *b*. A current will pass from the top disk to the vessel if the two are linked by a conductor. Two piles may be linked together by a metal strip, as at *c c*, and the effect doubled.

The 'crown of cups' is a series of vessels of salt water or dilute acid in which are pairs of plates of different metals, connected by metal strips *a a*. The action is as with the pile.

In England water was decomposed by current in the very year of Volta's publication. It was a generally held view that the chemical changes in the pile were the source of the electric current. Thus chemical affinity began to be correlated with electricity, which Franklin and others after him had come to regard as related to 'fire' or heat.

We may note that the 'crown of cups', each cup containing two plates of different metals steeped in salt water or a dilute acid, is the direct ancestor of the various forms of electric 'cell'.

Transformations of Forces

The voltaic pile or the crown of cups provided an entirely new means for the decomposition of certain substances. In the decomposition of water by the current so produced, very great interest was aroused by the sight of oxygen and hydrogen bubbling off from the separate plates. Humphry Davy was among the first to develop this most fruitful mode of analysis, from which he had very great hopes, believing that it must 'carry with it perfectly new views of corpuscular action'. He himself showed by its means that in the decomposition of water the volume of hydrogen is double that of oxygen. Before many years electrical decomposition in his hands had yielded a whole series of new elements, notably sodium and potassium (1807–8).

The nature of the process of electrical decomposition and the cause of migration of its products to the two poles of the electric cell gave rise to much speculation. Davy developed or adapted a theory that the electric pile breaks the particles near it into two factors. Thus in decomposition with a zinc-copper couple the copper repels and the zinc attracts the oxygen. Oxygen being given off, the hydrogen is thereby set free and attracts oxygen from the nearest particle. Thus again hydrogen is released and again attracts the nearest oxygen. A chain of decomposition is formed resulting in the discharge of hydrogen at the zinc pole and oxygen at the copper.

The process of electrical decomposition was given quantitative expression by Faraday (1833, p. 310). Its two primary laws, still known by his name, are:

(a) The mass of the product liberated by electrical decomposition is proportional to the quantity of electricity passed.

(b) When the same current is passed through solutions of different substances, the masses of the liberated products are proportional to the chemical equivalents of those products.

Thus a definite relation was established between electrical and chemical action.

(vi) Electromagnetism.

Soon after the completion of Davy's electro-chemical researches a new orientation of electrical science set in. The year 1820 was especially eventful. In that year the Dane, HANS CHRISTIAN

OERSTED (1777–1851), demonstrated exactly the long-suspected connexion of electricity with magnetism. He found that if a wire carrying an electric current was placed near and parallel to a magnetic needle it deflected it (Fig. 78), but not if the wire carrying the current was at right angles to the needle. The direction in which the needle turns depends on whether the wire carrying the current is above or below the needle, and on the direction of the current. The significance of this linking of electricity with magnetism

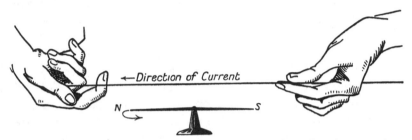

FIG. 78. Oersted's experiment on the effect of an electric current on a magnetic needle.

was at once recognized by the French investigator, FRANCOIS ARAGO (1786–1853), who showed (1820) that a spiral of copper wire, through which a current was passed, attracted previously unmagnetized iron filings, which clung to the wire as long as the current flowed, but dropped off when the circuit was broken. Such a coil, in fact, acts like a magnet. In 1824 he found that rotation of a copper disk produced rotation of a magnetic needle supported above it (Fig. 79). This phenomenon was rendered intelligible by Faraday in 1831 (p. 314).

ANDRE MARIE AMPERE (1775–1836), very soon after Oersted's publication, revealed the laws governing the deflection of the magnetic needle by the electric current and the mutual attractions and repulsions of electric currents. He showed that two parallel wires carrying currents attract each other if the currents flow in the same direction, and repel each other if the currents flow in opposite directions and he showed, as Arago had already done, that a cylindrical coil behaves like a magnet when a current is passed through it. He proceeded to a mathematical analysis of these phenomena (1822–7) and showed that an electric current

Transformations of Forces

is equivalent in its external effects to a magnetic shell. He propounded the theory that magnetism is the result of molecular electric currents. His memory is perpetuated in the well-known 'Ampere's Rule', formulated by him for determining the deflec-

FIG. 79. Arago's experiment of rotating a copper disk below a magnetic needle.

tion of a magnet by an electric current, and in the *ampère*, the practical unit of electric current, which is named after him.

The work of these investigators, especially of Ampère, provided a means of detecting a current and of measuring it on some arbi-

FIG. 80. The simplest form of galvanometer or apparatus for measuring electric current. It consists of a magnetic needle set in a non-conducting rectangular framework around which are wound many turns of wire through which passes the current, the effect of which is to be measured.

trarily chosen scale by means of its magnetic effect. Instruments devised for this purpose, *galvanometers*, appeared in 1821 from the hands of several inventors. In their simplest form, they consist of a coil of many turns of wire carrying the current and deflecting a magnetic needle suspended on a pivot at the centre of the coil (Fig. 80).

The Mechanical World

(vii) *The Dynamo.*

A main achievement of MICHAEL FARADAY (1791–1867), one of the greatest of scientific geniuses, was the demonstration that an electric current can be used as a source of power. From the experiments of Oersted and from his own Faraday realized that a current traversing wire creates a magnetic 'field of force'. Any part of this may be presented graphically, in any plane at right angles

FIG. 81. Faraday's apparatus for demonstrating how an electric current can be disposed so as to produce a continuous rotational movement.

to the wire, by a series of circles concentric to the wire (Fig. 84). Faraday thought that this force might cause a magnet to move round the wire. Moreover he argued that if a magnetic pole can be made to rotate round a current it should be possible to cause a wire carrying current to rotate round a magnetic pole.

A circuit, consisting of two vessels of mercury and connecting wires, was arranged by Faraday so that in one vessel there was a fixed magnet and a wire free to rotate, while in the other the wire was fixed and the magnet movable (Fig. 81). Electric current passed from the wire through the mercury in the left-hand cup to a copper rod running into the base of the vessel. The magnet in this cup was fastened to the copper rod by a thread. In the right-hand vessel the fixed magnet was placed in a socket in the stem of the vessel, and the wire which dipped into the mercury was able to move freely. As soon as the circuit was completed, the magnet in the first vessel and the wire in the second commenced to rotate, and continued to do so while the current was passing. Faraday had thus transformed electrical current into continuous movement (1821).

Faraday knew of the demonstration by Ampère that a cylin-

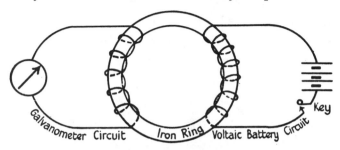

FIG. 82. Faraday's ring.

drical coil of wire behaves like a magnet when a current is passed through it. The converse that a magnet could produce a current was shown by him to be equally true. The experiments that led him to this conclusion have become classics.

Around an iron ring he wound two separate coils of wire. One was connected with a voltaic battery, the other with a galvanometer. A key made it possible to break or make circuit. On making or breaking the current in the voltaic circuit, the galvanometer showed that a current also flowed for an instant in its circuit, but the currents on making and breaking were in the opposite direction (1831, Fig. 82).

But if a circuit can act as a magnet, as Arago had shown, cannot a magnet produce this same result with an iron ring? Is not the battery unnecessary? The testing of this point was a critical experiment for the whole future of electrical science. Faraday wound a coil of wire round a bar of iron and completed the circuit so as to include a galvanometer. He then placed the bar between the north pole of one bar magnet and the south pole of another, the other ends of the magnets being in contact. Whenever contact between the magnets was made or broken the galvanometer indicated the momentary passage of a current (Fig. 83).

311

The Mechanical World

On this discovery a wit of the time wrote:

> Around the magnet Faraday
> Was sure that Volta's lightnings play:
> But how to draw them from the wire?
> He took a lesson from the heart:
> 'Tis when we meet, 'tis when we part,
> Breaks forth the electric fire.

Faraday had dispensed with a battery. Could he, by retaining the battery, dispense with a magnet, substituting for it a current?

FIG. 83. Production of momentary electric current by magnetic 'make' and 'break'.

Using a wooden bobbin for the iron ring, Faraday wound a coil of wire round it, and connected it to a voltaic coil. Round this 'primary' coil was wound another and much longer coil, the 'secondary', its ends being joined to a galvanometer. As before, both on make and break, momentary currents were indicated by the galvanometer. Faraday had revealed the process of 'inducing' a current, and with the knowledge of induction currents a new era in the application of electricity had opened (1831).

It was now clear that the essential factor in the production of the magneto-electric effects was change, movement of the magnet or of the coil, or making and breaking of the current or the contact. Magneto-electric effects are related somehow to 'fields of force' which fade out as we pass farther from the site of the change. These fields of force can be arranged or mapped in lines as indicated by the behaviour of iron filings placed on cards within their area.

In seeking a general explanation of these phenomena Faraday

312

was thinking much about the lines of magnetic force which came to play a very important part in electrical sciences. They were by no means a new conception. Gilbert (p. 188) had a clear idea of them, Descartes had seen in them evidence for his hypothetical vortices, and certain eighteenth-century physicists had even mapped them, but it was reserved for Faraday to indicate their

FIG. 84. Lines of force due to current in a straight conductor.

significance. Throughout the rest of his career he continued to speculate and experiment on these lines of force which are now a familiar scientific conception.

The general character of the lines of force due to a current can be easily demonstrated either by manipulating a small compass needle in the neighbourhood of a current or by running an electric wire carrying a current through a card on which iron filings are spread. These filings take the position of curves in the neighbourhood of the wire and the lines of force can similarly be represented as concentric circles at right angles to the current (Fig. 84).

Faraday had already succeeded in making a magnet rotate round a wire carrying a current, and a wire carrying a current rotate round a magnet. Such movements are related to the distribution of the lines of force due to current or magnet setting up certain stresses in the medium. This wire or magnet is continually urged away from the strong part of the field. Ampère had shown that parallel wires carrying current attract one another if the currents are in the same direction and repel one another if the currents are in opposite directions. This fact Faraday was able easily to fit into his conception of lines of force. If currents

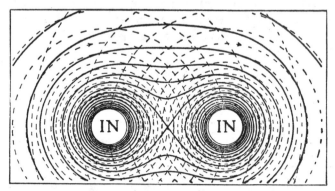

FIG. 85. Field due to currents in the same direction.

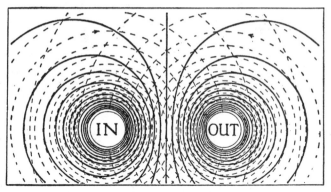

FIG. 86. Field due to currents in the opposite direction.

run in the same direction in two neighbouring wires, the resultant field of lines of force will be such that they will be driven from the strong parts of the field to the weaker and so drawn together (Fig. 85). If the currents run in opposite direction they will again be driven to the weaker parts of the field and so driven apart (Fig. 86).

Arago's demonstration of the effect of a rotating copper disk on a magnet suspended over it (p. 308) was now explicable in terms of lines of force. As the disk moves, it cuts through the

Transformations of Forces

lines of force of the magnet. Induced currents are therefore set up. The movement of the magnet is simply the result of the mutual action of the magnet and of the magnetic fields due to the induced current. By visualizing the lines of force as endowed with certain physical properties, it is possible to link together many otherwise disconnected phenomena.

Faraday in his very fruitful year 1831 provided also the converse to Arago's experiment (Fig. 79). He made a copper disk rotate between the two poles of a horseshoe magnet. The axis and the edges of the disk were connected with a galvanometer. As the disk turned, the galvanometer showed that an induced current was produced. This was the first magneto-electric machine or dynamo. This discovery of electro-magnetic induction was thus the starting-point for the utilization of electricity on a large scale, and for the application of such power for lighting and traction.

A dynamo consists essentially of a suitable conductor, built up of many coils, which rotates in a magnetic field. The rotating conductor cuts through the lines of force of the magnetic field and an induced current is thereby set up in the coils of the rotating conductor. In each coil the induced current changes its direction during each revolution. Such a current is said to *alternate*. By means of a well-known device the alternating current may be made *direct* by reversing the current in each coil of the armature each time it passes a pair of conductors.

With Faraday's ring with two coils of wire (Fig. 82) it is possible to obtain a high electromotive force from a current given by a very few cells. Many experimenters after him sought to construct apparatus which should give a high electromotive force by inductive action of one circuit on another. It was H. D. RUHM-KORFF (1803–77), a Parisian instrument maker, who in 1851 produced the type of coil still known by his name and so rendered practical the development of the electric motor.

About this time, when Faraday's researches were thus assuming practical significance, scientific men began to appreciate the exactness and preciseness behind much of his simple language. It is astonishing how many general theorems, the methodical deduction of which require the highest mathematical powers, Faraday attained by some sort of intuition without the help of mathematical formulae. Thus the first important scientific contribution of JAMES CLERK-MAXWELL (1831–79) was *On Faraday's Lines of Force* (1856). In it he sought 'to show how by a strict application of the ideas and methods of Faraday, the connexion of the very different orders of phenomena which he discovered, may be placed before the mathematical mind'. He followed the suggestion of William Thomson, later LORD KELVIN (1824–1907), who had been working at the subject since 1849. The analogies that Clerk-Maxwell worked out were those of heat and of hydrodynamics. These gave rise to his conception of electric and magnetic effects as due to changes in the ether (1862) and to his great contribution *On a Dynamical Theory of the Electro-magnetic Field* (1864). In the latter he showed that electro-magnetic action travels through space at a definite rate, in waves, and that these waves are, like those of light (pp. 316 ff.), transverse to the direction in which the waves are propagated. Since he was able also to prove that the velocity of these waves is the same as that of light (1867), an electro-magnetic theory of light thereby became possible.

(viii) *Undulatory Theory.*

At the end of the eighteenth century there were in the field two rival conceptions of the nature of light, the emission theory and the undulatory theory.

The emission theory is of great antiquity but was given modern scientific form by Newton. He treated a luminous body as emitting streams of minute corpuscles moving progressively in a straight line corresponding to the direction of the ray. Vision was supposed to be produced by the impact of these streams on the eye. The bending of the ray as it passes from air into a denser medium—as, for example, into glass or into water—is explained by assuming that as each corpuscle approaches the denser surface of the medium at any given angle it begins to be attracted towards it.

The undulatory theory of Christian Huygens, put forward in 1678 and especially in his famous *Treatise on Light* (1690), treated all space as pervaded by a subtle and elastic medium, the *ether*, through which waves are propagated in all directions from a light-source. These undulations spread in a regular spherical form from

Transformations of Forces

the point of origin, just as waves produced by a stone dropped into water spread in circles.

Huygens applied this theory to explain the phenomena of refraction. A source of light may be regarded as emitting a series of spherical waves in the ether. Any point A on the surface of such a wave (Fig. 87) may in its turn be regarded as a source of light. Every other point on the surface of the same wave, as for example B, C, or D, emits similarly its own spherical wave. At

any distance from the original source the surface of all these waves can be regarded as combining together to form what is called a 'wave-front'. If the source of light be sufficiently distant the wavefront is on so large a sphere that a small part of it may be treated as flat (or, in section, linear), while the lines radiating to it from the source of light may be treated as parallel.

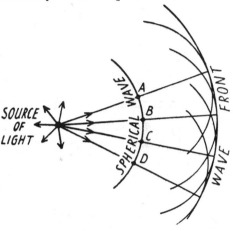

FIG. 87. Huygens's conception of 'wave-fronts'.

We have now to consider, as did Huygens, the application of this wave theory to the known facts of refraction and notably to Snell's law (p. 194). Those facts require (as we shall presently see) that the velocity of propagation of light should be less in a denser than in a rarer medium. The change in the rate of propagation will produce a change in the direction of the wave-front.

In the diagram (Fig. 88) A and C are parallel rays derived from a distant source of light with wave-front a plane surface at right angles to their line of advance. They strike the surface of a denser medium obliquely, A reaching it at A_1, along wave-front $A_1 M$, before C reaches it at C_1. Suppose the velocity in the denser to be $\frac{2}{3}$ of that in the rarer medium. While C advances from M to C_1, A will reach a point A_2 which is $\frac{2}{3}$ as far from A_1 as M is from C_1. For another ray B, that strikes the

317

The Mechanical World

surface at B_1 midway between A and C, the wave-front $B_1 N$ may be considered. Now C gets to C_1 at the moment when B having struck the surface at B_1 reaches a point B_2 at $\frac{2}{3}$ the distance from B_1 that N is from C_1, or $\frac{1}{2}$ of $\frac{2}{3}$ the distance that M is from C_1.

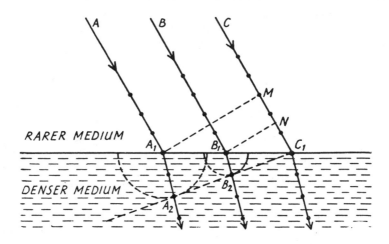

FIG. 88   Refraction in terms of Wave Theory. Beads on the lines mark equal intervals. Velocity of light in denser is represented as two-thirds of that in the rarer medium.

Thus $A_1 A_2$ is $\frac{2}{3} MC_1$ and $B_1 B_2$ is $\frac{1}{3} MC_1$. A_2 and B_2 are located on wave-front circles with centres A_1 and B_1 respectively and with radii equal respectively to $\frac{2}{3}$ and $\frac{1}{3}$ of MC_1. Thus the wave-front when C arrives at C_1 will be a straight line $A_2 B_2 C_1$, which only touches but cannot cut either of the two circles. That straight line is, in fact, a common tangent to all circles formed on the proportionate construction here considered. The angular change of direction of wave-front from $A_1 M$ to $A_2 C_1$ corresponds to the necessities of Snell's law.

The wave theory of light that prevailed in the nineteenth century was propounded at its dawn by THOMAS YOUNG (1773–1829). In two communications (1801), which place him in the forefront of scientific investigators, he set out his wave theory and its essential principle of *interference*.

318

Transformations of Forces

'Suppose', he said, 'a number of equal waves of water to move upon the surface of a stagnant lake with a certain constant velocity, and to enter a narrow channel leading out of the lake; suppose then, another similar cause to have excited another equal series of waves, which arrive at the same channel with the same velocity and at the same time as the first. One series of waves will not destroy the other, but their effects will be combined. If they enter the channel in such a manner that the elevations of the one series coincide with those of the other, they must together produce a

FIG. 89. To illustrate the principle of interference.

series of greater joint elevations; but if the elevations of one series are so situated as to correspond to the depressions of the other, they must exactly fill up those depressions, and the surface of the water must remain smooth—at least, I can discover no alternative, either from theory or experiment. Now, I maintain that similar effects take place whenever two portions of light are thus mixed, and this I call *the general law of the interference of light.*'

This view of interference is perhaps most simply presented if we picture waves from one centre of disturbance entering two channels of unequal length which subsequently meet. If at the meeting-point the waves are in opposite phases they will evidently neutralize each other (Fig. 89).

Newton had himself discussed the wave theory, and had dismissed it, saying:

'If light consisted in motion, *it would bend into the shadow*, for motion cannot be propagated in a fluid in right lines beyond an obstacle which stops part of the motion, but will bend and spread every way into the quiescent medium beyond the obstacle. . . . A bell may be heard beyond a hill which intercepts the sounding body . . . but light is never known to bend into the shadow.'

Young proved, however, that light does bend. The bend is extremely small, owing to the minuteness and immense speed of the waves, but it is greater in some mediums than in others, in water for example than in air (Fig. 90).

Young demonstrated this bending of light-rays by a simple

experiment. Light reflected from the sun was admitted through a pin-hole in the side of a dark chamber, making a cone of light. In the pathway of this cone was interposed a narrow strip of card. Faint fringes of colour were seen on either side of the shadow thus cast on the opposite wall, while in the shadow itself was a sequence of faint dark and light upright bands, finishing off in a faint light band in the middle of the shadow. Since light normally travels equally in all directions, a part of it, passing on each side of the strip of card, must spread out behind it. But why should the light arrange itself in strips, and not fall equally all over the shadow? When an opaque object was placed so as to prevent the light from passing one of the edges

FIG. 90. Waves diverging from centre A, pass through aperture BC. They extend themselves on each side—that is, they 'bend into the shade'— so as to fill the space $BCDE$ while affecting the parts outside this area much less or not at all. (Young's diagram.)

of the card the fringes disappeared. Therefore, so long as the light passes in one direction behind the card it spreads itself out equally, and only when two sets of rays from the two sides of the card meet do 'interference' bands appear. This is a close analogy to what happens in the case of water-waves.

Light does not, however, always travel through a transparent medium equally in all directions. Thus it had long been known that light traversing two crystals of Iceland spar in any but one direction gives two streams of (usually) unequal brightness. The relative intensity of the two streams was known to depend on the relative positions of the crystals. In certain positions one stream disappears entirely. The French mathematician ÉTIENNE LOUIS MALUS (1775–1812) found that he could elicit results comparable

Transformations of Forces

to those of Iceland spar by light *reflected* from transparent surfaces. Misunderstanding the nature of the process, he called it *polarization* (1805), a misleading title which it still bears. Such phenomena as those investigated by Malus were inexplicable on the wave theory until it was given its modern form by the French experimenter AUGUSTE JEAN FRESNEL (1782–1827), in correspondence with Young.

In sound-waves, from which Young had drawn his picture of light-waves, the vibrating particles move in a direction parallel to the propagation of the wave. This is 'longitudinal' vibration. In water-waves the water particles move up and down at *right angles* to the forward direction of the wave. This is 'transverse' vibration. The ether vibrations of a light-wave are transverse. They have, however, this complication, that the plane of vibration is not restricted, so that a ray of light may consist of waves vibrating in any plane at right angles to the direction of the ray.

Graphically represented (Fig. 91), looking 'end-on' at a wave, we can

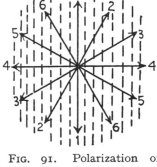

FIG. 91. Polarization of Light. The observer is supposed to view a light ray end-on as it advances toward him. The line of advance is along an axis represented by the central point. Vibrations in the ether take place in all planes through which the axis passes. These planes, from the observer's point of view, are seen as straight lines, of which six are represented. If the ray encounters a medium which acts as a grating (such as is represented by the dotted lines), permitting the passage of vibrations in only one plane (1 in diagram), the light is 'polarized'.

visualize a series of short straight lines signifying the extremes between which the 'ether particles' vibrate. Although vibrating in planes at all angles to the line of the advancing light, yet all vibrations are at right angles to the direction in which the wave advances, that is, for the purposes of our diagram, they vibrate in the plane of the paper. The action of Iceland spar upon the light-waves impinging on it may be compared to a set of railings with vertical chinks. Vibrations parallel to the rails will pass on between its chinks, but the remainder will be stopped. The light that passes on is said to be 'polarized'.

321

The Mechanical World

Fresnel also used the conception of interference to bring the undulatory view of light within mathematical range by making possible a quantitative estimate of the length of light-waves (1821). Two metal mirrors in almost the same plane (Fig. 92) reflect light from a pin-hole in the wall of a dark chamber on to a white screen. Looking into the mirrors from the screen the observer would see two 'virtual images' of the Whole, as at A and B, and the optical effects are as though the light really proceeds

FIG. 92. Fresnel's Interference Experiment.

from those points. By rotating the mirrors, A and B can be made to approach each other until, when the mirrors are in the same plane, the points coincide into a single virtual image. A line drawn from this, vertical to the screen, meets it at C. Consider any point P on the screen in the area that receives light from both A and B. PA is longer than PB, but the difference becomes less the nearer P is to C. This difference, PA minus PB, can be calculated from the known conditions of the experiment. Now P sometimes shows a dark, sometimes a light, band. This will be according as the difference between PA and PB approximates to an odd or even multiple of a half wave-length; whether, in fact, the waves of light as from A and B strike the screen in the same or in opposite phases (Fig. 93).

We have seen (p. 317) that the wave theory requires that the velocity of light in rarer media should be greater than in denser, the opposite being demanded by the emission theory. Thus a direct proof that light passes more rapidly through air than water would be a further confirmation of the wave theory. This was achieved by JEAN LEON FOUCAULT (1819–68) in a very well-known series of experiments begun in Paris in 1850, and described in full detail in 1862. He had already done work with his exact contemporary, HIPPOLYTE LOUIS FIZEAU (1819–96), on allied themes, such as chromatic polarization of light and interference of heat-rays and of light-rays of greatly differing length of path. He is also well remembered for his invention of the gyroscope (1852) and for his method of giving the reflectors of optical instruments a spheroid or paraboloid form (1857). His name is, moreover, attached to several electrical devices. Fizeau had made determinations of the absolute velocity of light in 1849. These deter-

FIG. 93. *ABCD* represents a system of transverse waves propagated toward the right. At *P* it is joined by a second system, of equal amplitude and wave-length, but with oscillations half a wave-length later. There results a state of oscillatory rest or 'interference'. Should the waves of the second system have the same wave-length but unequal amplitude, the amplitude of the first would be reduced. If the second has a different wave-length, a more complex system will arise.

minations of Foucault and Fizeau—in the neighbourhood of 300,000 kilometres per second—open the modern classical period of optics. Fizeau introduced certain conceptions of the relative motion of matter and ether that were later developed by Clerk-Maxwell.

(ix) *Doctrine of Energy.*

The History of Science submits, no more easily than the history of other subjects, to arbitrary time divisions. Nevertheless there are certain seminal scientific ideas, the appearance of which makes it possible for the historian to establish time boundaries sufficient for the division of his narrative. Such a one is the doctrine that any form of measurable physical activity is convertible into any other form, and that the total amount of such activity in the world is limited and remains the same. This Doctrine of Energy became accepted about the middle of the nineteenth century and opened a new era in the history of scientific ideas.

An important advance in this direction was made by the brilliant

The Mechanical World

young Frenchman, SADI CARNOT (1796–1831), in his only publication, *Reflexions on the Motive Power of Fire* (1824). Carnot measured and defined work as 'weight lifted through a certain height'. He established quite clearly the principle that heat and work are reversible conditions and that the efficiency of a reversible engine depends on the temperatures between which it works. The work of Carnot attracted little attention during his lifetime. The principles involved were grasped some twenty years later by the Englishman J. P. JOULE (1818–89), a pupil of Dalton, who developed the subject with great experimental skill.

Joule's work began to assume significance in 1840, when he was emphasizing the idea of the importance of physical units. Those which he then adopted involved the conception of the transference of chemical into electrical activity in a measurable way. His unit of static electricity was the quantity needed to decompose 9 grammes of water, and his degree of current electricity the same amount propagated in an hour. He regarded the consumption of the metal in the electric battery as a source of energy analogous to that of the coal that drives the steam-engine.

In considering the electric motor invented by Faraday, Joule was able to demonstrate a numerical relation between the chemical effect in the battery, the mechanical effect in the motor, and the electrical effect in the circuit. Thus, if a given weight of zinc be dissolved in acid, a certain measurable amount of heat is given off. Make the zinc an element in a battery and a measurably *less* amount of heat is produced in the course of its solution. If the current passes through a wire, it heats the wire. This amount of heat corresponds, he showed, to the difference between the heat produced by the simple solution of zinc in acid and that produced when it is dissolved as an element in a battery. Moreover, if the current drives a motor yet more heat is missing. The amount missing is proportional to the work done by the motor.

Joule's historic paper of 1843 *On the calorific effects of Magneto-Electricity and on the mechanical value of Heat* brings out very clearly the relation between work and heat. It sets forth 'Joule's Equivalent', as it is now called, that is, the amount of work which must be transformed in order to give one unit of heat. This unit of heat was the amount needed to raise one pound of water one degree Fahrenheit. His unit of work was the amount required to raise one pound weight a height of one foot. His equivalent, as he then determined it, was 838 foot-pounds.

In the years that followed, Joule pursued his idea with many refinements. Thus he measured the work required and the heat produced when water is driven through fine tubes, when air is compressed or allowed to expand, when a paddle-wheel is driven through water or through more viscous fluids, and so on. But not until 1847 did he give the first full and clear exposition of that principle now called *energy*, a term first applied in that capacity by William Thomson (Lord Kelvin) in his great paper, *Dissipation of Mechanical Energy* (1852).[1] Joule's superb exposition of 1847 had been given in the form of a popular lecture in a church reading room! This great scientific pronouncement, after rejection by several journals, appeared in a Manchester weekly paper with the title: *Matter, Living Force and Heat*. His *living force* is, of course, what we call 'Energy'. He said:

'Living force (*vis viva*) is one of the most important qualities with which matter can be endowed, and as such it would be absurd to suppose that it can be destroyed. . . . Experiment has shown that wherever living force is *apparently* destroyed, whether by percussion, friction, or any similar means, an exact equivalent of heat is restored. The converse is also true, namely, that heat cannot be lessened or absorbed without the production of *living force* or its equivalent attraction through space. . . . Heat, living force and attraction through space (to which I might also add *light*, were it consistent with the scope of the present lecture) are mutually convertible. In these conversions nothing is ever lost.'

In the same year appeared the little book of HERMANN HELMHOLTZ (1821–94), *Erhaltung der Kraft* ('Conservation of Energy') In the same year, too, Joule came in contact with William Thomson, afterwards LORD KELVIN (1824–1907), who had long been interested in the transformation of heat. Helmholtz, in his famous pamphlet, in rejecting the possibility of perpetual motion, sought to establish the doctrine that through all transformations of energy the sum total of all energies in the universe remains constant. Thomson accepted the conclusions of Joule and Helmholtz and

[1] Thomas Young had used the word in an analogous sense in 1807.

applied himself from 1848 onward to the mathematical implications of these doctrines. 'The first step toward numerical reckoning of the properties of matter', he wrote, 'is the discovery of a continuously varying action of some kind, and the means of measuring it in terms of some arbitrary unit or scale division. But more is necessary to complete the science of measurement in my department, and that is the fixing on something absolutely definite, as the unit of reckoning.'

Thomson reached his conception of a fixed point. He was familiar with Carnot's view of a reversible cycle and was one of the first to draw attention to it (1848), in illustration of the fact that the melting-point of ice is lowered by pressure. He saw clearly that the amount of work performed by an engine does not depend directly on the thermometer scale value of the temperatures between which it is working. Thus, to take a simple example, the work done between 100° and 150° is not the same as the work done between 150° and 200°. Therefore, before it is possible to reach a clear conception of the interchange of forces it is necessary to find some absolute scale which is not arbitrarily determined by the changes of state of a single substance as is that of the ordinary thermometer by the freezing-point and boiling-point of water. Now for an engine to be theoretically perfect, that is, for all its heat to be converted to work, it is necessary that the lower of the temperatures between which it works be the minimum possible. This minimum point Thomson called 'the absolute zero of temperature'.

Working between the temperatures 0° and 100° Thomson found that for every 373 parts of heat put in at 100° the engine will return 273 parts into the receiver, converting 100 parts into mechanical work. In other words, if boiling-point under the stated conditions be taken as one fixed point and freezing-point be taken as another, then—treating the working range between these two points as 100—the lowest conceivable temperature, the zero of this absolute scale, would be −273°. This is the zero of an absolute thermometric scale'. That scale is concerned solely with the work done by the substance employed and has nothing to do with its physical properties.

The recognition of an absolute scale and of its implications with

Transformations of Forces

the doctrines of energy, of the transformation of forces, of the ether, and of atoms provides the foundations on which was built the impressive structure of classical physics during the second half of the nineteenth century.

6. Multiplicity of Organic Forms.

(i) Early Classificatory Systems.

As the exploration of the globe proceeded, the number of kinds of living organisms known to science rapidly increased and became very large. Some system of codification and standardized description became an urgent need. Many attempts were made in this direction, but the successful and accepted scheme was that of the Swede KARL LINNAEUS (1707–78). Its pre-eminent convenience led to its rapid adoption to the exclusion of all other systems.

Linnaeus took the parts of a plant or animal in regular sequence and described them according to a recognized rule. This introduced what was almost a new international language, very condensed, very clear, and very easily learned. As 'botanical Latin' it has survived and maintained its usefulness. The method was a great improvement on the verbose and confused accounts usual till that time. It is best expounded in his *Philosophia botanica* (1751).

Linnaeus also constructed a system of arrangement in which every known species of animal and plant had a position assigned to it. This involved grouping the *Species* into *Genera*, the Genera into *Orders*, and the Orders into *Classes*.

For plants the Classes and Orders were based on the number and arrangement of the parts in the flower. Linnaeus had a clear though not very accurate or searching conception of the sexual character of the floral elements. The number of 'stamens' or free male parts was his first consideration. Thus Linnaeus grouped plants with one stamen in the Class *Monandria*, plants with two in the Class *Diandria*, plants with three in the Class *Triandria*, and so on. Each Class was then divided into Orders, according to the number of 'styles', or free female parts, in the flower. Thus, the Class *Monandria* was divided into the Orders *Monandria*

Monogynia with one style, *Monandria Digynia* with two, and *Monandria Trigynia* with three, and so on.

For animals, Linnaeus distinguished the Classes of Mammals, Birds, Reptiles, Fishes, Insects, and Vermes. The first four had already been grouped together by Aristotle as 'Animals with red blood' or, as we now call them, 'Vertebrata' or backboned animals. The remaining Classes, Insects and Vermes, contain, bundled together, all the Orders of animals without vertebrae or backbones. Here Linnaeus was behind Aristotle, who had broken up these groups more effectively (p. 41).

The contribution through which the name of Linnaeus will, however, always be remembered and is daily recalled by naturalists is his 'binomial nomenclature', the system of defining every known living thing by two Latin names, the first being that of its *genus* and the second that of its *species*. It will naturally be asked what is meant by these words. To this no one can give a clear or even an intelligible answer, though there is evidence that an answer is slowly emerging from certain current work. Naturalists have been occupied for over two centuries with the more exact individual application of these terms without reaching any general definition of them. It is unparalleled in scientific history that undefined and undefinable terms should remain indispensable for so long and so active a period.

But although no one can, even now, define species in general terms, Linnaeus had certain ideas concerning their nature which are of great historical importance. He held that species are constant and invariable, a view in which he differed from John Ray. 'There are just as many species as there were created in the beginning', wrote Linnaeus, and again, 'There is no such thing as a new species'. In this matter we have departed completely from his standpoint.

The *Systema Naturae* of Linnaeus is nevertheless a permanent contribution. It was first drafted in 1735, and he modified it and amplified it in its many editions. Of these, biologists have agreed on the tenth, which appeared in 1758, as the permanent basis for the scientific names of living things. If a species is given its 'Linnaean name' by a modern naturalist, it means that adopted in this tenth edition.

328

Multiplicity of Organic Forms

Linnaeus was an extremely stimulating teacher. He had a great number of enthusiastic pupils, many of whom went on expeditions to distant lands and discovered and described multitudes of species. He and his disciples, by concentrating their interest on external parts, which are specially valuable for purposes of classification, withdrew attention from the intimate structure and working of the living organism. The search for new species thus remained for long the chief aim of most naturalists, to the neglect both of anatomical and of physiological studies.

Much of the immense appeal of Linnaeus to his generation and those which followed was due to his appreciation of wild life. There have been few greater nature lovers. His tradition can be traced especially in Britain. It is commemorated in the 'Linnean Society' (established 1790), and its impact happened to coincide in time and was ancillary to the literary movement known as the 'Romantic Revolt'. Natural History in Britain had long interested the country gentry and clergy. There came a time when these were reinforced by the scientific tastes of the rising and wealthy industrial class. Thus the study of Nature became 'fashionable'. Societies for it were founded in every major centre of population in Britain, from Kirkwall in the Orkneys to Penzance in Cornwall. Darwin, who approached his great task in the dual capacity of systematist and observer of wild life, was a typical product of this dual Linnaean tradition. Among its gifted literary exponents were the Rev. Gilbert White of Selborne (1789) and Charles Waterton of the *Wanderings* (1823). Other eminent and typical figures associated with the movement were Banks (p. 340), Lyell (p. 281), Murchison (p. 283), T. A. Knight, and the entomologist, the Rev. W. Kirby (1759–1850).

Since the time of Linnaeus almost every important biological movement has left its mark on the system of classification current in its day. The classification of living things adopted by a biological writer may often be treated as an epitome of his views on many important biological problems, and especially on 'comparative' studies. This was notably the case with the system of Cuvier.

The French naturalist GEORGES CUVIER (1769–1832) wielded great authority and determined the general direction of biological,

329

and especially of zoological, activity in the first half of the nineteenth century. His general approach, unlike that of Linnaeus, was analytic. He laid stress on the structure and relations of the inner parts rather than on their external characters. Cuvier divided the animal kingdom into four great divisions, each of which, in his view (1817), is built on its own peculiar and definite plan.

I. VERTEBRATA, with a backbone.
II. MOLLUSCA, slugs, oysters, snails, &c.
III. ARTICULATA, jointed animals, insects, spiders, lobsters, &c.
IV. RADIATA, all remaining animals.

In drawing up this scheme Cuvier was guided by his analysis of two main sets of functions. The heart and circulation provide, he considered, a centre for the 'vegetative functions' of growth, reproduction, &c., to which the breathing apparatus is accessory. The brain and spinal cord he regarded as presiding over the 'animal functions' which are associated with active movement and are served by the muscular system. We are here reminded of the 'vegetative' and 'animal soul' of Aristotle. The thought of Cuvier is, in fact, infused with that of his great predecessor. Though the conception of vegetative and animal functions have, since Cuvier's time, changed beyond recognition, much of our modern classificatory system is based on him and through him on Aristotle.

The Genevan botanist, AUGUSTIN PYRAMUS DE CANDOLLE (1778–1841), did for plants similar service to that of Cuvier for animals. He was a searching and patient investigator. Much of his classification of the higher plants (1824) survives in the systems developed by modern botanists.

(ii) *Main Subdivisions of Biological Study.*

A feature in the biological outlook of the early nineteenth century was the slowness with which microscopical research took an important place. The great microscopists of the seventeenth century had singularly few successors in the eighteenth. Thus, when Humphry Davy needed descriptions of the microscopic structure of stems and leaves for his *Agricultural Chemistry* (1813)

Multiplicity of Organic Forms

he had to rely on Grew's *Anatomy of Plants* (1682) of 130 years earlier. In the third decade of the nineteenth century certain improvements in the microscope began to be available, and awoke the interest of biologists. A number of observers were soon devoting themselves to the intensive study of microscopic organisms, and to the microscopic analysis of larger forms. From now on the microscope became the essential instrument of the biological sciences. Microscopic observations have since provided the building material for a dozen separate departments of biology.

In the first half of the nineteenth century philosophical naturalists were largely occupied in establishing 'affinities' between different types of organisms. These workers may be divided into fairly definite groups according to the character of the problems that they set themselves to solve. Of these groups five may be distinguished as of particular historical importance.

(*a*) Those concerned with comparing external characters of living forms nearly allied to each other; that is, in the work of establishing the nature and limits of species, genera, and families, and of their degrees of affinity. These are the systematists or 'taxonomists' (Greek *taxis*, 'arrangement'; *nomia*, distribution. The word was introduced by de Candolle 1813). A very great exponent of this study was Darwin. His concentration on its problems led him to his historic consideration of the *origin* of species. An investigator who worked on comparable lines but came to different conclusions was Louis Agassiz. Important botanical taxonomists were A. P. de Candolle and J. D. Hooker.

(*b*) Those occupied in investigating the inner structure of contrasted forms, that is of forms belonging to widely separated groups as Orders and Classes. These are the *comparative anatomists* or *morphologists*. (The term 'Comparative Anatomy' was introduced in 1672 by Grew. Morphology, Greek *morphé*, 'form', was introduced by Goethe about 1817.) Typical exponents of this method were the versatile JOHANNES MÜLLER (1801–58) in Germany, RICHARD OWEN (1804–92), opponent of Darwin, and first director of the British Museum of Natural History, ROBERT BROWN (1773–1858), 'botanicorum facile princeps', and ÉTIENNE GEOFFROY ST. HILAIRE (1772–1844), opponent of Cuvier.

(*c*) Workers engaged in the comparative anatomy of fossil forms

331

The Mechanical World

are known as *palaeontologists*. Among the greatest of these were RICHARD OWEN (1804–92), and the palaeobotanist W. C. WILLIAMSON (1816–95). The word 'Palaeontology' was introduced into English by Sir Charles Lyell (1838).

(*d*) It was early realized that the structure of embryos revealed affinities that are less apparent in adults. Moreover, in certain respects, the knowledge of the formation of the parts in the embryo was found to make the structure of adult forms more intelligible. The beginnings of life had always excited wonder and curiosity. The investigation of embryos required, however, unusual kinds of skill, and 'embryologists' were early differentiated. (The term *embryologie* was admitted into the French language by the *Académie* in 1762. It did not enter English till the nineteenth century.) Important early embryologists were the Germans KARL ERNST VON BAER (1792–1876) and ROBERT REMAK (1815–65), the Swiss ALBRECHT KÖLLIKER (1817–1905), and the Swiss American LOUIS AGASSIZ (1807–73).

(*e*) Quite apart from the schools of 'naturalists' and 'biologists', the first half of the nineteenth century saw a great extension of scientific interest in the analytical study of animal function by means of physical and chemical experiment. The exponents of this science of 'physiology' were mainly preoccupied with its medical applications. Such physiologists were not usually concerned to compare different forms. Choosing for preference those likest to man—the 'higher' animals—they devoted themselves rather to the examination of the parts and functions in their developed state. The results have been portentous in bulk, complexity, and interest, and have given rise to a picture of the animal machine which has deeply influenced the current conception of the nature of Man, and of his place in Nature. Among the greatest exponents of this department of science were SIR CHARLES BELL (1774–1842), JOHANNES MÜLLER (1801–58), and CLAUDE BERNARD (1813–78).

(iii) *Naturphilosophie*.

The startling revelations of the microscopists and the 'mechanist' physiologists of the seventeenth century induced, especially in German thought, an era of speculative activity. The conception

332

of the 'ladder of Nature' assumed a new importance. Aristotle had been content with its formal projection (p. 41). During the eighteenth century it took the form of a rigid framework into which observations were to be fitted.

Certain microscopic observations had given rise to the false idea that the organism is already fully formed in the germinal original, that is in the ovum of the female or, alternatively, in the spermatozoon of the male. The Genevan CHARLES BONNET (1720–93) raised this idea of 'preformation' to the rank of a scientific and philosophic doctrine (1762). Both this conception and the process of reproduction without fertilization (parthenogenesis, Greek, = virgin birth), which he rediscovered (1745), he made to serve theological ends.

In this peculiar intellectual atmosphere Bonnet and his followers developed a rigid interpretation of the conception of a 'ladder of nature'. Passing from the most subtle of the elements, fire, through air, water, and the densest, earth, this 'scala naturae' ascended through the finer minerals, such as crystals, to living things, proceeding through what were then regarded as the lowest of these, namely the moulds, via plants, insects, and worms, upward to fish, birds, mammals, and finally to man. The medieval and Christian view of 'man as the measure of all things' was thus given a new significance by Bonnet and his school. 'All beings,' he wrote, 'have been conceived and formed on one single plan, of which they are the endlessly graded variants. This prototype is man, whose stages of development are so many steps toward the highest form of being.' Each being was believed to be 'preformed' in the male or female 'primordium' or germ, the spermatozoon or ovum.

Such views pass insensibly into the attitude, known later as *Naturphilosophie*, which became especially popular in Germany. Some of its developments in that country became fantastic to the verge of insanity. Yet there were several effective thinkers for whom this attitude became a useful approach to natural knowledge. Among these were the two loftiest intellects that Germany has produced, Kant and Goethe.

The thought of the age was given a new direction by the Königsberg philosopher, IMMANUEL KANT (1724–1804) in his famous *Critique of Pure Reason* (1771). He had begun as a man of science, and it was from his treatment of scientific problems that his philosophical interest emerged. Beginning with a world of phenomena, of nature, of experience—the determinate world of the man of science—he gradually passed into the world of the intelligible, of ends, of the philosopher.

To most men then, and to most men still, these two worlds seem to confront one another. Men of science affirm this when they say that 'the study of purpose in Nature is inconsistent with the scientific aim, which is the adequate description of phenomena'. It was Kant's thought that the two attitudes are neither opposite nor irreconcilable. He reduces the problem to the discussion of the relation between our perception of things and their real nature. Our perceptions, Kant held, come into relation with the real nature of things through the character of our processes of thought. In other words, our thoughts work along Nature's own lines. Kant pointed out that, if we consider living organisms, we perceive that they are composed of parts which are comprehensible only as conditions for the existence of the whole. The very existence of the whole implies an end. True, says Kant, Nature exhibits to us nothing in the way of purpose. Nevertheless we can only understand an organism if we regard it as though produced under the guidance of thought for the end. The naturalist tacitly admits this when he considers the different organs or parts in relation to their function in the whole living organism.

The opposition, so familiar to the biologist, between the mechanist and the teleological (p. 42) or vitalist view, is, Kant held, due to the nature of our knowledge, that is of our experience. But our thoughts must be distinguished from our experience. In thought we pass constantly from the view of the *part as mechanism* to a view of the *whole as purpose,* and back again. Nor do we separate these two views unless deflected by some specific doctrine that the parts are really separate. There is, Kant believes, a hidden basic principle of Nature which unites the mechanical and teleological. That principle is none the less real because our reason fails to grasp it or our powers to formulate it. So far as actual practice and use of language go, such a principle is, in fact, accepted by every biologist, the most convinced 'mechanist' no less than the most extreme 'teleologist'.

Multiplicity of Organic Forms

Kant's scientific influence is to be traced especially in JOHANN WOLFGANG VON GOETHE (1749–1832), whose pre-eminence as a poet and writer must not obscure his importance for science. Goethe did great service in emphasizing the fact that all organisms accord in structure to a certain quite limited number of patterns or plans. These represented for him 'ideas' in the mind of God. By searching for these ideas—'plans' or 'types' as they came later to be called—Goethe and his followers did much to stimulate the systematic comparison of diverse living things. Not the least of their services was that they thus persuaded biologists to abandon the point of view, derived from medical applications, that regards the structure of man as the type to which that of all other creatures must be referred.

Goethe expounded several doctrines of great importance, some of which are still of value. His most valuable scientific conceptions were the following:

(*a*) The genera of a larger group (Family, Order, Class, or Phylum) present something in the nature of variants on a common plan. These are all expressions of the same 'idea' or 'type'.

(*b*) The various parts of the flower are but modifications of leaves. The 'cotyledons' of germinating seeds (cf. the terms 'Monocotyledons' and 'Dicotyledons') are but the first leaves borne by the infant shoot.

(*c*) Similarly all the parts of living beings are referable to one original model or 'primordium'. Thus not only is there a primordial species of animal and a primordial species of plant, corresponding to the animal 'idea' and the plant 'idea', but also there is a primordial part of each animal and plant. The bones of the spine provide a good illustration of this conception. These 'vertebrae', fundamentally of the same origin and structure, have different forms, and perform different functions in different parts of the backbone. All are variants on the 'primordial' vertebra. Goethe believed that a like story might be told of other organs. This position is now indefensible in its original form, but it has in it an element of truth which provided a basis for much research and a good framework for the classification of observations.

Bonnet, Kant, Goethe, and their followers, the 'Nature

Philosophers', were thinkers rather than observers, though their observational activities certainly cannot be despised. Some of their ideas, strange and strained as they now seem, and repellent as they are to many men of science, recur again and again during the centuries.

Whence comes the continued fascination of thoughts so little related to the daily task of the scientific observer? The essence of the thought of such men is that the processes of the mind reflect the processes of Nature. In this there is surely a truth, though it is presumptuous to suppose that we have any deep understanding of this parallelism. To say that 'the burnt child dreads the fire' is but to give a special instance of the wider statement that reason is generalized experience. Our minds, as much the product of evolution as our bodies, have in the ages developed as mirrors of the world in which we dwell; they are attuned to Nature. The mathematical thought of ages on the nature of certain curves elaborated a knowledge which Kepler and Newton fitted into the phenomena of planetary movements. The minds of the pre-Keplerian mathematicians were attuned to Nature. They were working on Nature's lines, though they knew it not. To say that we live in a rational world is but to say that by reasoning aright we may learn something about that world. This is as true for biology as for astronomy, though no such diagrammatic illustration is to hand in the biological realm. Yet by those whose minds were specially attuned to biological studies, truths have often been discerned which were verified later by experience. This in itself is sufficient justification for that speculative attitude which is more productive during some scientific episodes than during others, but is never without its value.

(iv) *Correlation of Parts.*

GEORGES CUVIER (1769–1832), 'the dictator of biology', was especially interested in structure rather than function. He was essentially a 'morphologist'. The main conception that guided him was that of the 'correlation of parts', the nature of which must be discussed.

Organs do not exist or function separately in nature, but only as parts of complete living things. In these living things certain relations are observed which are fundamental to their mode of life. Thus feathers are always found in birds, and never in other creatures. The presence of feathers is related to a certain formation of the forelimb, with reference to its action as wing. Feathers are never found without wings, and in other winged animals than birds the structure of the wing is very different from that of birds, and never has feathers. But the wing structure peculiar to the bird is in turn related to certain formations of the collar-bone and breast-bone, with reference to the function of flight; these, again, to the form and movement of the chest; these, again, to the function of breathing, and so on throughout the entire body of the bird.

This principle of 'correlation' is traceable in the structure and working of each and every organ and probably in every part of every organ of the bird. Thus, given a feather, it is possible to infer that its owner had a particular form of collar-bone, a particular kind of skeleton, a particular type of mouth, a particular structure of lung, a particular method of breathing, excretion, digestion, a particular temperature and heart-beat, even a particular kind of mind. Again, given a particular form of collar-bone, skeleton, mouth, lung, &c., we can infer a feather. If enough be known of the comparative morphology of the bird group, it is possible by the use of this principle to make astonishingly sweeping and accurate inferences.

Cuvier was far from being the first to apply his principle. In a sense it is obvious. If anyone were to find a severed hand, he would know that it had once been attached to the body of a human being, and not to that of an animal. He could make a very likely guess at sex, occupation, age, state of health, and the social position of the owner of the hand. This is nothing but the 'principle of correlation' which is the theme of most detective stories. Aristotle had, to some extent, been able to act upon this principle, but Cuvier, out of the great stores of his knowledge of organic forms, refined and extended the application of it far beyond any of his predecessors. In Cuvier's hands the principle of correlation could often be brought to bear upon the merest fragment. From a little bit of leg bone, for example, even the 'leggy' nature of which no one but a trained naturalist could guess, he succeeded in reconstructing an entire giant bird of a very aberrant type. His reconstruction was proved to be accurate by subsequent discoveries.

The principle of correlation has been of special value in the study of fossils, since these are usually fragmentary. Cuvier was therefore in a good position to elucidate the relationship between the living and the extinct forms. Thus arose the modern science of 'palaeontology' which owes to none so great a debt as to him. In his time large numbers of very strange fossil forms were being discovered.

The effect on the mind of Cuvier of these strange discoveries may itself seem strange. He realized that the evidence of geology showed that there had been a succession of different types of animal population, and he recognized that vast numbers of species, many no longer existing, had appeared upon the earth at different periods. Following Linnaeus, he was a firm believer in the fixity and unalterability of species. He had, however, to account for the extinction of many forms of life, and the new appearance of many other forms. His explanation was that the earth had been the scene of a series of great catastrophes, of which the last was the Flood recorded in Genesis! He expressly denied the existence of fossil man.

Cuvier did not commit himself to the doctrine of a special creation following each catastrophe. He suggested that on each occasion the earth was repeopled from the remnant that survived. This did not explain the regular succession of new species in geological time. He believed that these came from parts of the world still inadequately explored by geologists. His followers carried the matter farther, and elevated his teaching into a doctrine of successive creations. This came to assume fantastic forms even in the hands of serious scientific exponents, one of whom, as late as 1849, expounded the science of palaeontology on the basis of twenty-seven successive creations.

Cuvier's great work *Le Règne Animal* appeared in 1819. With various enlargements, modifications, and improvements by his pupils it remained standard for many years. To him and through his disciples Comparative Anatomy owes so much that the work may be said to be still standard. His personality lit up a zeal for comparative anatomy and palaeontology which lasted throughout the nineteenth century. Of the many inspired by this movement a typical representative was RICHARD OWEN (1804–92). He was also influenced by *Naturphilosophie* and was an obstinate opponent of Darwinian evolution.

Owen embarked on an immense investigation of the teeth of mammals set forth in his *Odontography* (1840–5). Teeth, being the hardest parts of the body, are found fossilized more often than any others. Thus his investigations led him into palaeontology, of which he became an admitted master. Among his best-known works in that department are those on the giant bird, the recent but extinct Dinornis of New Zealand (1846), and the much more ancient giant walking sloth, the fossil Mylodon of South America (1842).

In 1856 Owen became Director of the Natural History Department of the British Museum, and his activity and industry rose to the occasion. His great *Anatomy and Physiology of the Vertebrates* (1866–8) was based entirely on personal observation, and was the most important of its kind since Cuvier. The system of classification he adopted has not won favour, but as a record of facts the book was of very great value.

The activity of the comparative anatomists during the nineteenth century was immense. Many new Classes were described on the basis of fossil material. The teaching of Darwin, providing a framework into which comparative studies could be fitted, gave the effective stimulus to such work. The alliance of comparative studies with evolutionary doctrine had the effect of focusing attention on structure as distinct from function. Comparative physiology almost ceased to be studied in the later nineteenth century, and is only now reviving. Comparative anatomy in its turn became largely a study of developmental stages, and embryology became the comparative study *par excellence*.

(v) *Biological Exploration.*

In the exploratory voyages of the eighteenth century the practice was begun of carrying naturalists with equipment for observing and collecting. One of the earliest and most important of the expeditions thus provided sailed the Pacific between 1768 and

The Mechanical World

1776 under Captain JAMES COOK (1728–79). JOSEPH BANKS (1745–1820), a young amateur of great wealth and scientific competence, accompanied him and provided equipment. The staff included several artists, and a pupil of Linnaeus went as botanist. The voyage yielded many plants and animals new to science. Cook's two other voyages were also very productive.

Among such expeditions a most important place is taken by the voyage of the *Beagle* in 1831–4, which carried as naturalist the youthful CHARLES DARWIN (1809–82). His name is so associated with the evolutionary idea through which he profoundly influenced scientific, philosophical, political, religious, and ethical thought, that certain of his other claims are often forgotten. To appreciate his distinction, it is necessary to recall that, had he never written on evolution, he would still stand in the front rank among naturalists, and would have to be included in any history of science. Thus, as a single example, even during the voyage in the *Beagle* he reached conclusions that modified and extended the fundamental working principles of geology and palaeontology.

In Darwin's record of experience in the *Beagle* in the famous *Journal of Researches* (1839) a special interest attaches to his observations on the highly peculiar animals and plants connected with oceanic islands. The Galapagos and St. Helena are good examples. Their extraordinary wealth of peculiar forms and the difference of these from those of the nearest neighbouring land—either continental or insular—are among the most striking phenomena in the distribution of living things. They, more perhaps than any other, suggested to Darwin his solution of the problem of the origin of species.

Second only in importance to the voyage of the *Beagle* was that of the *Erebus* and *Terror* (1839–43) which explored the Antarctic under the command of Sir James Ross (1800–62). As naturalist there accompanied him JOSEPH DALTON HOOKER (1817–1911), afterwards in charge of the Botanic Gardens at Kew.

Hooker was an industrious collector and skilled systematist. None of his numerous writings is of more weight than those (published 1844–60) on the flora encountered in this voyage. They include accounts of the plants of the Antarctic area. as well as those of Tasmania and New Zealand, and laid the foundation of the systematic study of plant geography. Further Hooker showed the vast importance in the economy of Nature of the minute marine plants known as 'diatoms'.

The expedition was also important for its revelation of a very varied fauna in a region hitherto unexplored, namely the depths of the sea. Four hundred fathoms were sounded by Ross, and life was proved to be abundant there. We now know that there is life in the open sea at every depth with a great concentration near the surface and at the bottom. Until about 1869, however, with the laying of the first Atlantic cable, it was not realized how vast and varied a fauna and flora there is, and how different are the conditions of life at the two levels. The effective knowledge of the ocean fauna dates from the work of the *Challenger* naturalists. They showed that most of the living matter in the world is contained in the microscopic plant forms that float at and near the surface.

The greatest of all biological explorations was that undertaken by this British Admiralty vessel *Challenger* in 1872–6. She carried full equipment for six naturalists under CHARLES WYVILLE THOMSON (1830–82). She travelled 69,000 nautical miles in the course of which every ocean and the least frequented parts of the world were visited, and hundreds of deep-sea soundings were taken.

The vast collections of the *Challenger* were investigated by a whole army of naturalists under JOHN MURRAY (1841–1914). The results were issued by the British Government in fifty large volumes. These provide the best-worked-out account of any biological expedition, and form especially the solid bases of a science of Oceanography. They made it evident that, for any understanding of the life of our planet as a whole, an exact knowledge of the physical conditions of the sea is essential. Oceanography has since developed in a manner which demonstrates the interdependence of the biological and the physical sciences. A study which involves more than two-thirds of the earth's surface, and implicates the whole past and future history of the other third, is of primary importance to our conception of life as a whole.

The voyage of the *Challenger* was succeeded by that of the United States Government steamer *Tuscarora*, whose scientific staff investigated the floor of the Pacific. Other American and Norwegian expeditions followed in rapid succession. ALEXANDER AGASSIZ (1835–1910) was especially prominent in this work. Trained as an engineer, he was able greatly to improve the apparatus of oceanic investigation. Among his most remarkable results was his demonstration that the deep-water animals of the Caribbean Sea are more nearly related to those of the Pacific depths than they are to those of the Atlantic. He concluded that the Caribbean was once a bay of the Pacific, and that it has been cut off from the Pacific by the uprise of the Isthmus of Panama.

(vi) *Distribution of Living Things.*

Many facts significant of the past or present configuration of the earth's surface have been revealed by the study of oceanography, and it may be said that the subject has greatly modified our general conception of the world of life. Nor is this remarkable, since the major part of the earth's surface is covered by sea, and the general level of depression of the sea is much greater than the general level of elevation of the land.

Oceanic plants dwelling near the surface were studied on the *Challenger* in conjunction with the floating fauna with which they dwell. The name *plankton* (Greek 'drifting') was invented for this whole community by VICTOR HENSEN (1835–1924) of Kiel (1888). The study of plankton has become of great importance. Hensen, primarily a physiologist, began it while considering the production of nutritive substances under different meteorological conditions. He thus laid the foundations of the systematic study of the economics of the life of the ocean—*oceanic bionomics*, as we may call it. The subject is fundamental for our conception of the course of life as a whole upon this planet.

The circumstances of life on the ocean floor, as revealed by the *Challenger*, and by later expeditions, are entirely different from those at the surface. The pressure at 5,000 fathoms is about 5 tons to the square inch as against 15 lb. at the surface. No sunlight penetrates there; below 200 fathoms all is dark. The temperature in the depths is uniform, and not much above freezing. There are no currents, and no seasons. Conditions are substantially uniform the world over, on the equator and at the poles. There is no vegetable life to build up the bodies of the animals that dwell there, and thus the animals prey only on one another, drawing their ultimate supplies from the dead matter that rains down from above.

The results of the deep-sea dredging have been in certain respects disappointing. Specimens of numerous new genera, and species of known families have been brought up. Many are interestingly specialized, but few are widely different in essential structure from more familiar forms. No 'missing links' have been discovered, no new Classes or Orders found.

The *Challenger* found, and further exploration has confirmed, that the species of plants and animals of the open ocean, whether on the surface or at the bottom, are mostly very widespread. An exception must be made for the inhabitants of the extremest depths. The distribution of oceanic forms is determined by such factors as temperature, degrees of saltness, intensity of light, pressure, &c.

The extension of the knowledge of the conditions that prevail in the ocean and in its superincumbent atmosphere is leading to a new range of scientific ideas. As the laws of oceanic life were seen to come into relation with those of physical conditions, a most impressive physico-biological parallelism was distinguished which may one day provide a real 'physiology' of the ocean. The word *physiologia* was, in fact, originally applied to the material working of the world as a whole, and not to the individual organism. Thus Gilbert ushered in the modern scientific era with his *Earth as a Magnet, a New Physiology* (1600).

For a philosophical view of our planet as a whole a knowledge of the distribution of the life on land as well as in the sea is necessary. That different countries had different kinds of living forms was always obvious. In the eighteenth century Buffon (pp. 278–9) drew attention to 'natural barriers' delimiting flora and fauna. Lyell (1834, p. 287) convinced his readers that the present distribution of life is determined by past changes involving the major land-masses. The materials obtained by Darwin in the *Beagle* (published 1839–63) brought out striking facts in the geographical distribution of animals, both living and extinct. The peculiar way in which existing species were placed on the earth's surface was, however, a special object of interest to the traveller and collector ALFRED RUSSEL WALLACE (1823–1913), known for writings on South America and the Eastern Archipelago.

Wallace produced in 1876 his *Geographical Distribution of Animals*, still the most important work on the subject. He based his discussion on mammals, dividing the land-surface of the earth into six zoogeographical regions. These he named *Palaearctic, Nearctic, Ethiopian, Oriental, Australian*, and *Neotropical*. These have been retained in great part by more modern workers. The most important changes since his time are (*a*) the separation of Madagascar (Malagasy) from the Ethiopian region; (*b*) the general recognition that the Palaearctic and Nearctic regions are more nearly allied to each other than to any other region, and their union into a Holarctic region; and (*c*) the subdivision of the 'Australian' or Pacific region.

Wallace. demonstrated many remarkable faunal contrasts. None is more striking than that between the islands of Bali and Lombok, near Java. These islands are separated by a deep strait which at its narrowest is but fifteen miles. Yet, as Wallace remarked, they 'differ far more in their birds and quadrupeds than do England and Japan'. This strait, known as the 'Wallace Line', has been generally regarded as delimiting the Oriental from the highly peculiar Australian zoogeographical region.

The zoogeographical regions into which the earth's surface can be divided must obviously depend upon the particular group of animals chosen, since different groups are of different geological age and have different modes of dispersal. It happens, however, that the division of geographical regions based on mammals accords closely with that based on perching birds, and is not vastly different from that based on certain invertebrate groups, e.g. the spiders, earthworms, &c. Very different from these, on the other hand, is the division based on such very ancient groups as reptiles or molluscs.

The general principles that determine plant regions are similar to those of animals, but their application is considerably different. The subject has been broached mainly in connexion with the flowering plants. These are geologically younger than the groups on which zoogeographical regions are based. Moreover, tempera-

Multiplicity of Organic Forms

ture and moisture are of overwhelming importance in the life of plants. Even between countries in the same regions of plant geography which present but slight differences of climate certain notable floristic differences may occur. Further, the means of dispersal of flowering plants are more effective than those of most animal groups. The effects of this are sufficiently evident on oceanic islands.

A pioneer plant geographer was the German philosopher and traveller, Alexander von Humboldt (p. 282). Von Humboldt began

Fig. 94. The main zoogeographical regions. The 'Australian' region includes an immense number of islands, too small to appear on the map.

his *Kosmos* (1845–7) when he was seventy-six, and he completed it in what he called the 'improbable years' which followed. This great book, now seldom read, did good service in emphasizing the relations between the forms and habits of plants and the character and soil of their habitat.

Certain resemblances between the flora of Africa, South America, and Australia had impressed Humboldt and other naturalists. In 1847 J. D. Hooker (p. 340) suggested in explanation a land connexion between South America and Australia as late as Jurassic times. Various names, forms, and areas have been ascribed to this now fragmented continent.

Attempts to delimit definite plant regions have been less successful than those of the zoogeographers. A simple scheme is to divide the earth's flora into three primary areas: (*a*) the North

345

Temperate Zone, (*b*) the Tropical Zone, and (*c*) the South Temperate Zone. The northern tropic cuts off (*a*) from (*b*) with considerable accuracy. The southern tropic separates (*b*) from (*c*) with less precision.

(*a*) The North Temperate Zone contains most land. It is continuous save for the geologically recent break at the Behring Straits. It is characterized (i) by needle-leaved cone-bearing trees; (ii) by catkin-bearing and other trees that lose their leaves in winter; and (iii) by a great number of herbaceous plants that die down annually.

(*b*) The Tropical Region occupies areas widely separated by intervening ocean. It is characterized (i) by giant Monocotyledons, notably the palms, by the Banana family, and by the enormous grasses known as 'bamboos'; (ii) by evergreen polypetalous trees, and by figs; (iii) by the rarity of herbaceous plants which, in this region, are mostly parasitic on other plants.

(*c*) The South Temperate Zone occupies very widely separate areas of South Africa, South America, Australia, and New Zealand. It is characterized by a number of peculiar Natural Orders, mostly of shrub-like habit. Many are intolerant of moisture. Individual species are very numerous and often very restricted in area of distribution.

Geographical regions are biologically interesting not so much in themselves, but as revealing or summarizing the history of the various groups from which they are constructed. Thus the distribution in space of living forms is ultimately referable to their distribution in time. The discussion of the one is of little profit without the other. The first systematic efforts to correlate the two sets of facts for plants were made by WILLIAM CRAWFORD WILLIAMSON (1816–95) of Manchester, who came early under the influence of William Smith (p. 280) and began his work on plants in 1858. Williamson demonstrated that in coal are to be found gigantic woody forms similar to the higher existing flowerless plants, such as horse-tails, ferns, and club-mosses.

Knowledge of the geological succession of plant forms became astonishingly detailed, and the floristic landscape at various periods and in various parts of the world was confidently restored. Moreover, owing to the fact that plant cells have definite and thick walls which may be preserved in fossils, it is sometimes possible to examine the minute structure of fossil forms. In many cases even the reproductive processes are susceptible of close examination. Such studies have produced remarkably definite theories of the line of descent of plant forms, and of the interrelation between the great groups.

7. Physical Interpretation of the Living Organism.

(i) Beginnings of Modern Physiology.

Throughout all history there has been an opposition, alike in philosophy and in science, between the interpretation of the nature of life in terms of mechanism and that in terms of some other entity. In the first quarter of the eighteenth century this conflict came into very clear view.

GEORGE ERNST STAHL (1660–1734), professor at Halle and a fashionable physician, was an extremely voluminous writer, especially on chemistry. He saddled that science with the unfortunate theory of 'phlogiston' (p. 288) which held its ground until Lavoisier. In physiology he set himself especially against the mechanism of Descartes. To the French philosopher the animal body was a machine. To the German physician the word *machine* expressed exactly what the animal body was not. The phenomena characteristic of the living body are, Stahl considered, governed not by physical laws but by laws of a wholly different kind. These are the laws of the *sensitive soul* which, in its ultimate analysis, is not dissimilar to the *psyche* of Aristotle. Stahl held that the immediate instruments, the natural slaves, of this sensitive soul, are chemical processes (1708).

Almost exactly contemporary with Stahl was his rival at Halle FRIEDRICH HOFFMANN (1660–1742), who was no less skilled a chemist and at least as verbose a writer. In Hoffmann's view the body is like a machine. Nevertheless he separated himself, on the one hand from the pure mechanists of the school of Descartes and Boerhaave by claiming that bodily movements are the exhibition of properties peculiar to organic matter and, on the other, from the Stahlian vitalists by denying the need to invoke a sensitive soul. 'Life', he wrote, 'consists in the movements of the blood. This circular movement maintains the integrity of that complex which makes up the body. The vital spirits which come from the blood are prepared in the brain and released therefrom to the nerves. Through them come the acts of organic life which can be reduced to the mechanical effects of contraction and expansion' (1718).

An important participator in the controversy was HERMANN BOERHAAVE (1668–1738), professor of medicine at Leyden, and one of the greatest physicians of all time. He, too, was skilled in chemistry. His admirable *Institutiones medicae* (1708) remained the standard account of physiology for half a century. In this work Boerhaave goes systematically through the functions and actions of the body, seeking to ascribe chemical and physical laws to each. He does lip-service to the influence of mind on body, but in practice is as completely mechanist as Descartes. Thus he still believed that something material passes down the nerves to cause movement by distending the muscles. He set the tone to physiological thought for at least a century.

Throughout the eighteenth and nineteenth centuries nearly all important physiological investigators were medical men. An exception was the exemplary parish priest, the Rev. STEPHEN HALES (1677–1761), who made many important advances in both animal and vegetable physiology. His work on the functional activity of plants was the most important until the nineteenth century. His *Vegetable Staticks* (1727) contains the record of a great number of experiments on living plants, devised to interpret their activity in terms of recognized physical forces. Thus, measuring the amounts of water taken in by the roots and that given off by the leaves, he estimated what botanists now call 'transpiration'. He compared this with the amount of moisture in the earth, and showed the relationship of the one to the other. He calculated the rate at which the water rises in the stems, and showed that this has a relation to the rate at which it enters by the roots and is transpired through the leaves. He measured the force of the upward sap-current in the stems. He sought to show that these activities of living plants might be explained in mechanical terms with reference to their structure.

An interesting contribution by Hales was his demonstration that the air supplies something material to the substance of

348

Physical Interpretation of the Living Organism

plants. This we now know to be carbon dioxide. Following upon this he showed with the aid of the air-pump that air enters the plant, not only through the leaves, but also through the rind.

Hales endeavoured to give a quantitative expression to the conception of the circulation of the blood in animals. He showed that just as there is a 'sap-pressure' that can be measured, so there is a 'blood-pressure' that can be measured. Moreover he perceived that this pressure in the vessels varies according to circumstances. It is different in the arteries and the veins; different during contraction of the heart from what it is during its dilation; different with a failing and with an active heart; different in large and in small animals. All these differences Hales measured. He measured, too, the rate of flow in the capillaries of the frog. These experiments and conclusions of Hales initiated the quantitative phase of the science of animal physiology.

In contrast to the secluded career of Hales is that of ALBRECHT VON HALLER (1708–77), a Swiss of noble birth and ample means who, after many active years in Dutch and German universities, retired to his native Berne. He exhibited literary and scientific activity almost unparalleled in range and volume. His great *Elementa Physiologiae* (1759–66) set forth his conceptions of the nature of living substance and of the action of the nervous system. These formed the main background of physiological thinking for a hundred years after his time, and are still integral parts of physiological teaching.

Associating life with movement and muscular contraction, Haller concentrated on an investigation of the muscle-fibres. A muscle-fibre, he pointed out, has in itself a tendency to shorten with any stimulus, and afterward to expand again to its normal length. This capacity for contraction Haller called 'irritability'. He recognized irritability as an element in the movement of various organs, and notably of the heart and of the intestines. The salient features of irritability are (*a*) that a very slight stimulus produces a movement altogether out of proportion to the original disturbance, and (*b*) that it will continue to do this repeatedly, so long as the fibre remains alive. We now recognize irritability as a property of all living matter.

349

Besides its own inherent force of irritability, Haller showed that a muscle-fibre can develop another force which (a) comes to it from without, (b) is carried from the central nervous system by a nerve, and (c) is that by which muscles are normally called into action after the death of the organism as a whole. This is the 'nerve force' which he thus distinguished from irritability. It provides one way of arousing irritability.

Having dealt with movement, Haller turned to feeling. He showed that the tissues are not themselves capable of sensation, but that the nerves are the channels or instruments of this process, and that all the nerves are gathered together into the brain. These views he supported by experiments involving lesions or stimulation of the nerves and of different parts of the brain. He ascribed special importance to the outer part or cortex, but the central parts of the brain he regarded as the essential seat of the living principle, the soul. Although his view on the nature of the soul lacks clarity, he separates such conceptions sharply from those which he is able to deduce from actual experience. His work has, throughout, a modern ring, and he may reasonably be regarded as the father of modern physiology.

(ii) *Foundations of Bionomics.*

Light was thrown on the vital activities of plants by the chemist JOSEPH PRIESTLEY (p. 288). In his *Experiments and Observations on Different Kinds of Air* (1774) he demonstrated that plants immersed in water give off the gas which we term 'Oxygen'. He observed, too, that this gas is necessary for the support of animal life. His contemporary the French chemist, Lavoisier (p. 289), made quantitative examinations of the changes during breathing (1774, p. 289). These displayed the true nature of animal respiration, and proved that carbon dioxide and water are the normal products of the act of breathing.

In the meantime JAN INGENHOUSZ (1730–99) was introducing the highly important concept of the balance of animal and vegetable life. He was a Dutch engineer who worked in London with Hunter, and in 1779 published his *Experiments upon Vegetables, discovering their great power of purifying the common air in the sunshine and of injuring it in the shade and at night.* It contains

350

a demonstration that the green parts of plants, when exposed to light, fix the free carbon dioxide of the atmosphere. He showed that plants have no such power in darkness, but that they give off, on the contrary, a little carbon dioxide. This most significant discovery is the foundation of our whole conception of the economy of the world of living things. Animal life is ultimately dependent on plant life. Plants build up their substance from the carbon dioxide of the atmosphere together with the products of decomposition of dead animals and plants. Thus a balance is kept between the animal and the plant world. The balance can be observed in the isolated world of an aquarium.

The biological contribution of JOHN HUNTER (1728–93), whose life was closely contemporary with his pupil Ingenhousz, is peculiarly elusive and difficult to present. His older contemporary Linnaeus and his young contemporary Cuvier were both occupied in classifying organisms. To do this they sought always differences. It was similarities, however, that attracted Hunter. He experimented on and anatomized over 500 species. He designed to trace systematically through all these the different phases of life, as exhibited by their organs, their structure, and their activities. But his main work was his museum. A spirit informs it which is as different as possible from the 'magpie instinct' which has been the motive of many great collections. Here every object has its place and its reason for being included. Hunter created the modern idea of a museum by his conception of a collection to illustrate the varieties of structure and function right through the organic series.

Hunter was ever seeking the general principles that underlie the dissimilarities in organic forms. The most general of all is that mysterious thing called life. Life is never exhibited by itself, but is seen in the various activities of living things. As a surgeon Hunter naturally stressed, among these, the power of healing and repair. This power is peculiar to living things, and cannot be paralleled in the non-living world. He considered that, whatever life may be, it is something held most tenaciously by the least organized being. It must therefore be independent of structure and must be somehow an attribute of a substance which all organic forms contain. These ideas lead to the conception of

351

protoplasm, the substance, simple in appearance, yet inconceivably complex in ultimate structure and composition, without which life is never found. Hunter did not use the word 'protoplasm', which was invented (in 1846) fifty years after his death. But he was reaching out toward the conception of a common material basis of life (pp. 358–61).

The orderly observations of vital phenomena by naturalists such as Hunter, Linnaeus, and Cuvier were given an entirely new direction by the chemical workers of the next generation. Respiration had already been made chemically intelligible by Priestley, Lavoisier, and Ingenhousz. Many other processes of the living organism were now chemically interpreted by Liebig and his school.

JUSTUS VON LIEBIG (1802–73), professor of chemistry at Giessen, was an exceedingly stimulating teacher who had an immense following and did much to introduce laboratory teaching. He greatly improved the methods of organic analysis and, notably, he introduced a method for determining the amount of *urea* in a solution. This substance is found in blood and urine of mammals, and was the first organic compound to be prepared from what were then regarded as inorganic materials. It is of very great physiological importance, for it is regularly formed in the animal body in the process of breaking down the nitrogenous substances, known as 'proteins' (p. 360), characteristically found in association with all living substances.

With his colleague, FRIEDRICH WÖHLER (1800–82), who had already synthetized urea (1828), Liebig showed that a complex organic group of atoms—a 'radicle' as it is now called—is capable of forming an unchanging constituent which can be traced through a long series of compounds. A radicle may behave throughout as though it were an element (1832). The discovery is of primary importance for our conception of the chemical changes in the living body.

From 1838 onwards Liebig devoted himself to attempting a chemical elucidation of living processes. In the course of his investigations he did pioneer work along many lines that have since become well recognized. Thus he classified articles of food with reference to the functions that they fulfilled in the animal economy (fats, carbohydrates, proteins), and he taught the true doctrine, then little recognized, that all animal heat is the result of combustion, and is not 'innate'.

Very important was Liebig's teaching that plants derive the constituents of their substance, their carbon and nitrogen, from the carbon dioxide and ammonia in the atmosphere, and that these compounds are returned by the plants to the atmosphere in the process of putrefaction. This development of the work of Ingenhousz made possible a conception of a sort of 'circulation' in Nature. That which is broken down is constantly built up, to be later broken down again. Thus the wheel of life turns on, the motor power being energy from without, derived ultimately from the heat of the sun.

By far the major part of existing living matter is contained in green plants. These also provide the ultimate source of aliment for the entire animal kingdom. The economic significance of the sources from which the substance of plants is replenished cannot, therefore, be exaggerated. A most important source is carbohydrate, especially in the form of starch, the formation of which is associated with the green matter itself.

We now know that starch is built up in the plant from the carbon dioxide absorbed from the atmosphere (p. 349); that starch formation is a function of the living plant-cell, intimately connected with the green substance; and that the process is active only in the presence of light. (The name 'Chlorophyll', Greek = 'leaf green', was coined in 1817.) Steps toward the modern position were made by the French botanical experimenter HENRI DUTROCHET (1776–1847). A key to the working of the living organism is the process by which the gases of the atmosphere come into contact with the tissues. In animals the general character of this is fairly evident, especially in such as breathe actively. Plants, however, were long in giving up their secret. Dutrochet showed (1832) that little openings on the surface of leaves— 'stomata' (Greek, plural of *stoma*, 'mouth') as he called them— communicate with spaces in the substance of the leaf, but it was sixty years before the stomata were generally recognized as the normal channel of gaseous interchange. Dutrochet also knew from Ingenhousz that the plant as a whole gave off oxygen and absorbed

353

carbon dioxide, and he showed that only those cells that contain green matter are capable of absorbing the carbon dioxide (1837).

From carbon dioxide assimilation and carbohydrate formation we turn to a consideration of the origin and fate of nitrogenous substances in living things. Davy, Liebig, and others were well aware of the importance of nitrogen in the substance of plants. Liebig showed that nitrogen is taken into the plant by the roots in the form of ammonia compounds and nitrates. He made the general process of nutrition intelligible by a wide generalization of the utmost importance. Rejecting the old idea that plants grow by the absorption of humus, he claimed that carbon dioxide, ammonia, and water contain in themselves all the necessary elements for the production of vegetable matter and that these substances are also the ultimate products of their processes of putrefaction and decay (1840).

JULIUS SACHS (1832–97) of Würzburg was immersed from 1857 onward in problems of plant nutrition. He demonstrated that the green matter of plants, chlorophyll, is not diffused in tissues but contained in certain special bodies—'chloroplasts' as they were later (1883) named. He showed also that sunlight plays the decisive part in determining the activity of chloroplasts in absorption of carbon dioxide. Further, chlorophyll is formed in them only in the light. Moreover, in different kinds of light the process of carbon dioxide assimilation goes on with different degrees of activity. The views and discoveries of Sachs were brought together in his treatise on botanical physiology (1865).

The French mining engineer, JEAN BAPTISTE BOUSSINGAULT (1802–87), applied himself persistently, and, in the end, successfully to the nitrogen problem. During the fifties he succeeded in proving that plants absorb their nitrogen not from the nitrogen of the atmosphere but from the nitrates of the soil. He showed further that plants can grow in soil devoid of organic or carbon-containing matter, provided that nitrate be present, and that therefore the carbon in plants must be derived from the carbon dioxide of the atmosphere.

Thus was built up a definite economic picture of the world of life, plants drawing their substance from the inorganic world, animals drawing their substance from plants, and the decomposition of both going back into the inorganic world to be reabsorbed by plants.

(iii) *Cell Theory*.

While chemists were interpreting in their own terms the processes by which living things build up the substance of their bodies, microscopists were investigating the details of those bodies that were invisible to the naked eye. The mystery of the unexplored lay still over the world of microscopic beings with their bizarre forms and entrancing strangeness. By some they were fancifully endowed with complex organs that they do not possess, but the 'minima naturae' were more generally regarded as the 'simplest' and 'most primitive' of beings wherein the secrets of life might most hopefully be sought.

Such inquiries were prosecuted especially with those minute creatures—'animalcula' was the old name for them—that appeared, seemingly 'spontaneously', in infusions of various kinds. The term *Infusoria* soon, however, came to include certain other minute organisms that present superficial resemblances to the animalcula of infusions (1764). The limits and definition of the *Infusoria* were long disputed.

As so often, the discussion was barren until directed along lines which corresponded to a concrete and intelligible theory. It came gradually to be realized that all non-microscopic and certain microscopic organisms are aggregates, each unit (cell) of which has some degree of individual life. Not until this position was reached could the Infusoria be properly definable and the term restricted to unicellular forms to the exclusion of cell aggregntes (1841). Again, as so often in scientific history, this position was repeatedly approached and even temporarily occupied before it was actually won. Such a pioneer attempt was that of the wildly speculative *Naturphilosoph*, LORENZ OKEN (1779–1851), who in 1805 compared Infusoria to the 'mucous vesicles (cells) of which all larger organisms are composed', spoke of 'the infusorial mass or *Urschleim* (protoplasm) of which larger organisms fashion themselves', and claimed that such organisms are equivalent to 'agglomerations of Infusoria'.

The conception that 'cells' of various forms and functions, but variants of a common plan with a greater or less degree of independent life, form the basis of larger organisms came slowly to be accepted doctrine. The progress occupied the first half of the nineteenth century. The nomenclature of the earlier part of this period is naturally confused. The term 'cellula' dates back to Hooke (1664), who, however, applied it only to the cell-walls of plant-cells. The word 'cell' is frequently used by late eighteenth- and early nineteenth-century writers to describe the microscopic divisions perceptible in most tissues with suitable treatment. The central body of the 'cell' substance, its controller, is the *nucleus*. This term was first applied to the important structure now known by that name in 1823. Robert Brown (p. 331) in 1831 realized that the nucleus was a regular feature of plant-cells and he normalized the use of the word.

The great Czech naturalist JOHANNES EVANGELISTA PURKINJE (1787–1869) in 1835 drew attention to the close analogy of the packed masses of cells in certain parts of animals with those in plants. FELIX DUJARDIN (1801–62) of Toulouse, a most penetrating observer, entered in that year upon a critical examination of microscopic forms. Two conceptions of primary importance emerged from his researches. First, he clearly distinguished unicellular organisms as such, and adequately delimited the Infusoria. Secondly, he discerned that life is always associated with a substance of mucilaginous consistence with certain very definite optical, chemical, and physical characteristics. Purkinje, who worked on comparable lines, gave to it the name *protoplasm* (1839; Greek = first formed). It became recognized that the living parts of all cells were composed of protoplasm.

The first adequate presentation of the knowledge of the cell as a body of doctrine (1839) was made by THEODOR SCHWANN (1810–82), a pupil of Johannes Müller. He extended the discussion to the ovum or egg which is the beginning of the animal or plant body. In some animals, as the hen, the egg is very large, being distended with food substance—the yolk—and surrounded by a larger and protective substance—the white or albumen. In other eggs, as the frog's, the amount of yolk and albumen is much less. In yet others yolk and albumen are reduced to a minimum, as in the microscopic eggs of mammals then recently discovered

Physical Interpretation of the Living Organism

(1828) by von Baer (p. 332). Schwann discerned that all these are essentially cells and exhibit the characteristic elements of cells—nucleus, protoplasm, cell membrane, &c.

The development of the egg into the young animal (or plant) proceeds by division of the egg cell. This process of 'segmentation' is particularly evident in the earliest stages of development, and had been casually noted in a variety of organisms by several early naturalists. Schwann treated the process as a normal part of embryonic development. He showed that the continued division of the egg or 'germ-cell' gives rise to the organs and tissues, and he distinguished on a cellular basis five classes of tissues:

(a) Tissues in which the cells are independent, isolated, and separate. Such is the blood.

(b) Tissues in which the cells are independent but pressed together. Such is the skin.

(c) Tissues in which the cells have well-developed walls that have coalesced to a greater or less degree. Such are cartilage, teeth, and bones.

(d) Tissues in which the cells are elongated into fibres. Such are tendons, ligaments, and fibrous tissue.

(e) Tissues 'generated by the coalescence of the walls and cavities of cells'. Here he included muscles and nerves.

Schwann now passed to a general statement of his belief as to the cellular origin and structure of animals and plants. His conclusion may be expressed thus:

(a) The entire animal or plant is composed either of cells or of substance thrown off by cells.

(b) The cells have a life that is to some extent their own.

(c) This individual life of all the cells is subject to that of the organism as a whole.

This general attitude is still valid.

The synthesis of the ideas of protoplasm, unicellular organisms or 'protozoa', and egg or germ-cell was made by MAX SCHULTZE (1825–74). He devoted himself to a study of tissues—'histology' —in a wide range of animals. In 1861 he gave the definition of a cell as 'a lump of nucleated protoplasm', and in 1863 defined protoplasm as 'the physical basis of life'. He showed that protoplasm presents essential physiological and structural similarities

357

in plants and animals, in lower and higher forms, in all tissues wherever encountered.

Great influence on the biology of the second half of the nineteenth century was exercised by the liberal Berlin professor RUDOLF VIRCHOW (1821–1902). His main contributions are set forth in his *Cellular Pathology* (1858), in which he analyses diseased tissue from the point of view of cell-formation and cell-structure, and enunciates the now familiar idea that the body may be regarded 'as a state in which every cell is a citizen. Disease is a civil war, a conflict of citizens brought about by external forces.' Further: 'Where a cell arises, there a cell must have been before, even as an animal can come from nothing but an animal, a plant from nothing but a plant. Thus in the whole series of living things there rules an eternal law of continuous development, nor can any developed tissue be traced back to anything but a cell.'

Virchow crystallized the matter in his famous aphorism, *Omnis cellula e cellula* ('Every cell from a cell'), to be placed beside *Omne vivum ex ovo* ('Every living thing from an egg') of Harvey, and *Omne vivum e vivo* ('Every living thing from a living thing') of Pasteur. These are three of the widest generalizations to which biology has attained. They were all reached within the ten years around the middle of the nineteenth century, for though Harvey's was stated much earlier, he had not the evidence on which to base it.

(iv) *Protoplasm.*

From a time when it was first recognized that a similar substance 'protoplasm' underlies all vital phenomena there has been much interest in its chemical and physical composition. Strictly the subject is insoluble since protoplasm can only be adequately investigated when it has ceased to be the basis of life. We may learn what protoplasm takes in and what it throws out. We may gain some idea of its local reactions to ingested or applied substances. But living protoplasm is beyond the reach of the chemist's activities. It is protoplasmic products and dead protoplasm that have been the subject of most of his researches.

Dead protoplasm consists of a very complex mixture of numerous substances. Of these the bulkiest is water. The others are

358

Physical Interpretation of the Living Organism

largely made up of the complex nitrogenous groups known as *proteins* and their derivatives, of the *lipoids* or fats, and of the *carbohydrates* or starchy substances. The general significance of these three types was first made definite by Justus von Liebig about 1840 (p. 352).

Living protoplasm is liquid. Nevertheless, an elementary acquaintance with its behaviour shows that it exhibits a considerable degree of 'viscosity', that is it has some of the properties of a sticky or of a jelly-like substance. Modern views of the intimate structure or composition of living protoplasm have become closely linked with a comparison of its behaviour with that of other substances in the *colloid* ('glue-like') state. The study of the colloid state, one of the many areas in which the old sciences of chemistry and physics have become merged, was initiated by THOMAS GRAHAM (1805–69) in 1850 while Master of the Mint in London. The term was already in use, but he applied it to a particular state of matter. He divided soluble substances in general into the two great classes, *colloids* and *crystalloids*. He observed that certain substances (*a*) pass very slowly into solution, (*b*) do not crystallize, and (*c*) cannot diffuse or diffuse very slowly through organic membranes. Of these substances glue is the type, hence the name *colloid*. In this class are starch (compare starch paste), white of egg, gelatine (the basis of most table jellies). Opposed to these in all three respects are the crystalloids.

Graham was aware that certain substances—silica for instance —could exist as either colloid or crystalloid. He recognized, too, that instability was a characteristic of colloids. Moreover, he perceived that most colloids are of organic origin. He foresaw certain modern views of the nature of vital activity in his conception that the surface energy of colloids 'may be looked upon as the probable primary source of the force appearing in the phenomena of vitality'.

The knowledge of the essential nature of colloids was but little extended until the twentieth century. Investigators of our own generation have given a physical interpretation to the differences between the colloid and crystalloid states.

Among the colloids, biologically the most important is the vast and varied class known as proteins. They are absolutely necessary

359

to the building up of protoplasm. Dead protoplasm largely consists of them. They are not only essential for growth and repair of living substance, but they can be used by the living organism as a source of energy and of heat, though the carbohydrates and fats share this function with them. Chemically the proteins are all built up of very large molecules.

The modern chemistry of the proteins is based on the work of the great German chemist EMIL FISCHER (1852–1919) from 1882 onwards. Fischer demonstrated that proteins are built up of linkages or condensations of numbers of molecules of the substances known as amino-acids. The members of this very peculiar class are characterized by the presence in each molecule of one or more NH_2 ('amino') groups and one or more COOH ('carboxyl') groups. The former gives them basic qualities, the latter acid. According as one or the other predominates, the amino-acid acts as a base or as an acid.

A favourite theory of the nature of protoplasm regards it as a mixture of amino-acids. These can become immeasurably complex by associating with each other in varyingly intimate ways. A modern mechanist view of life pictures all vital activity as a continuous change and interchange of the conditions and relations of amino-acids. These, it is held, act through local changes in the degree of viscosity. Many other phenomena of the living cell have been interpreted as due to changes in degree of viscosity.

Another aspect of protoplasmic activity is that of enzyme action. The word 'enzyme' (Greek 'in yeast') was introduced by Willy Kühne (1878) to distinguish a class of organic substance which activates chemical change. Such an enzyme can act on an indefinite amount of material without losing its activating power. The living body produces a large number of enzymes. These are remarkably specific in their action.

Within the protoplasm, though not of it, are numerous materials, the so-called 'food substances', which are often of relatively simple composition. Under this heading are to be included sugars and their derivatives, fats, and the 'reserve' proteins. The problem of the nature of protoplasm thus resolves itself into that of the nature of the matrix in which a vast variety of controlled reactions are taking place, and the ways in which the matrix can influence these reactions. The chemical processes at any moment within a single cell are of many and varied types. In spite of the smallness of cellular dimensions, these must some-how be spatially separated from one another.

(v) *Physiological Synthesis.*

With the parallel development of a knowledge of the vital processes as a continuous elaboration and breaking down of living substance, of the living body as a structure composed of cells, and of the physical basis of life as protoplasm, there developed new views of the organism as a physico-chemical mechanism. Mechanist views have never lacked critics, and it is significant that the most effective critic of mid-nineteenth-century mechan-ism, JOHANNES MÜLLER (1801–58), was himself an experimental physiologist of genius who is largely responsible for the picture of the body as a machine. In Müller's *Handbook of Physiology* (1834–40) the results alike of microscopic and of comparative anatomy, of physics and of chemistry were, for the first time, systematically brought to bear on physiological problems. His researches on the chemistry of the animal body touch on those of Liebig at many points. His most important physiological investi-gations, however, dealt with the action and mechanism of the senses, and were important starting-points for modern research.

The doctrine specially associated with Müller's name is the 'principle of specific nerve energies'. This teaches that the general character of the sensation, following the stimulation of a sensory nerve, depends not on the mode of stimulation, but on the nature of the sense organ with which the nerve is linked. Thus mechani-cal stimulation of the nerve of vision produces luminous impres-sions, and no other; stimulation of the nerve of hearing gives rise only to an auditory impulse, and so on. This doctrine is of such importance that it is well to consider some of its implica-tions.

What do we know of the world in which we live? Only what our senses tell us. But how do our senses convey anything to us? That no man can answer. All we know is that certain external events somehow initiate specific disturbances in certain nerves, that these nerves convey the disturbances to the brain or central nervous system, and that a sensation then arises. We can dimly picture a mechanism by which the external event may elicit a specific nerve-impulse, and we know a little about the nature of the impulse and how it travels up the nerve. But how that impulse becomes a sensation, which is what we experience, and how experience gives rise to something which so alters a nerve or series of nerves that it induces action—of these things we are not only completely ignorant but it is difficult to believe that we can ever be other than ignorant. Indeed, there are reasons to believe that here is a veil which never can be rent by mortal man.

But consider further. External events are known to us only through our senses. Nevertheless from one and the same event we may receive completely different sensations. Thus, an electric stimulation of the optic nerve will give rise to a visual sensation; the same stimulation of the olfactory nerve yields a sensation of smell; of the auditory nerve a sensation of sound. Further, different events may give rise to the same order of sensation. Thus it matters not whether the optic nerve be stimulated by electricity, by heat, or mechanically, the sensation aroused will be visual. If our optic nerve were grafted to our auditory organ and our auditory nerve to our optic organ we should find ourselves transported to a world so strange that we cannot form the remotest conception of it. (Such an operation may actually be practicable in certain organisms.) To beings with senses different from ours the world would be utterly different.

The law of specific nerve energies is thus fundamental for our view as to the range of validity of scientific method, and indeed of experience as a whole. That law is a standing criticism of the 'common-sense' view that the world is as we see it, and that its contents, and particularly the living things in it, can be completely understood by us.

Müller was a convinced vitalist. He laid emphasis on the existence of something in the vital process that was, and must remain, insusceptible of mechanical explanation or physical measurement. This doctrine, however, occasionally misled him. Thus he held it impossible to measure the velocity of the nervous impulse. Yet that velocity was measured by his own pupil, Helmholtz, some ten years later. Vitalistic views are useful to the philosopher, but the working man of science had best, with Claude Bernard, forget them while he is at his appointed task.

The French physiologist CLAUDE BERNARD (1813–78), one of the greatest of all biological thinkers and experimenters, was the most effective contributor to the presentation of a concert of all the bodily processes as a chemico-physical mechanism. Perhaps his greatest discovery was that the liver builds up, from the nutriment brought to it by the blood, certain highly complex substances which it stores against future need, and that these substances, and notably that known as *glycogen*, it subsequently modifies for distribution to the body according to its requirements.

It was already recognized that the source of bodily energy is the breaking down of nitrogenous substances, of which the final degradation product is urea (p. 352). Bernard, by his work on glycogen, demonstrated that the body not only can *break down* but also can *build up* complex chemical substances. This it does according to the requirements of its various parts.

Bernard thus destroyed the conception, then still dominant, that the body could be regarded as a bundle of organs, each with its appropriate and separate functions. He introduced a conception that the various forms of functional activity are inter-related and subordinate to the physiological needs of the body as a whole.

No less important, as bearing on this conception, was Bernard's work on digestion. Up to his time, an elementary knowledge of the facts of digestion in the stomach constituted the whole of digestive physiology. Bernard showed that this digestion is 'only a preparatory act' and that numerous other processes are involved. Thus the juice of the 'pancreas' or sweetbread, poured into the intestine near the lower opening of the stomach, emulsifies the fatty food substances as they leave the stomach and splits them into fatty acids and glycerin. He showed further that the pancreatic juice has the power to convert insoluble starch into soluble sugar for distribution to the body in the blood, and that it has a solvent action on such proteins as have not been dissolved in the stomach.

A third great synthetic achievement of Bernard was his exposition of the manner of regulation of the blood-supply to the different parts of the body. This we now call the 'vaso-motor mechanism'. In 1840 the existence of muscle-fibres in the coats of the smaller arteries was discovered. Bernard showed that these small vessels contract and expand, thereby regulating the amount of blood supplied to the part to which they are distributed. This variation in calibre of the blood-vessels is, he showed, associated with a complex nervous apparatus. The reactions of the apparatus depend upon a variety of circumstances in a variety of other organs. Thus he provided another illustration of the close and complex interdependence of the various functions of the body upon each other.

Bernard's clear conception of the reciprocal relations of the organic functions led him to a very valuable generalization. He perceived that the characteristic of living things, indeed the test of life, is the preservation of internal conditions despite external change. 'All the vital mechanisms', he held, 'varied as they are, have only one object, that of preserving constant the conditions of life in the internal environment.' This phrase is the seal on Bernard's belief that the living organism is something *sui generis*, something quite different from everything in nature that is not living. The organism has an object, and it uses a mechanism for attaining that object. Is this conception infinitely removed from that of Aristotle?

What is the internal environment of an organism? Bernard was thinking chiefly of the blood. But if we think of a part in terms of cells we see the environment of the cell made up of four main factors:

(*a*) The neighbouring cells and cell products.

(*b*) The substances that are brought to it by the blood.

(*c*) The substances that it throws off and that are removed from it by the blood.

(*d*) The nervous impulses that come to it.

The whole vast mass of physiological research since Bernard's time may be regarded as a commentary on these four factors of the internal environment.

(vi) *Supremacy of Nervous System.*

It will be impossible to follow further all the factors of

Physical Interpretation of the Living Organism

internal environment, but there is one group upon which it is well to enlarge since the whole standpoint with regard to it has altered fundamentally since the time of Bernard. It is the consideration of the nervous system and its relation to the body as a whole. By the time of Haller (pp. 349–50), the naked-eye anatomy of the nervous system had become quite familiar. A new physiological phase was opened by LUIGI GALVANI (1737–98) of Bologna, who showed (1791) that if a nerve be subjected to a certain method of stimulation, the muscle to which it leads will contract. The electric nature of Galvani's method was revealed by ALESSANDRO VOLTA (1745–1827) of Pavia. In the fifth decade of the nineteenth century the Berlin professor, EMIL DU BOIS-REYMOND (1818–96), pupil and successor of Johannes Müller, showed that a nervous impulse is always accompanied by the passage along the nerve of a change of electrical state. He and other investigators demonstrated, moreover, that chemical changes in the muscle accompany contraction. These chemical changes are initiated—'lit up', we might say—by the nervous impulse.

In the meantime SIR CHARLES BELL (1774–1842) had been at work on the double spinal roots from which most of the nerves of the body arise. He showed that of these roots, one conveys only sensory elements, while the other conveys only motor elements. Thus the investigation of the action of individual nerves became possible.

In the first half of the nineteenth century there appeared many comparative studies on the nervous system. Cuvier based his classificatory system in part upon the nervous reactions (p. 330). He had himself explored the nervous system of Molluscs, Starfish, and Crustaceans. His influence may be traced in many works on the anatomy of the vertebrate nervous system prepared in the first half of the nineteenth century, but it was not until the appearance of T. H. Huxley's *Manual of the Anatomy of the Invertebrated Animals* (1877) that full stress came to be laid on the ascendancy of the nervous system in all members of the animal series.

Despite the lead of Huxley, the nervous physiology of invertebrates remained neglected. But the internal structure of the nervous system of mammals has been investigated with very

365

great detail. It has been found to be almost inconceivably complex. The investigations have been greatly helped by the introduction of new technique, at which we may now glance.

The early anatomists, from Vesalius onward, recognized that the central nervous system consists of two main parts—the grey, and the white matter. It was perceived that in the brain the grey matter is mostly on the surface, while in the spinal cord it is mainly central in position.

Soon after the foundation of histology as a special science it was observed that white matter consists of masses of enormous numbers of fibres while grey matter contains also numerous cells. These facts were known to Purkinje (1835, p. 356) and were formally set forth by JACOB HENLE (1809–85). It was, however, more than forty years before the Swiss ALBRECHT KÖLLIKER proved that all nerve-fibres are nothing more than enormously elongated processes given off from nerve-cells with which they retain continuity (1889). These nerve-cells are to be found either in the central nervous system itself or in the various ganglia.

In 1873 the Pavia professor, CAMILLO GOLGI (1844–1926), introduced a method of depositing metallic salts within various cell structures. These deposits are very evident under the microscope, and Golgi succeeded in applying this method to the central nervous system. He showed that the cells in that system tend to resemble irregular polygons from the angles of which project processes, *axons*, the essential parts of the nerve-fibres which ultimately end in a complicated system of branches, *dendrites*. The dendrites form twig-like 'arborizations' round other dendrites linked to other cells. Ultimately the system ends in terminal cells associated with sense organs, glands, or muscles.

The method of Golgi has been developed especially by RAMON Y CAJAL (1852–95) of Madrid, almost the only important scientific investigator that Spain has hitherto produced. His researches stamped upon biology the conception of an immensely complex series of systems for the transport of nervous impulses. These systems, if intact and working well, determine the activities, the reactions, the whole life of the organism. Most significant work has been done during the last half-century in the light of this conception.

Physical Interpretation of the Living Organism

While the various nervous tracts were thus being traced, much work was in progress in the localization of the functions of the different parts of the nervous system.

In 1861 the French surgeon, PAUL BROCA (1824–80), demonstrated in a post-mortem room at Paris a relationship between loss of speech and injury to a definite area of the cortex. Broca made many contributions to the knowledge of the brains of men and of apes.

Others soon continued his work in the experimental field. Thus in 1870 a very versatile naturalist, GUSTAV FRITSCH (1838–91), and a student of insanity, EDUARD HITZIG (1838–1907), working together at Berlin, found that stimulation of certain parts of the cortex regularly produced contraction of certain muscles. The Englishman DAVID FERRIER (1843–1928), followed this up by demonstrating that other areas of the cortex, which do not evoke muscular activity, are nevertheless functionally differentiated (1876).

In the half-century that has since elapsed the surface of the brain has been mapped in great detail. Special areas have been associated with movements of different parts and different organs. Others are related to various forms of sensory discrimination such as sight, sense of position, weight, taste, and the like. Yet others are involved in the use of language, both in written and spoken form.

Influential in determining modern views of the action of the nervous system have been researches on the nature of 'reflex action', that is, non-voluntary movement in response to a sensory stimulus. The conception may be traced in physiological writings from Descartes onwards. The term 'reflex action' was invented (1833) by the English physiologist MARSHALL HALL (1790–1857). The study of reflexes has resulted in the localization of functions in the grey matter of the spinal cord much as with the grey matter of the cortex.

Since Hall's time there has been vast extension of the conception of reflexes. In addition to the simple nervous arc there are also more complex arcs which depend for their action on an elaborate mechanism. Beside 'spasmodic' events, as sneezing, coughing, scratching, &c., many of the ordinary acts of life, standing, walking, breathing, &c., are expressible as reflexes. The attempt has also been made by Pavlov (1849–1936) and others to press even the 'instincts' into the same category, and the cortex has been shown to have the power of establishing new reflexes. The school that has been thus occupied seeks to explain all the reactions, and indeed the whole life, of the higher organisms on a purely objective basis without reference to volitional elements.

If the simple reflexes of animal bodies are tested, it will be found that they clearly serve certain ends. Lightly touch the foot of a sleeping child and it will withdraw it. Tickle the ear of a cat and it will shake it. Exhibit savoury food to a hungry man and his digestive process will at once get to work, his mouth will 'water'. These instances might be multiplied a hundredfold. Such reflexes are admirably adapted to their ends. Many will continue in an animal in which the brain has been removed, provided that the spinal cord be still intact. Nevertheless, in the higher animals, and especially in man, the reflexes are controllable to a greater or less extent by the will.

But to leave the question at that would give a false idea of the extremely complex functions performed by the central nervous system. Thus, the spinal cord which, to the naked eye, is a longitudinal and little differentiated nervous mass, is, in fact, a collection of nerve-centres which have historically, both in the individual and in the race, been formed by the union of a series of separate segments. Each segment in this system governs certain functions or movements of the body, and the activity of each segment is related in various ways to the activity of the other segments. There is thus a very complex process of 'integration' which runs right through the nervous system.

The growing knowledge of the bodily functions of chemical and physical nature gradually revealed that these activities are far more largely under nervous control and discipline than was formerly conceived. Thus, the main factor in the activity of any part is its blood-supply, but the blood-supply is determined, as Bernard showed (p. 364), by the state of contraction of the vessels of supply which are in their turn under nervous control. Similar relations prevail for the state of nutrition of muscles, for the action of the sweat glands of the skin, for the mechanism of childbirth, and for a thousand bodily states. The regulation and control of all these events, processes, and states by the nervous system has since come to be called *nervous integration*.

(vii) *Mind as Condition of Life*.

During the.nineteenth century there was an enormous extension of scientific interest in the analytical study of animal function through physical experiment. The exponents of this science of physiology applied themselves mainly to the higher animals. They devoted themselves to an examination of the parts or functions in the adult or developed state. The results were portentous in bulk, complexity, and interest, yet they went a very little way to help us in considering the organism as a whole.

The animal body is, as it were, a vast and complex maze. The physiologist enters it, and he wanders there as long as he will. But his close and detailed report on its paths and walls helps but little toward the exposition of the design as a whole, for the physiologist, in his special studies, is well nigh bound to consider isolated functions wall by wall, path by path. He selects respiration, nutrition, muscular movement, the action of the nervous system, or the like. But the performance of each of the functions of each of these systems is inextricably linked with the performance of the functions of all the other systems.

We are always looking for metaphors in which to express our idea of life, for our language is inadequate for all its complexities. Life is a labyrinth. But a labyrinth is a static thing, and life.is not static. Life is a machine. But machines do not repair themselves, nor do they reproduce themselves. Life is a laboratory, a workshop. But it is a workshop in which a thousand processes go on within a single microscopic cell, all crossing and intercrossing and influencing each other, and it is a workshop which is constantly multiplying itself and producing its like.

Life is a dance. There was a 'dance of death', and there is a dance of life. It is but a metaphor. When we speak of the ultimate things we can, maybe, speak only in metaphors. Life is a dance, a very elaborate and complex dance! The physiologist cannot consider the dance as a whole. That is beyond his experimental power. Rather he isolates a particular corner or a particular figure. His conception of the dance, as thus derived, is imperfect in itself and, moreover, in obtaining it he has disturbed the very pattern of the dance. The shortcoming of his method becomes fairly evident when he seeks to relate his corner to another in a far distant part of the dance.

Moreover, even should he seek to treat the organism as a whole, he is still almost bound to consider it as an 'individual' complete and separate in itself, shut off from its environment and its history, born, as was Minerva, armed and fully equipped from the head of Jove. But in fact living beings are not so. There is every degree of independence of their fellows among organisms. 'Individuality' comes into prominence only in the more differentiated groups. The term is almost inapplicable to plants, in which physiology is, in effect, of a community, and that is a study not far, in its conceptions, from that of bionomics. The very idea of the 'individual' involves a historical record which the science of physiology has hitherto almost ignored.

Physiology alone is of its nature incapable of presenting any picture of the mode of action of the organism as a whole, though modern doctrines of the workings of the nervous system have given some explanation of certain forms of animal behaviour. Yet the functions of the nervous system, like those of other systems, are relative to the other functions of the body. Not only is respiration, for example, regulated by the nervous system, but the nervous system itself is regulated by the character of the respiration. Raise the amount of carbon dioxide in the blood, and the respiratory movements are first stimulated and finally diminished via action on the respiratory centres. It would be possible to show that the same is true of any system or part of a system in relation to any other. What picture, then, can physiological processes give us of the interrelated complex of activities that we call an organism?

The physiologist has found that his science can be best prosecuted on the higher animals. Why? Because the functions of these creatures are best differentiated. If he wishes to study movement, respiration, nutrition, nervous action, he finds in the higher animals separate organs devoted to these processes. Such organs he cannot so easily, or cannot at all, find in the lower organisms. In the lowest of all, the Protozoa, every process is carried on in a minute single cell.

But the most distinctly and clearly developed characteristics of the highest animals are their mental powers. To discuss these in the mechanistic nomenclature adopted by physiology is merely contradiction in terms. The one thing that we really know is our own thoughts, and external things—including the science of physiology—we know only in relation to these. How then can external things be said in any sense 'to explain' our thoughts? It is more intelligible to invert the process and to say that phenomena—including those of physiology—are parts of our thinking, than to say that our thinking can be built up of phenomena.

But if we emphasize the conception of science as dealing with phenomena—'things which appear'—we reach a *modus vivendi* both for a conception of mind, and for the findings of science. Having agreed that science shall deal only with phenomena, we expressly exclude our own mind, which is not an appearance at all, but that to which appearances happen. Science must keep to the phenomenal level. On that level she may prosecute physiological study. But no amount of that study will truly represent an entity in which any element of mind exists. Is that element of mind found in other organisms than myself? Unless the solipsist view be taken, this question must be answered in the affirmative. The man who answers it in the affirmative is a vitalist.

8. *Evolution.*

(i) *The Word.*

The leading contributions of the nineteenth century to the conception of a mechanical world are the twin doctrines of Energy and Evolution. As with most important scientific ideas, the enunciation of neither can be dated exactly or placed to one man's credit. To the doctrine of Energy it is convenient to attach the name of Joule, and the date 1842 (pp. 324–5). The doctrine of Evolution has become so closely linked with the name of Darwin that 'Darwinism' is often taken as a synonym of this doctrine which is dated to 1859, the year of publication of the *Origin of Species*. The term 'Evolution' should, however, be retained for the philosophical view that the world attained its present form not by a single creative act, but by a slow process over long ages. This was held by a number of ancient thinkers such as Plato (pp. 32 7), and by several unorthodox medieval thinkers, such as Averroes (p. 139). Of this view, the doctrine of Evolution of Organic Forms, or Darwinism proper, is a special case.

The Latin word *evolvere* means to unroll, to roll forth, to revolve. In classical usage its noun *evolutio* acquired the special meaning of the unrolling of a scroll in order to read it, 'the opening of the records' as we might say. In the Vulgate version of the Scriptures, *evolvere* is used either in its literal sense or, most often, to designate passage of time as marked by the revolving heavens. Derivatives of *evolvere* had little application in the Middle Ages, since scrolls had been replaced by books with leaves, and no form of it occurs in the Authorized Version of the English Bible (1611). The word *Evolution* was given currency in modern literature by the group of seventeenth-century philosophers known as the 'Cambridge Neoplatonists'. They employed it to describe the unrolling, as of a scroll, of vast records of Time (cf. *Revelation* vi, 14; *Isaiah* xxxiv, 4). 'The whole Evolution of ages, from everlasting to everlasting, is represented to God at once', wrote (1667) their founder Henry More (1614–87), paraphrasing 'a thousand years in Thy sight are but as yesterday when it is past' (*Psalm* xc, 4).

Search of the writings of many philosophers of the eighteenth century, notably those of Leibnitz (1646–1714), Diderot (1713–84), and Kant (1724–1804), reveals uses of the word evolution extended from that of the Cambridge Neoplatonists, and even adumbrations of the modern philosophical sense considered under heading (*e*) below. During the same century the word 'Evolution' was developed on lines comparable to those of the Cambridge Neoplatonists by the 'Naturphilosophen', and notably by Oken (p. 355), in connexion with their doctrine of 'ideas'. In this sense it was reimported into nineteenth-century English, probably by Samuel Taylor Coleridge (1772–1834). 'The sensible world', he wrote, 'is but the evolution of Truth, Love and Life or their opposites in Man' (1820).

In the course of its varied and adventurous career the word 'Evolution' thus acquired many different meanings and shades of meaning. It entered into the technical vocabulary of biological

Evolution

science—where we are chiefly concerned with it—in at least five clearly distinguishable senses.

(a) Evolution naturally and conveniently designated the process, mainly an unfolding, of the parts of a bud opening into a flower; or again of the imago of an insect, such as a butterfly, in its final transformation from the pupa.

(b) There were two rival theories as to how living things develop. One held that the germ contained the living organism in a substantially complete state, folded on itself. This had to *unfold* in order to pass from the embryonic stage. The other held that the germ was at first uniform, and that the form of the embryo was later generated in it. The philosophical biologist Bonnet (p. 333) gave wide currency to the former view under the name *Evolution* (1762), while the latter came to be known as *Epigenesis*. It is usually said that it is the epigenetic view that has prevailed. In the literal sense, but not in certain other senses, this is the case (p. 356).

(c) There has always been a philosophical problem of the relation of Being to Becoming. We need not follow this discussion in its vast divarications. St. Augustine posed the problem for the next millennium and a half: 'In the beginning God made Heaven and Earth, that is the *seeds* of Heaven and Earth, for the material of Heaven and Earth was yet in confusion; but since it was inevitable that from these seeds Heaven and Earth would be, therefore the material is thus called' (*De genesi contra Manichaeos*). These are the *seminales rationales* of the great medieval Christian thinkers who stressed *being* rather than *becoming*. These *seminales* in the mind of God were for them the ultimate reality. Bonnet is, in this sense at least, a belated medieval, insisting that every being already is, and only seems to become. Seventeenth-century thinkers, startled by the changes newly revealed by the telescope in the heavens, and by the extraordinarily complex processes discerned by means of the microscope in the development of individuals on earth, directed attention to *becoming*. This was expressed by the scientific dilettante Matthew Hale (1609–76), for example, who writes of an 'ideal principle in the *evolution* whereof Humane Nature must consist'. Several eighteenth-century authors treat in a

373

similar manner of the 'evolution of ideas', including 'ideas' in the technical sense of the *Naturphilosophie*.

(*d*) Great confusion has been caused by an early and still current misapplication of this last use of the word in biology. The process of *development* of the organism (*not* its unfolding) became called its 'evolution'! Thus Erasmus Darwin, grandfather of Charles, wrote of 'the gradual *evolution* of the young animal or plant from the egg or seed' (*Botanic Garden*, 1791), meaning its epigenetic development, and *not* its evolution in the sense of Bonnet. This confusing usage has persisted to our time.

(*e*) Finally the word is used for a process (or the result of a process) by which, in long stretches of time, organic types develop (or have developed) from other types. More or less definite expressions of this view can be traced very far back, but no earlier use has been found of the word 'Evolution' to designate it than that of Lyell (1797–1875) in his *Principles*. There he discusses in detail the biological theories of Lamarck, and notably the view of that naturalist that 'certain organisms of the ocean existed first, until some of them by gradual *evolution*, were improved into those inhabiting the land' (1831).

The word 'Evolution' has been awarded numerous other technical meanings in departments other than biology, as for instance in mathematics, and in military tactics, where it is not our quarry. It is necessary, however, to remind the reader that the biological meanings of the word all interdigitate, and that this fact is not without significance in the development of the philosophical conception of evolution. The word, in fact, carries with it all the trailing clouds of a confused and intricate past.

(ii) *Eighteenth-Century Evolutionists.*

Among naturalists, the idea of the transformation of species was more or less overtly expressed by Hooke (1635–1703), Ray (1627–1705), Goethe (1748–1832), Oken (1779–1851), and many others. That it was much in the air is shown by the repeated insistence by Linnaeus, Haller, Bonnet, and many orthodox biological thinkers that species are not transformed from other species but exist in the form in which they were first created. (The difficulties that arose from the geological record and the way in which they were met are reviewed on p. 338.) The whole direction of biological activity in the period of Linnaean dominance was against discussion of variation or transformation, and in favour of the treatment of the world of life as something static. Nevertheless, a few eighteenth-century naturalists were able to break away from this view. We discuss two of these.

The first naturalist to give both form and substance to a conception of evolution of living things was GEORGES LOUIS LECLERC, COMTE DE BUFFON (1707–88). He was an attractive writer, and perhaps the ablest scientific popularizer that has ever lived. His great *Natural History* (1749–1804), in forty-four volumes which took fifty-five years to publish, sought to cover the whole area of natural knowledge, and was the first modern work of its kind. He himself regarded it as a sort of commentary on Newton's conception of a mechanical world. A new element in Buffon's work was its inclusion of living Nature which Newton had disregarded.

Buffon paid little attention to minor differences between organisms on which biological classificatory systems must necessarily be based. For that reason the Linnaean system did not appeal to him. He was interested rather in features that can be traced through very long series of organic forms. As regards the fixity of species he expressed himself variously, but he settled gradually into opposition to that view. Particularly he noted that animals possess parts which have no function as, for example, the lateral toes of the pig which, though perfectly formed, can never come into action. To explain these, he conceived that a species may alter in type from time to time, but retain marks of its previous form, as the pig retains its disused toes. Then, moving a little further, he concluded that some species are degenerate forms of others. Thus the ape is a degraded man, the ass a degraded horse, and so on. (We have already discussed his views of the history of the earth, and its relation to organic forms, on pp. 278–9.)

The ideas of Buffon were examined by ERASMUS DARWIN (1731–1802), grandfather of Charles Darwin. He, like Buffon, was anxious to show that living phenomena fitted in with those of the inorganic and mechanical world. With this in view, he sought

The Mechanical World

some way of showing how living things had naturally acquired their manifest adaptations to their environment. His solution of this problem is buried in the verbiage of several of his bulky works. In the best of these, his *Zoonomia; or the Laws of Organic Life* (1794–6), he sums up the general nature of the difficulties among which Buffon had been groping. For his solution he gathers together precisely those classes of facts that were most to impress his grandson.

'When we revolve', writes Erasmus Darwin, 'first the *changes which we see naturally produced in animals after their birth*, as in the butterfly with painted wings from the crawling caterpillar, or the [air-breathing] frog from the [water-breathing] tadpole; secondly the *changes by artificial cultivation*, as in horses exercised for strength and swiftness, or dogs for strength, courage, or acuteness of smell, or swiftness; thirdly, the *changes produced by climate*, the sheep of warm climates being covered with hair instead of wool, and the hares and partridges which are long buried in the snow becoming white during the winter months; fourthly, the *changes produced before birth by crossing or mutilation*; fifthly, the *similarity of structure in all the warm-blooded animals, including mankind*, one is led to conclude that they have alike been produced from a similar living filament.'[1] (Very greatly abbreviated.)

Erasmus Darwin held that similar changes in nature produce species in the course of time. These changes, he held, were passed on to the offspring. The process is epigrammatically pictured in his conspicuously bad poem *The Temple of Nature*:

> Organic life beneath the shoreless waves
> Was born and nurs'd in ocean's pearly caves;
> First forms minute, unseen by spheric glass,
> Move on the mud, or pierce the watery mass;
> These, as successive generations bloom,
> New powers acquire and larger limbs assume;
> Whence countless groups of vegetation spring,
> And breathing realms of fin and feet and wing.

The mechanism by which such changes come about is, he believed, the transmission of character acquired sometimes at least as an act of will. 'All animals undergo perpetual transformations;

[1] This 'filament' is a spermatozoon which he regarded, following Buffon, as a sort of biological unit.

which are in part produced by their own exertions . . . and many of these acquired forms or propensities are transmitted to their posterity' (*Zoonomia*).

(iii) '*Transformism.*'

JEAN BAPTISTE DE MONET DE LAMARCK (1744–1829), unquestionably the greatest systematist of his age, was unfortunate in the simultaneous possession of too arid a style and too fertile an imagination. Many of his views were so fanciful that he was lightly esteemed by most of his contemporaries. Cuvier, who adhered to the fixity of species, formed a low opinion of his abilities. Charles Darwin, among his successors, held him almost in contempt. The interest of the theory by which Lamarck is remembered was not fully realized until after his death. It was discussed in great detail by Lyell (1831).

Lamarck held that no frontiers can ultimately be found between species. It seemed to him, therefore, intrinsically improbable that they are permanently fixed. In reaching this conclusion he also laid stress on the domesticated animals, which vary greatly from their wild originals. Who, seeing for the first time a greyhound, a spaniel, and bulldog, would not think of them as different species? Yet all have a common ancestor. Their different characters have been produced by man's selective breeding. In Nature, too, variations comparable to these in kind are occasionally found within the same species. The agent that produces them is, according to Lamarck, the environment. Species, he thought, maintain their constancy only so long as their environment remains unchanged.

Lamarck, having decided on the importance of variation in the production of new species, had to consider its mechanism. How do changes of environment give rise to variation and so to production of species? In answer, he enunciated the 'law of use and disuse', inseparably connected with his name. He supposed that changes of environment lead to special demands on certain organs. These, being specially exercised, become specially developed. Such development, or some degree of it, is transmitted to the offspring. Thus a deer-like animal, finding herbage scanty, took to feeding on leaves of trees. It needed a longer neck to reach the leaves. In the course of generations, during which the poor creatures were always straining their necks to reach their food, long necks became an ever more accentuated feature of their anatomy. Thus emerged a beast recognizable as a giraffe. Conversely, useless organs, such as the eyes of animals that live in darkness, being unexercised, gradually became functionless and finally disappeared. The character of a longer neck or of defective eyes was acquired by the individual in its lifetime and transmitted, in some degree at least, to its descendants.

The great assumption is that acquired characters are inherited. Whether and in what sense acquired characters can be inherited is a matter of current discussion, but it is certain that in the sense suggested by Lamarck they are not inherited. Nevertheless Lamarck's work was of value in directing attention to one of the most important problems in the whole range of biological thought. Unfortunately some of his early supporters set forth evolutionary schemes that were fantastic to the last degree. This resulted in biological speculation falling into disrepute for the first half of the nineteenth century.

Yet there was one writer of the time, whose work bore upon the subject, against whom the charge of reckless speculation could most certainly not be made. The Rev. T. R. MALTHUS (1766–1834) was a cautious and somewhat formal writer on mathematical and economic subjects. He produced anonymously in 1798 his *Essay on Population*. At that time political theory was a matter of acute controversy in connexion with the French Revolution. Such topics as the 'rights of man', 'natural justice', and the like were in the public mind. The most flourishing school of thought in England was the 'utilitarian', which was the direct ancestor of that liberal philosophy on which Britain rose to industrial and imperial greatness during the nineteenth century. Adam Smith (1723–90), Joseph Priestley (1733–1804), and Jeremy Bentham (1748–1832) were the chief early spokesmen in England of this great movement. Many believed that a day was dawning when, amidst universal peace, all men would enjoy complete liberty combined with complete equality. Malthus, who followed in general the line of utilitarian thought, brought out the difficulties that must arise in such a state from over-population, by his famous

378

(but fallacious) principle that populations increase in geometric, but subsistence at best only in arithmetic ratio. He argued that a stage must be reached at which increase in population will be limited by sheer want. Thus he held that 'checks' on population are a necessity in order to avoid vice and misery.

Darwin read the *Essay* of Malthus in 1838, and the *Principles* of Lyell in 1831. The one suggested to him the idea of the Struggle for Existence and the Survival of the Fittest, the other the general doctrine of Evolution. In the first half of the nineteenth century both these ideas were discussed by a number of writers, and notably by several English amateur naturalists accessible to Darwin. None put the two ideas together, or at least none put them together adequately. Darwin derived nothing or next to nothing from such predecessors.

(iv) *'The Origin of Species.'*

It is the great achievement of CHARLES DARWIN (1809–82) that he persuaded the scientific world, once and for all, that many diverse organic forms are of common descent, that species are inconstant and in some cases impossible of definition, and that some mechanism must be sought to explain their evolution. In search of this mechanism, he directed attention to the occurrence of variation, to its persistence, and to the question of its origin and its fate.

In 1859 appeared Darwin's classic *Origin of Species*. He had opened a note-book on the subject in 1837, made a first draft of it in 1842, a second in 1844, and in 1858 published, simultaneously with ALFRED RUSSEL WALLACE (1823–1913), a preliminary sketch of his views. It is an interesting fact that Wallace, like Darwin, seems to have caught his idea immediately from Malthus.

The *Origin* is one of the world's great books, and has proved significant for almost every human activity. It is unnecessary to discuss its greatness or its importance. But despite the conviction that it carried, and despite the fact that for the half-century after its publication its ideas provided the main stimulus for biological research, its arguments are frequently defective.

Darwin's basic claim is that organs and instincts have been 'perfected by the accumulation of innumerable slight variations, each good for the individual'. For this, he says, it is necessary to admit only three propositions. (*a*) 'That gradations in the perfection of any organ or instinct, either do now exist or could have existed, each good of its kind.' (*b*) 'That all organs and instincts are, in ever so slight a degree, variable.' (*c*) 'That there is a struggle for existence leading to the preservation of each profitable deviation of structure or instinct.' But this assumes that the 'profitable deviations' are inherited. Thus not three but four propositions are needed.

Again, after discussing our knowledge of the distribution of species in time and space—which carries irresistible conviction of organic evolution as an historical process—he turns to discuss conditions under which a variation is perpetuated.

'Man does not produce variability [in domestic animals]; he only exposes beings to new conditions, and then nature acts on the organisation, and causes variability. But man can select variations, and accumulate them in any desired manner. He thus adapts animals and plants for his own benefit. He can influence the character of a breed by selecting, in each successive generation, individual differences so slight as to be quite inappreciable by an uneducated eye. That many of the breeds produced by man have to a large extent the character of natural species, is shown by the doubts whether many are variations or aboriginal species.

'In the preservation of favoured individuals and races, during the Struggle for Existence, we see the most powerful means of selection. More individuals are born than can survive. A grain in the balance will determine which shall live and which die— which variety or species shall increase in number, and which shall decrease, or finally become extinct.

'There will in most cases be a struggle between the males for possession of the females. The most vigorous individuals will generally leave most progeny. But success will often depend on special weapons or means of defence, or on the charms of the males; and the slightest advantage will lead to victory.'

There are here, as we can now see, certain fallacies and erroneous assumptions.

(*a*) All domestic breeds have not been produced by selecting very slight individual differences. Some domestic breeds have certainly been produced by breeding from individuals which presented great deviations from the normal.

Evolution

(b) That a natural variation should confer an advantage is not enough to secure its perpetuation. The advantage must be effective, and it must be transmissible. Now it is difficult to believe that the earlier stages of some developments are effective as, for example, a wing so little developed as to give no power of flight or of gliding.

(c) Darwin assumes that species differ from their nearer relatives in having some special advantages that enable them to adapt themselves to slightly different conditions. Closely allied species are, however, often found living in identical areas and under identical conditions. There are very few cases indeed in which the characters by which such species differ from their fellow species can be shown to be advantageous, and there are some cases in which they can, perhaps, be shown not to be advantageous.

Darwin's presentation of Natural Selection as an effective agent is probably at its weakest in dealing with the problem of disuse. Here he assumes the inheritance of acquired characters in a form hardly differing from that of the despised Lamarck.

'Disuse, aided sometimes by natural selection, will often tend to reduce an organ, when it has become useless under changed conditions of life; and we can clearly understand on this view the meaning of rudimentary organs. But disuse and selection will generally act on each creature, when it has come to maturity and has to play its full part in the struggle for existence, and will thus have little power of acting on an organ during early life; hence the organ will not be much reduced or rendered rudimentary at this early age. The calf, for instance, has inherited teeth, which never cut through the gums of the upper jaw, from an early progenitor having well-developed teeth; and we may believe that the teeth in the mature animal were reduced, during successive generations, by disuse or by the tongue and palate having been better fitted by natural selection to browse without their aid; whereas in the calf, the teeth have been left untouched by selection or disuse, and on the principle of inheritance at corresponding ages have been inherited from a remote period to the present day.'

The full title of Darwin's book was *The Origin of Species by means of Natural Selection, or the Preservation of Favoured Races in the Struggle for Life.* Darwin himself compared the action of natural selection to that of a man building a house from stones of all shapes. The shapes of these stones, he says, would be due to definite causes, but the uses to which the stones were put in the building would not be explicable by those causes. The conception reveals the general weakness of Darwinistic thought which treats natural selection as though it were an active and directive agent. For when a man builds a house, there is the intervention of a definite purpose, directed towards a fixed end and governed by a clearly conceived idea. The builder, in the proper sense of the word, *selects*. But the acts of selection—mental events in the builder's mind—have no relation to the 'causes' which produced the stones. They cannot be compared with the action of Natural Selection. If a metaphor be sought for the action of Natural Selection, a better one might be the arrangement of stones on a sandy shore. Large stones are found high up on the beach. The stones become smaller as we descend toward the sea. On approaching the brink, we come upon a zone of sand. This arrangement is due to the forces of winds, waves, and tides acting, according to their nature, and according to the nature of the rocks of which the cliffs are built, over a long period of time. Provided that it be kept well in mind that it is a metaphor, and provided that no teleological view is implied, there can be no harm (and not very much good) in calling this a 'selective action' of the forces of wind, waves, and tides upon the disintegrated rocks.

Darwin repudiated teleology, but in his title, almost as though wishing to emphasize it, he repeats the teleological metaphor and speaks of the *Preservation of Favoured Races*. But how do we know that races are favoured? By their preservation! And what is preservation? A favour! And what is a favour? Preservation!

So, too, with the phrase *Survival of the Fittest*. In the sense in which the Darwinians used the word, fittest was often and naïvely confused with physical or even athletic fitness, and to it an ethical corollary was sometimes forcibly adjusted. But the only kind of fitness implied in the Darwinian phrase was fitness for survival. It is doubtless a good thing, on an ethical level, to be brave as a lion, and a bad thing, on an ethical level, to be timid as a rabbit. But, on a biological level, either quality may indicate fitness. Lions survive because of their courage in seeking their prey. Rabbits survive because of their cowardice in fleeing from those that prey upon them. Courage and cowardice are alike tests of fitness. Those that survive are fit, and those that are fit survive; and survival is the test of fitness, and fitness the test of survival!

Thus these phrases are, on analysis, devoid of ultimate meaning. This is very far from saying that no meaning can be extracted from the history of their use. Darwin was an investigator of the very first rank, but he was inexpert in the exact use of language and had little philosophical insight. His biological discovery, though of the highest scientific importance, was not quite of the nature that many of his followers thought it to be.

(v) *Doctrine of Descent of Man.*

There is one species whose origin raised acute controversy. Ancient and modern anatomists had drawn attention to the likeness of the anatomy of man to that of the apes. Darwin at first expressed no opinion on this point. Several of his supporters, notably T. H. HUXLEY (1825–95), devoted attention to it. The formal expression of Darwin's views was reserved till 1871, when at the opening of *The Descent of Man* he wrote: 'Huxley has conclusively shown that in every visible character man differs less from the higher apes than those do from the lower members of the same order of Primates.' This was very different from a demonstration of any intermediate form between man and the higher man-like apes. Nevertheless, evidence of this sort was gradually accumulating.

In 1856, three years before the publication of the *Origin*, the long bones and part of the skull of a man-like being had been unearthed in the small ravine of Neanderthal in Rhenish Prussia. They were all at first misinterpreted as pathological. Huxley ultimately recognized them as those of a human being, but the most ape-like yet found. He held that man is 'more nearly allied to the higher apes than the latter are to the lower'. The species to which these bones belong is now entitled *Homo Neanderthalensis*.[1] The remains of about a hundred individuals of this species are now known.

Since the discovery of Neanderthal man, a number of other

---

[1] A Neanderthal skull had been found at Gibraltar as early as 1848, but had not been brought to scientific notice.

species of fossil man have been discovered. On the other hand, several fossil species of apes approaching nearer than living forms to the human stem have also been found. The ape-man series is now probably more complete than that of most comparable mammalian groups. About two hundred fossil individuals are known, distributed over eleven or more species of quaternary and late tertiary measures.

Even before Darwin, and still more after him, evolutionary doctrine was applied to human habits, language, social organization, and psychology. Thus arose a science of Anthropology, which owes a deep debt to the French investigator, JACQUES BOUCHER DE PERTHES (1788–1868). As early as 1830 de Perthes discovered in the gravels of the Somme certain flints which he believed bore evidence of very ancient human workmanship. In 1846 he demonstrated the existence of such flints in company with the remains of elephant, rhinoceros, and other tropical or extinct forms. In his great *Antiquités celtiques et antédiluviennes* (1847–64), he established the existence of man from human products in Pleistocene and early Quaternary times. In 1863 de Perthes clinched this view by discovering near Abbeville, in a Pleistocene deposit, a human jaw associated with worked flints.

These conclusions were accepted, though with caution, by Lyell in his *Antiquity of Man* (1863). Since that time the study of the works and arts of Stone-Age man has developed parallel with the study of his physical structure. The succession of the cultures, crafts, and art of Palaeolithic Man and their emergence into those of Modern Man and notably into the culture known as Neolithic have now become familiar. It has been equated with geological and geographical change.

The subject of Organic Evolution has been pursued along many paths which pass the frontiers of biology and indeed of the sciences in the limited sense, and enter into many departments where we cannot here follow. Evolution illumines the whole history of life, the life of Man in all its manifold variety as well as the lives of organisms in all their manifold variety.

(vi) *Reception of the Doctrine of Evolution.*

In 1852—seven years before the publication of the *Origin*—the

Evolution

philosopher, HERBERT SPENCER (1820–1903), expounded doctrines of Evolution in a work where that word was used to describe a general process of production of higher from lower forms. He devoted the remainder of his long life to a highly elaborate exposition of what he regarded as the implications of evolution in every department of the inorganic and the organic world, in the structure of human society, and in the human mind. He eagerly adopted Darwinian principles as soon as the opportunity arose. Since his political philosophy of extreme 'individualism' fitted well the feeling of the age, his works were very widely read. They were translated into many languages, occidental and oriental, and thus did more, perhaps, than those of any other man to spread evolutionary views. The phrase 'Survival of the Fittest' was coined by him (1864).

That the evolutionary philosophical system of Spencer is an object of derision is one of the few points on which all philosophers seem now to agree. There are few living who can claim to have studied all his works. That the many who have done so are dead is a cause for reflection rather on their number than their state. But despite his extreme dryness as a writer, Spencer was a very great phrase-maker. A surprising number of his dicta have obtained currency. A selection of passages from one section of one chapter of his first independent work *Social Statics* (1850) will suffice to indicate not only his general attitude, which altered but little in later years, but also the philosophical atmosphere of the scientific public to which the *Origin* was delivered, nine years later.

'Progress is not an accident but a necessity. It is part of nature.' 'All perfection is a fitness to the condition of existence.' 'Evil tends perpetually to disappear.' 'Nature's rules have no exceptions.' 'In virtue of an essential principle of life, non-adaptation of an organism to its conditions is ever being rectified. Whatever possesses vitality obeys this law. We see it illustrated in the acclimatisation of plants, in the altered habits of domestic animals, in the varying characteristics of our own race. . . . *Such changes are towards fitness for surrounding conditions.*' 'Civilisation instead of being artificial is a part of nature ; all of a piece with the development of the embryo or the unfolding of a flower. . . . Man needed one moral constitution to fit him for his original state ; he needs another to fit him for his present state; and he has been, is, and will long continue to be, in process of adaptation. By *civilisation* we signify the adaptation that has already taken place. In virtue of this process man will eventually become completely suited to his mode of life.'

The *Origin* came thus to a world well prepared. It had been drafted as early as 1842, and Darwin himself had used the phrase 'Natural Selection' in a letter as far back as 1837. The central idea of the work was far from being new or even modern. Nevertheless, it created a revolution in biology, and indeed in almost every department of thought. It was the first work by a cautious, penetrating, highly competent, and experienced investigator that set forth a large and carefully sifted body of evidence on the subject of Evolution. Darwin himself was not very fond of using this word, but usually refers to it, in his modest way, as 'the species question'. His great book, however, was the first that suggested a simple and apparently universally acting biological mechanism producing changes of form. The struggle of living forms, presented as natural selection by the survival of the fittest, as set forth by him, proved an extremely stimulating suggestion.

The story of the rise of Darwinism has been so well and so often told that it is unnecessary to repeat it. It is probably the most familiar incident in the history of science. Among the opponents of Darwin were Owen, who occupied a very important scientific position and was the leading comparative anatomist in Europe, and Agassiz of Harvard, a very accomplished naturalist and the leading comparative anatomist in America. Both were still bemused by *Naturphilosophie*, as was also von Baer (p. 332), now in extreme old age. All opposed to evolution the '*idea*' or '*type*' of Goethe and Cuvier, a metaphysical conception and, of its nature, insusceptible of demonstration.

In Germany, then swept by 'liberal' ideas, Darwinism made rapid progress and gave rise to something that was very near a religion. The ablest continental critic of Darwinism was the Swiss professor at Würzburg, ALBRECHT KÖLLIKER (1817–1905). Without denying the inconstancy of specific forms, and while fully accepting evolution within the limits of certain wider groups, he

Evolution

indicates several real weaknesses in the Darwinian position, namely:

(a) Absence of any experience of the formation of a species.

(b) Absence of any evidence that unions of different varieties (i.e. of *incipient species* on Darwin's view) are relatively more sterile than unions of the same variety.

(c) Extreme rarity of true intermediate forms between known species, whether living or fossil.

Kölliker and other critics claimed that the 'chance' element in Darwin's scheme was but a veiled teleology. Natural selection had been elevated to the rank of a 'cause' leading to an 'effect' and science has to deal not with causes but with conditions. In Kölliker's view, Darwin was dealing with the 'might' and 'may be' and not with any theory that could be tested by experience. Here Kölliker was right. Evolution is perhaps unique among major scientific theories in that the appeal for its acceptance is not that there is evidence for it, but that any other proposed interpretation of the data is wholly incredible.

In France the reception of Darwinism was on the whole hostile and its advance slow. The influence of Cuvier was still paramount. The ultimate victory was complete, though several very able biologists, such as Bernard (p. 363), remained unconvinced to the end. The movement led to a revival of interest in Lamarck, and *transformisme*, as evolution was called, received in France a Lamarckian tinge.

The battle of evolution is now a stricken field, and the whole of modern biology has been called 'a commentary on the *Origin of Species*'. Biologists are now at one in the view that living forms correspond to a limited number of common stocks, and tolerable agreement has been reached as to the evolutionary history of these stocks. It was not many decades, however, before doubt began to dawn as to the mechanism of evolution. Even during Darwin's active period, Gregor Mendel (1822–84) was at his unnoticed work (1857–69), the rediscovery of which (1900) introduced a particulate view of inheritance, a view of which Darwin and the generation after him knew nothing.

ENVOI

Without as within the realm of biology, the leading feature of later nineteenth-century thought was its occupation with the conception of evolution. By unphilosophic minds and by the public generally evolution was, erroneously, elevated from a process into a 'cause' and from a law into a force. Further, by constant association with the conceptions of Natural Selection and Survival of the fittest, Evolution was frequently confused with them. The varied uses of these stock phrases provide good illustrations of the control of ideas by words.

It fell out that the rise of evolutionary theory coincided with a period of industrial expansion and also with a period of social change to which the much abused term 'progress' may reasonably be attached. Naturalists discerned in the vast ranges of geological time a process of development of living forms to which the term 'progress' might also reasonably be attached. The conditions of human life in England of the mid-nineteenth century were, on the whole, much better than those of the pre-industrial age. The adaptations of the living forms of our world to their environment are, on the whole, much better than those of earlier geological ages. The two processes were often equated and, for various reasons, a belief in 'evolutionary progress' conquered the imagination of the generation. At first the fact was missed, even by many naturalists, that adaptation to environment might lead to loss of 'higher' qualities. Darwin himself placed opposite the title-page of the *Origin* a passage from Bacon's *Advancement of Learning* 'Let no man think that a man can search too far . . . in God's word or God's works, divinity or philosophy [that is science]; but rather let men endeavour an endless *progress* in both.' But the poet who wrote:

I dipt into the future, far as human eyes could see
Saw the vision of the world, and all the wonder that would be;
Till the war-drum throbbed no longer, and the battle-flags were furl'd
In the Parliament of man, the Federation of the world.

(*Locksley Hall*, 1832.)

had to write, sixty years later:

388

Envoi

Is there evil but on earth? or pain in every peopled sphere?
Well, be grateful for the sounding watchword 'Evolution' here;
Evolution ever climbing after some ideal good,
And Reversion ever dragging Evolution in the mud.

(Locksley Hall, Sixty Years After, 1886.)

The fallacies of the nineteenth-century evolutionists were from the first clearly discerned by professed philosophers. But in those days and in this country, professed philosophers dwelt securely and apart in scientifically constructed ivory towers, erected in and protected by ancient universities. There they spoke (to each other) in the idiom of Plato. Such missives as they sent down to mortals (if they sent any) were incomprehensible to that considerable majority that did not understand the idiom. Thus the fallacies of Spencer and of the more optimistic Darwinians attained the widest vogue. Thus misunderstandings of Darwin's method and limitations were given a degree of notoriety that amounted to general acceptance.

It would be wrong to end this book with the impression of any desire to belittle a very great naturalist. Himself a modest man, he rated low—and rightly—his own philosophic powers. This estimate of himself is additional evidence of his greatness, and of the soundness of his judgement. He never permitted himself to be drawn into any discussion of the wider implications of his views. Despite and perhaps because of his helplessness in the niceties of language he has many claims to be regarded as a great writer as well as a great naturalist. His services to science were enormous, and among them his greatest was to have laid bare the process of formation of organic types. Any other view of the origin of species than the evolutionary is incredible. That his 'explanation' of organic evolution turns out to be rather a redescription, is a charge against his philosophic but not against his scientific powers. Such redescription is the normal process of advance of scientific theory.

Thus we part with our story at the dawn of modern classical science. The task of science in the age following Newton was to describe the world in mechanical terms in the hope of reaching a unitary view. The age closed with a considerable advance towards a unitary conception of Force and with a suggestion for a

389

unitary conception of Matter, while in the world of life continuity, at least, had been demonstrated. These successes found their most characteristic celebration in the Doctrine of Energy, in the Atomic View of Matter, and in the Theory of Evolution of Organic Forms.

Despite such triumphs there yet remained in the narrative inconsistencies so evident and breaks so definite that they could be ignored only by the most optimistic or the least philosophical. Thus, for example, the Doctrine of Ether remained highly metaphysical, and there were unbridged gulfs between Matter and Force on the one hand and between the Living and the Not-living on the other. Nevertheless there were those who naïvely presented a supposedly complete picture of a world built up of changeless, spherical atoms, often compared to billiard balls—hard, impenetrable, inelastic, devoid of all secondary qualities—between which was only the mysterious Ether with a wealth of endowments which seemed to come from more worlds than one:—

> The gift which is not to be given
> By all the blended powers of earth and heaven.

These billiard balls were compelled, for a reason as impenetrable as themselves, to perform an everlasting dance. They were constantly changing partners under the orders of a protean dance-director in his various characters of 'Heat', 'Chemical Affinity', 'Electricity', &c., and the more he changed the more he remained the same. His masterpiece was 'Living Matter' which had itself somehow created its own dance-director called 'Natural Selection'. He was sometimes nicknamed 'Survival of the Fittest'—and had developed various other characters. Under him there opened the dreary prospect of the dance becoming ever more complex, for was not life the passage from the less to the more highly organized?

This depressing picture made little appeal to the professed philosophers. They saw that the whole structure of science had been built—and necessarily built—on certain metaphysical foundations. These, for the science of that age, were the unquestioned data of the Newtonian world system. But during the later nineteenth century it became apparent that even were the scientific narrative sufficiently consistent and continuous it could not be integrated into a comprehensive system unless and until its metaphysical data were clearly displayed and recognized for what they were.

During the nineteenth century science was immensely successful in many and revolutionary directions. It had improved the human lot. It had provided an intellectual stimulus that was far more effective than those of some other and more fatigued disciplines. It had rendered many current philosophical and theological positions completely untenable. It had—despite modern misunderstanding—introduced a humaner spirit into human relations. It provided a new basis for education, and had made certain of the older bases more than a little ridiculous. Most of all, it had inseminated a hopeful and at least partially justified view of the possibility of human progress. Nevertheless, the method has its limits which were, in fact, more readily recognized by working scientific men than by some who assumed the task of interpreting them.

Science, of its nature, is incapable of accomplishing or even of attempting the task of resolving all the various discrepancies of thought into one whole. For this reason, among others, a history of science is, in the strict sense of the word, hardly possible. Science cannot deal with the whole at all, but only with abstractions, with 'Departments of Scientific Inquiry' as we are accustomed to call them. But though it must perforce work in departments, it is by no means pledged to keep the boundaries of those departments fixed; it is committed to no doctrine of *status quo* for the frontiers on its maps. In changing those frontiers science must, at need, go back to its beginnings and question its own primary data: it must revise its metaphysic. In doing so it may well presuppose a philosophy different from the classical material istic plan. The world of science may well come to be regarded as an evolutionary scheme in which will emerge patterns of *value*, precisely that type of pattern in fact that was so stoutly repudiated by the materialist philosophers of a previous generation.

The generation of philosophers that could ignore the great scientific conclusions is now at rest and is not likely to be disturbed. It seems probable that Science itself is now reaching a stage in which an adequate scientific equipment will involve some regard to the world as an interconnected whole, in other words, in which Science and Philosophy will dwell less apart. This does not mean that Science will abandon its method of abstraction—for then it would cease to be Science—nor does it mean that Science will seek a refuge in that tomb which has become the peaceful abode of an older philosophy based on ratiocination. But it does mean that the frontiers of scientific abstractions may be rendered more fluid and that the philosophical method may have a share in determining the nature of the change. Notably it seems probable that the conceptions of the separation of mind from mind and of mind from matter may need modification. There are many indications that the tendencies of science since the later nineteenth century have been working in these directions.

INDEX

393

Index

Index

Index

396

Index

Index

398

Index

A CATALOG OF SELECTED
DOVER BOOKS
IN ALL FIELDS OF INTEREST

A CATALOG OF SELECTED DOVER
BOOKS IN ALL FIELDS OF INTEREST

CONCERNING THE SPIRITUAL IN ART, Wassily Kandinsky. Pioneering work by father of abstract art. Thoughts on color theory, nature of art. Analysis of earlier masters. 12 illustrations. 80pp. of text. 5⅜ x 8½. 23411-8 Pa. $3.95

ANIMALS: 1,419 Copyright-Free Illustrations of Mammals, Birds, Fish, Insects, etc., Jim Harter (ed.). Clear wood engravings present, in extremely lifelike poses, over 1,000 species of animals. One of the most extensive pictorial sourcebooks of its kind. Captions. Index. 284pp. 9 x 12. 23766-4 Pa. $12.95

CELTIC ART: The Methods of Construction, George Bain. Simple geometric techniques for making Celtic interlacements, spirals, Kells-type initials, animals, humans, etc. Over 500 illustrations. 160pp. 9 x 12. (USO) 22923-8 Pa. $9.95

AN ATLAS OF ANATOMY FOR ARTISTS, Fritz Schider. Most thorough reference work on art anatomy in the world. Hundreds of illustrations, including selections from works by Vesalius, Leonardo, Goya, Ingres, Michelangelo, others. 593 illustrations. 192pp. 7⅛ x 10¼. 20241-0 Pa. $9 95

CELTIC HAND STROKE-BY-STROKE (Irish Half-Uncial from "The Book of Kells"): An Arthur Baker Calligraphy Manual, Arthur Baker. Complete guide to creating each letter of the alphabet in distinctive Celtic manner. Covers hand position, strokes, pens, inks, paper, more. Illustrated. 48pp. 8¼ x 11. 24336-2 Pa. $3.95

EASY ORIGAMI, John Montroll. Charming collection of 32 projects (hat, cup, pelican, piano, swan, many more) specially designed for the novice origami hobbyist. Clearly illustrated easy-to-follow instructions insure that even beginning papercrafters will achieve successful results. 48pp. 8¼ x 11. 27298-2 Pa. $2.95

THE COMPLETE BOOK OF BIRDHOUSE CONSTRUCTION FOR WOODWORKERS, Scott D. Campbell. Detailed instructions, illustrations, tables. Also data on bird habitat and instinct patterns. Bibliography. 3 tables. 63 illustrations in 15 figures. 48pp. 5¼ x 8½. 24407-5 Pa. $2.50

BLOOMINGDALE'S ILLUSTRATED 1886 CATALOG: Fashions, Dry Goods and Housewares, Bloomingdale Brothers. Famed merchants' extremely rare catalog depicting about 1,700 products: clothing, housewares, firearms, dry goods, jewelry, more. Invaluable for dating, identifying vintage items. Also, copyright-free graphics for artists, designers. Co-published with Henry Ford Museum & Greenfield Village. 160pp. 8¼ x 11. 25780-0 Pa. $9.95

HISTORIC COSTUME IN PICTURES, Braun & Schneider. Over 1,450 costumed figures in clearly detailed engravings–from dawn of civilization to end of 19th century. Captions. Many folk costumes. 256pp. 8⅜ x 11¾. 23150-X Pa. $12.95

CATALOG OF DOVER BOOKS

STICKLEY CRAFTSMAN FURNITURE CATALOGS, Gustav Stickley and L. & J. G. Stickley. Beautiful, functional furniture in two authentic catalogs from 1910. 594 illustrations, including 277 photos, show settles, rockers, armchairs, reclining chairs, bookcases, desks, tables. 183pp. 6½ x 9¼. 23838-5 Pa. $9.95

AMERICAN LOCOMOTIVES IN HISTORIC PHOTOGRAPHS: 1858 to 1949, Ron Ziel (ed.). A rare collection of 126 meticulously detailed official photographs, called "builder portraits," of American locomotives that majestically chronicle the rise of steam locomotive power in America. Introduction. Detailed captions. xi + 129pp. 9 x 12. 27393-8 Pa. $12.95

AMERICA'S LIGHTHOUSES: An Illustrated History, Francis Ross Holland, Jr. Delightfully written, profusely illustrated fact-filled survey of over 200 American lighthouses since 1716. History, anecdotes, technological advances, more. 240pp. 8 x 10¾. 25576-X Pa. $12.95

TOWARDS A NEW ARCHITECTURE, Le Corbusier. Pioneering manifesto by founder of "International School." Technical and aesthetic theories, views of industry, economics, relation of form to function, "mass-production split" and much more. Profusely illustrated. 320pp. 6⅛ x 9¼. (USO) 25023-7 Pa. $9.95

HOW THE OTHER HALF LIVES, Jacob Riis. Famous journalistic record, exposing poverty and degradation of New York slums around 1900, by major social reformer. 100 striking and influential photographs. 233pp. 10 x 7⅞. 22012-5 Pa. $10.95

FRUIT KEY AND TWIG KEY TO TREES AND SHRUBS, William M. Harlow. One of the handiest and most widely used identification aids. Fruit key covers 120 deciduous and evergreen species; twig key 160 deciduous species. Easily used. Over 300 photographs. 126pp. 5⅜ x 8½. 20511-8 Pa. $3.95

COMMON BIRD SONGS, Dr. Donald J. Borror. Songs of 60 most common U.S. birds: robins, sparrows, cardinals, bluejays, finches, more—arranged in order of increasing complexity. Up to 9 variations of songs of each species. Cassette and manual 99911-4 $8.95

ORCHIDS AS HOUSE PLANTS, Rebecca Tyson Northen. Grow cattleyas and many other kinds of orchids—in a window, in a case, or under artificial light. 63 illustrations. 148pp. 5⅜ x 8½. 23261-1 Pa. $4.95

MONSTER MAZES, Dave Phillips. Masterful mazes at four levels of difficulty. Avoid deadly perils and evil creatures to find magical treasures. Solutions for all 32 exciting illustrated puzzles. 48pp. 8¼ x 11. 26005-4 Pa. $2.95

MOZART'S DON GIOVANNI (DOVER OPERA LIBRETTO SERIES), Wolfgang Amadeus Mozart. Introduced and translated by Ellen H. Bleiler. Standard Italian libretto, with complete English translation. Convenient and thoroughly portable—an ideal companion for reading along with a recording or the performance itself. Introduction. List of characters. Plot summary. 121pp. 5¼ x 8½. 24944-1 Pa. $2.95

TECHNICAL MANUAL AND DICTIONARY OF CLASSICAL BALLET, Gail Grant. Defines, explains, comments on steps, movements, poses and concepts. 15-page pictorial section. Basic book for student, viewer. 127pp. 5⅜ x 8½. 21843-0 Pa. $4.95

BRASS INSTRUMENTS: Their History and Development, Anthony Baines. Authoritative, updated survey of the evolution of trumpets, trombones, bugles, cornets, French horns, tubas and other brass wind instruments. Over 140 illustrations and 48 music examples. Corrected and updated by author. New preface. Bibliography. 320pp. 5⅜ x 8½. 27574-4 Pa. $9.95

HOLLYWOOD GLAMOR PORTRAITS, John Kobal (ed.). 145 photos from 1926-49. Harlow, Gable, Bogart, Bacall; 94 stars in all. Full background on photographers, technical aspects. 160pp. 8⅜ x 11¼. 23352-9 Pa. $11.95

MAX AND MORITZ, Wilhelm Busch. Great humor classic in both German and English. Also 10 other works: "Cat and Mouse," "Plisch and Plumm," etc. 216pp. 5⅜ x 8½. 20181-3 Pa. $6.95

THE RAVEN AND OTHER FAVORITE POEMS, Edgar Allan Poe. Over 40 of the author's most memorable poems: "The Bells," "Ulalume," "Israfel," "To Helen," "The Conqueror Worm," "Eldorado," "Annabel Lee," many more. Alphabetic lists of titles and first lines. 64pp. 5³⁄₁₆ x 8¼. 26685-0 Pa. $1.00

PERSONAL MEMOIRS OF U. S. GRANT, Ulysses Simpson Grant. Intelligent, deeply moving firsthand account of Civil War campaigns, considered by many the finest military memoirs ever written. Includes letters, historic photographs, maps and more. 528pp. 6⅛ x 9¼. 28587-1 Pa. $11.95

AMULETS AND SUPERSTITIONS, E. A. Wallis Budge. Comprehensive discourse on origin, powers of amulets in many ancient cultures: Arab, Persian Babylonian, Assyrian, Egyptian, Gnostic, Hebrew, Phoenician, Syriac, etc. Covers cross, swastika, crucifix, seals, rings, stones, etc. 584pp. 5⅜ x 8½. 23573-4 Pa. $12.95

RUSSIAN STORIES/PYCCKNE PACCKA3bI: A Dual-Language Book, edited by Gleb Struve. Twelve tales by such masters as Chekhov, Tolstoy, Dostoevsky, Pushkin, others. Excellent word-for-word English translations on facing pages, plus teaching and study aids, Russian/English vocabulary, biographical/critical introductions, more. 416pp. 5⅜ x 8½. 26244-8 Pa. $8.95

PHILADELPHIA THEN AND NOW: 60 Sites Photographed in the Past and Present, Kenneth Finkel and Susan Oyama. Rare photographs of City Hall, Logan Square, Independence Hall, Betsy Ross House, other landmarks juxtaposed with contemporary views. Captures changing face of historic city. Introduction. Captions. 128pp. 8¼ x 11. 25790-8 Pa. $9.95

AIA ARCHITECTURAL GUIDE TO NASSAU AND SUFFOLK COUNTIES, LONG ISLAND, The American Institute of Architects, Long Island Chapter, and the Society for the Preservation of Long Island Antiquities. Comprehensive, well-researched and generously illustrated volume brings to life over three centuries of Long Island's great architectural heritage. More than 240 photographs with authoritative, extensively detailed captions. 176pp. 8¼ x 11. 26946-9 Pa. $14.95

NORTH AMERICAN INDIAN LIFE: Customs and Traditions of 23 Tribes, Elsie Clews Parsons (ed.). 27 fictionalized essays by noted anthropologists examine religion, customs, government, additional facets of life among the Winnebago, Crow, Zuni, Eskimo, other tribes. 480pp. 6⅛ x 9¼. 27377-6 Pa. $10.95

FRANK LLOYD WRIGHT'S HOLLYHOCK HOUSE, Donald Hoffmann. Lavishly illustrated, carefully documented study of one of Wright's most controversial residential designs. Over 120 photographs, floor plans, elevations, etc. Detailed perceptive text by noted Wright scholar. Index. 128pp. 9¼ x 10¾. 27133-1 Pa. $11.95

THE MALE AND FEMALE FIGURE IN MOTION: 60 Classic Photographic Sequences, Eadweard Muybridge. 60 true-action photographs of men and women walking, running, climbing, bending, turning, etc., reproduced from rare 19th-century masterpiece. vi + 121pp. 9 x 12. 24745-7 Pa. $10.95

1001 QUESTIONS ANSWERED ABOUT THE SEASHORE, N. J. Berrill and Jacquelyn Berrill. Queries answered about dolphins, sea snails, sponges, starfish, fishes, shore birds, many others. Covers appearance, breeding, growth, feeding, much more. 305pp. 5¼ x 8¼. 23366-9 Pa. $8.95

GUIDE TO OWL WATCHING IN NORTH AMERICA, Donald S. Heintzelman. Superb guide offers complete data and descriptions of 19 species: barn owl, screech owl, snowy owl, many more. Expert coverage of owl-watching equipment, conservation, migrations and invasions, etc. Guide to observing sites. 84 illustrations. xiii + 193pp. 5⅜ x 8½. 27344-X Pa. $8.95

MEDICINAL AND OTHER USES OF NORTH AMERICAN PLANTS: A Historical Survey with Special Reference to the Eastern Indian Tribes, Charlotte Erichsen-Brown. Chronological historical citations document 500 years of usage of plants, trees, shrubs native to eastern Canada, northeastern U.S. Also complete identifying information. 343 illustrations. 544pp. 6½ x 9¼. 25951-X Pa. $12.95

STORYBOOK MAZES, Dave Phillips. 23 stories and mazes on two-page spreads: Wizard of Oz, Treasure Island, Robin Hood, etc. Solutions. 64pp. 8¼ x 11. 23628-5 Pa. $2.95

NEGRO FOLK MUSIC, U.S.A., Harold Courlander. Noted folklorist's scholarly yet readable analysis of rich and varied musical tradition. Includes authentic versions of over 40 folk songs. Valuable bibliography and discography. xi + 324pp. 5⅜ x 8½. 27350-4 Pa. $7.95

MOVIE-STAR PORTRAITS OF THE FORTIES, John Kobal (ed.). 163 glamor, studio photos of 106 stars of the 1940s: Rita Hayworth, Ava Gardner, Marlon Brando, Clark Gable, many more. 176pp. 8⅜ x 11¼. 23546-7 Pa. $12.95

BENCHLEY LOST AND FOUND, Robert Benchley. Finest humor from early 30s, about pet peeves, child psychologists, post office and others. Mostly unavailable elsewhere. 73 illustrations by Peter Arno and others. 183pp. 5⅜ x 8½. 22410-4 Pa. $6.95

YEKL and THE IMPORTED BRIDEGROOM AND OTHER STORIES OF YIDDISH NEW YORK, Abraham Cahan. Film Hester Street based on Yekl (1896). Novel, other stories among first about Jewish immigrants on N.Y.'s East Side. 240pp. 5⅜ x 8½. 22427-9 Pa. $6.95

SELECTED POEMS, Walt Whitman. Generous sampling from *Leaves of Grass*. Twenty-four poems include "I Hear America Singing," "Song of the Open Road," "I Sing the Body Electric," "When Lilacs Last in the Dooryard Bloom'd," "O Captain! My Captain!"—all reprinted from an authoritative edition. Lists of titles and first lines. 128pp. 5³⁄₁₆ x 8¼. 26878-0 Pa. $1.00

THE BEST TALES OF HOFFMANN, E. T. A. Hoffmann. 10 of Hoffmann's most important stories: "Nutcracker and the King of Mice," "The Golden Flowerpot," etc. 458pp. 5⅜ x 8½. 21793-0 Pa. $9.95

FROM FETISH TO GOD IN ANCIENT EGYPT, E. A. Wallis Budge. Rich detailed survey of Egyptian conception of "God" and gods, magic, cult of animals, Osiris, more. Also, superb English translations of hymns and legends. 240 illustrations. 545pp. 5⅜ x 8½. 25803-3 Pa. $11.95

FRENCH STORIES/CONTES FRANÇAIS: A Dual-Language Book, Wallace Fowlie. Ten stories by French masters, Voltaire to Camus: "Micromegas" by Voltaire; "The Atheist's Mass" by Balzac; "Minuet" by de Maupassant; "The Guest" by Camus, six more. Excellent English translations on facing pages. Also French-English vocabulary list, exercises, more. 352pp. 5⅜ x 8½. 26443-2 Pa. $8.95

CHICAGO AT THE TURN OF THE CENTURY IN PHOTOGRAPHS: 122 Historic Views from the Collections of the Chicago Historical Society, Larry A. Viskochil. Rare large-format prints offer detailed views of City Hall, State Street, the Loop, Hull House, Union Station, many other landmarks, circa 1904-1913. Introduction. Captions. Maps. 144pp. 9⅜ x 12¼. 24656-6 Pa. $12.95

OLD BROOKLYN IN EARLY PHOTOGRAPHS, 1865-1929, William Lee Younger. Luna Park, Gravesend race track, construction of Grand Army Plaza, moving of Hotel Brighton, etc. 157 previously unpublished photographs. 165pp. 8⅞ x 11¾. 23587-4 Pa. $13.95

THE MYTHS OF THE NORTH AMERICAN INDIANS, Lewis Spence. Rich anthology of the myths and legends of the Algonquins, Iroquois, Pawnees and Sioux, prefaced by an extensive historical and ethnological commentary. 36 illustrations. 480pp. 5⅜ x 8½. 25967-6 Pa. $8.95

AN ENCYCLOPEDIA OF BATTLES: Accounts of Over 1,560 Battles from 1479 B.C. to the Present, David Eggenberger. Essential details of every major battle in recorded history from the first battle of Megiddo in 1479 B.C. to Grenada in 1984. List of Battle Maps. New Appendix covering the years 1967-1984. Index. 99 illustrations. 544pp. 6½ x 9¼. 24913-1 Pa. $14.95

SAILING ALONE AROUND THE WORLD, Captain Joshua Slocum. First man to sail around the world, alone, in small boat. One of great feats of seamanship told in delightful manner. 67 illustrations. 294pp. 5⅜ x 8½. 20326-3 Pa. $5.95

ANARCHISM AND OTHER ESSAYS, Emma Goldman. Powerful, penetrating, prophetic essays on direct action, role of minorities, prison reform, puritan hypocrisy, violence, etc. 271pp. 5⅜ x 8½. 22484-8 Pa. $6.95

MYTHS OF THE HINDUS AND BUDDHISTS, Ananda K. Coomaraswamy and Sister Nivedita. Great stories of the epics; deeds of Krishna, Shiva, taken from puranas, Vedas, folk tales; etc. 32 illustrations. 400pp. 5⅜ x 8½. 21759-0 Pa. $10.95

BEYOND PSYCHOLOGY, Otto Rank. Fear of death, desire of immortality, nature of sexuality, social organization, creativity, according to Rankian system. 291pp. 5⅜ x 8½. 20485-5 Pa. $8.95

A THEOLOGICO-POLITICAL TREATISE, Benedict Spinoza. Also contains unfinished Political Treatise. Great classic on religious liberty, theory of government on common consent. R. Elwes translation. Total of 421pp. 5⅜ x 8½. 20249-6 Pa. $9.95

MY BONDAGE AND MY FREEDOM, Frederick Douglass. Born a slave, Douglass became outspoken force in antislavery movement. The best of Douglass' autobiographies. Graphic description of slave life. 464pp. 5⅜ x 8½. 22457-0 Pa. $8.95

FOLLOWING THE EQUATOR: A Journey Around the World, Mark Twain. Fascinating humorous account of 1897 voyage to Hawaii, Australia, India, New Zealand, etc. Ironic, bemused reports on peoples, customs, climate, flora and fauna, politics, much more. 197 illustrations. 720pp. 5⅜ x 8½. 26113-1 Pa. $15.95

THE PEOPLE CALLED SHAKERS, Edward D. Andrews. Definitive study of Shakers: origins, beliefs, practices, dances, social organization, furniture and crafts, etc. 33 illustrations. 351pp. 5⅜ x 8½. 21081-2 Pa. $8.95

THE MYTHS OF GREECE AND ROME, H. A. Guerber. A classic of mythology, generously illustrated, long prized for its simple, graphic, accurate retelling of the principal myths of Greece and Rome, and for its commentary on their origins and significance. With 64 illustrations by Michelangelo, Raphael, Titian, Rubens, Canova, Bernini and others. 480pp. 5⅜ x 8½. 27584-1 Pa. $9.95

PSYCHOLOGY OF MUSIC, Carl E. Seashore. Classic work discusses music as a medium from psychological viewpoint. Clear treatment of physical acoustics, auditory apparatus, sound perception, development of musical skills, nature of musical feeling, host of other topics. 88 figures. 408pp. 5⅜ x 8½. 21851-1 Pa. $10.95

THE PHILOSOPHY OF HISTORY, Georg W. Hegel. Great classic of Western thought develops concept that history is not chance but rational process, the evolution of freedom. 457pp. 5⅜ x 8½. 20112-0 Pa. $9.95

THE BOOK OF TEA, Kakuzo Okakura. Minor classic of the Orient: entertaining, charming explanation, interpretation of traditional Japanese culture in terms of tea ceremony. 94pp. 5⅜ x 8½. 20070-1 Pa. $3.95

LIFE IN ANCIENT EGYPT, Adolf Erman. Fullest, most thorough, detailed older account with much not in more recent books, domestic life, religion, magic, medicine, commerce, much more. Many illustrations reproduce tomb paintings, carvings, hieroglyphs, etc. 597pp. 5⅜ x 8½. 22632-8 Pa. $11.95

SUNDIALS, Their Theory and Construction, Albert Waugh. Far and away the best, most thorough coverage of ideas, mathematics concerned, types, construction, adjusting anywhere. Simple, nontechnical treatment allows even children to build several of these dials. Over 100 illustrations. 230pp. 5⅜ x 8½. 22947-5 Pa. $7.95

DYNAMICS OF FLUIDS IN POROUS MEDIA, Jacob Bear. For advanced students of ground water hydrology, soil mechanics and physics, drainage and irrigation engineering, and more. 335 illustrations. Exercises, with answers. 784pp. 6⅛ x 9¼. 65675-6 Pa. $19.95

SONGS OF EXPERIENCE: Facsimile Reproduction with 26 Plates in Full Color, William Blake. 26 full-color plates from a rare 1826 edition. Includes "TheTyger," "London," "Holy Thursday," and other poems. Printed text of poems. 48pp. 5¼ x 7. 24636-1 Pa. $4.95

OLD-TIME VIGNETTES IN FULL COLOR, Carol Belanger Grafton (ed.). Over 390 charming, often sentimental illustrations, selected from archives of Victorian graphics—pretty women posing, children playing, food, flowers, kittens and puppies, smiling cherubs, birds and butterflies, much more. All copyright-free. 48pp. 9¼ x 12¼. 27269-9 Pa. $5.95

PERSPECTIVE FOR ARTISTS, Rex Vicat Cole. Depth, perspective of sky and sea, shadows, much more, not usually covered. 391 diagrams, 81 reproductions of drawings and paintings. 279pp. 5⅜ x 8½. 22487-2 Pa. $6.95

DRAWING THE LIVING FIGURE, Joseph Sheppard. Innovative approach to artistic anatomy focuses on specifics of surface anatomy, rather than muscles and bones. Over 170 drawings of live models in front, back and side views, and in widely varying poses. Accompanying diagrams. 177 illustrations. Introduction. Index. 144pp. 8⅜ x11¼. 26723-7 Pa. $8.95

GOTHIC AND OLD ENGLISH ALPHABETS: 100 Complete Fonts, Dan X. Solo. Add power, elegance to posters, signs, other graphics with 100 stunning copyright-free alphabets: Blackstone, Dolbey, Germania, 97 more—including many lower-case, numerals, punctuation marks. 104pp. 8¼ x 11. 24695-7 Pa. $8.95

HOW TO DO BEADWORK, Mary White. Fundamental book on craft from simple projects to five-bead chains and woven works. 106 illustrations. 142pp. 5⅜ x 8. 20697-1 Pa. $4.95

THE BOOK OF WOOD CARVING, Charles Marshall Sayers. Finest book for beginners discusses fundamentals and offers 34 designs. "Absolutely first rate . . . well thought out and well executed."–E. J. Tangerman. 118pp. 7¾ x 10⅝. 23654-4 Pa. $6.95

ILLUSTRATED CATALOG OF CIVIL WAR MILITARY GOODS: Union Army Weapons, Insignia, Uniform Accessories, and Other Equipment, Schuyler, Hartley, and Graham. Rare, profusely illustrated 1846 catalog includes Union Army uniform and dress regulations, arms and ammunition, coats, insignia, flags, swords, rifles, etc. 226 illustrations. 160pp. 9 x 12. 24939-5 Pa. $10.95

WOMEN'S FASHIONS OF THE EARLY 1900s: An Unabridged Republication of "New York Fashions, 1909," National Cloak & Suit Co. Rare catalog of mail-order fashions documents women's and children's clothing styles shortly after the turn of the century. Captions offer full descriptions, prices. Invaluable resource for fashion, costume historians. Approximately 725 illustrations. 128pp. 8⅜ x 11¼. 27276-1 Pa. $11.95

THE 1912 AND 1915 GUSTAV STICKLEY FURNITURE CATALOGS, Gustav Stickley. With over 200 detailed illustrations and descriptions, these two catalogs are essential reading and reference materials and identification guides for Stickley furniture. Captions cite materials, dimensions and prices. 112pp. 6½ x 9¼. 26676-1 Pa. $9.95

EARLY AMERICAN LOCOMOTIVES, John H. White, Jr. Finest locomotive engravings from early 19th century: historical (1804–74), main-line (after 1870), special, foreign, etc. 147 plates. 142pp. 11⅜ x 8¼. 22772-3 Pa. $10.95

THE TALL SHIPS OF TODAY IN PHOTOGRAPHS, Frank O. Braynard. Lavishly illustrated tribute to nearly 100 majestic contemporary sailing vessels: Amerigo Vespucci, Clearwater, Constitution, Eagle, Mayflower, Sea Cloud, Victory, many more. Authoritative captions provide statistics, background on each ship. 190 black-and-white photographs and illustrations. Introduction. 128pp. 8⅞ x 11¾. 27163-3 Pa. $13.95

EARLY NINETEENTH-CENTURY CRAFTS AND TRADES, Peter Stockham (ed.). Extremely rare 1807 volume describes to youngsters the crafts and trades of the day: brickmaker, weaver, dressmaker, bookbinder, ropemaker, saddler, many more. Quaint prose, charming illustrations for each craft. 20 black-and-white line illustrations. 192pp. 4⅜ x 6. 27293-1 Pa. $4.95

VICTORIAN FASHIONS AND COSTUMES FROM HARPER'S BAZAR, 1867–1898, Stella Blum (ed.). Day costumes, evening wear, sports clothes, shoes, hats, other accessories in over 1,000 detailed engravings. 320pp. 9⅜ x 12¼. 22990-4 Pa. $14.95

GUSTAV STICKLEY, THE CRAFTSMAN, Mary Ann Smith. Superb study surveys broad scope of Stickley's achievement, especially in architecture. Design philosophy, rise and fall of the Craftsman empire, descriptions and floor plans for many Craftsman houses, more. 86 black-and-white halftones. 31 line illustrations. Introduction 208pp. 6½ x 9¼. 27210-9 Pa. $9.95

THE LONG ISLAND RAIL ROAD IN EARLY PHOTOGRAPHS, Ron Ziel. Over 220 rare photos, informative text document origin (1844) and development of rail service on Long Island. Vintage views of early trains, locomotives, stations, passengers, crews, much more. Captions. 8⅞ x 11¾. 26301-0 Pa. $13.95

THE BOOK OF OLD SHIPS: From Egyptian Galleys to Clipper Ships, Henry B. Culver. Superb, authoritative history of sailing vessels, with 80 magnificent line illustrations. Galley, bark, caravel, longship, whaler, many more. Detailed, informative text on each vessel by noted naval historian. Introduction. 256pp. 5⅜ x 8½. 27332-6 Pa. $7.95

TEN BOOKS ON ARCHITECTURE, Vitruvius. The most important book ever written on architecture. Early Roman aesthetics, technology, classical orders, site selection, all other aspects. Morgan translation. 331pp. 5⅜ x 8½. 20645-9 Pa. $8.95

THE HUMAN FIGURE IN MOTION, Eadweard Muybridge. More than 4,500 stopped-action photos, in action series, showing undraped men, women, children jumping, lying down, throwing, sitting, wrestling, carrying, etc. 390pp. 7⅞ x 10⅝. 20204-6 Clothbd. $25.95

TREES OF THE EASTERN AND CENTRAL UNITED STATES AND CANADA, William M. Harlow. Best one-volume guide to 140 trees. Full descriptions, woodlore, range, etc. Over 600 illustrations. Handy size. 288pp. 4½ x 6⅜. 20395-6 Pa. $5.95

SONGS OF WESTERN BIRDS, Dr. Donald J. Borror. Complete song and call repertoire of 60 western species, including flycatchers, juncoes, cactus wrens, many more–includes fully illustrated booklet. Cassette and manual 99913-0 $8.95

GROWING AND USING HERBS AND SPICES, Milo Miloradovich. Versatile handbook provides all the information needed for cultivation and use of all the herbs and spices available in North America. 4 illustrations. Index. Glossary. 236pp. 5⅜ x 8½. 25058-X Pa. $6.95

BIG BOOK OF MAZES AND LABYRINTHS, Walter Shepherd. 50 mazes and labyrinths in all–classical, solid, ripple, and more–in one great volume. Perfect inexpensive puzzler for clever youngsters. Full solutions. 112pp. 8⅛ x 11. 22951-3 Pa. $4.95

PIANO TUNING, J. Cree Fischer. Clearest, best book for beginner, amateur. Simple repairs, raising dropped notes, tuning by easy method of flattened fifths. No previous skills needed. 4 illustrations. 201pp. 5⅜ x 8½. 23267-0 Pa. $6.95

A SOURCE BOOK IN THEATRICAL HISTORY, A. M. Nagler. Contemporary observers on acting, directing, make-up, costuming, stage props, machinery, scene design, from Ancient Greece to Chekhov. 611pp. 5⅜ x 8½. 20515-0 Pa. $12.95

THE COMPLETE NONSENSE OF EDWARD LEAR, Edward Lear. All nonsense limericks, zany alphabets, Owl and Pussycat, songs, nonsense botany, etc., illustrated by Lear. Total of 320pp. 5⅜ x 8½. (USO) 20167-8 Pa. $6.95

VICTORIAN PARLOUR POETRY: An Annotated Anthology, Michael R. Turner. 117 gems by Longfellow, Tennyson, Browning, many lesser-known poets. "The Village Blacksmith," "Curfew Must Not Ring Tonight," "Only a Baby Small," dozens more, often difficult to find elsewhere. Index of poets, titles, first lines. xxiii + 325pp. 5⅜ x 8¼. 27044-0 Pa. $8.95

DUBLINERS, James Joyce. Fifteen stories offer vivid, tightly focused observations of the lives of Dublin's poorer classes. At least one, "The Dead," is considered a masterpiece. Reprinted complete and unabridged from standard edition. 160pp. 5³⁄₁₆ x 8¼. 26870-5 Pa. $1.00

THE HAUNTED MONASTERY and THE CHINESE MAZE MURDERS, Robert van Gulik. Two full novels by van Gulik, set in 7th-century China, continue adventures of Judge Dee and his companions. An evil Taoist monastery, seemingly supernatural events; overgrown topiary maze hides strange crimes. 27 illustrations. 328pp. 5⅜ x 8½. 23502-5 Pa. $8.95

THE BOOK OF THE SACRED MAGIC OF ABRAMELIN THE MAGE, translated by S. MacGregor Mathers. Medieval manuscript of ceremonial magic. Basic document in Aleister Crowley, Golden Dawn groups. 268pp. 5⅜ x 8½. 23211-5 Pa. $8.95

NEW RUSSIAN-ENGLISH AND ENGLISH-RUSSIAN DICTIONARY, M. A. O'Brien. This is a remarkably handy Russian dictionary, containing a surprising amount of information, including over 70,000 entries. 366pp. 4½ x 6⅛. 20208-9 Pa. $9.95

HISTORIC HOMES OF THE AMERICAN PRESIDENTS, Second, Revised Edition, Irvin Haas. A traveler's guide to American Presidential homes, most open to the public, depicting and describing homes occupied by every American President from George Washington to George Bush. With visiting hours, admission charges, travel routes. 175 photographs. Index. 160pp. 8¼ x 11. 26751-2 Pa. $11.95

NEW YORK IN THE FORTIES, Andreas Feininger. 162 brilliant photographs by the well-known photographer, formerly with *Life* magazine. Commuters, shoppers, Times Square at night, much else from city at its peak. Captions by John von Hartz. 181pp. 9¼ x 10¾. 23585-8 Pa. $12.95

INDIAN SIGN LANGUAGE, William Tomkins. Over 525 signs developed by Sioux and other tribes. Written instructions and diagrams. Also 290 pictographs. 111pp. 6⅛ x 9¼. 22029-X Pa. $3.95

ANATOMY: A Complete Guide for Artists, Joseph Sheppard. A master of figure drawing shows artists how to render human anatomy convincingly. Over 460 illustrations. 224pp. 8⅜ x 11¼. 27279-6 Pa. $10.95

MEDIEVAL CALLIGRAPHY: Its History and Technique, Marc Drogin. Spirited history, comprehensive instruction manual covers 13 styles (ca. 4th century thru 15th). Excellent photographs; directions for duplicating medieval techniques with modern tools. 224pp. 8⅜ x 11¼. 26142-5 Pa. $11.95

DRIED FLOWERS: How to Prepare Them, Sarah Whitlock and Martha Rankin. Complete instructions on how to use silica gel, meal and borax, perlite aggregate, sand and borax, glycerine and water to create attractive permanent flower arrangements. 12 illustrations. 32pp. 5⅜ x 8½. 21802-3 Pa. $1.00

EASY-TO-MAKE BIRD FEEDERS FOR WOODWORKERS, Scott D. Campbell. Detailed, simple-to-use guide for designing, constructing, caring for and using feeders. Text, illustrations for 12 classic and contemporary designs. 96pp. 5⅜ x 8¼. 25847-5 Pa. $2.95

SCOTTISH WONDER TALES FROM MYTH AND LEGEND, Donald A. Mackenzie. 16 lively tales tell of giants rumbling down mountainsides, of a magic wand that turns stone pillars into warriors, of gods and goddesses, evil hags, powerful forces and more. 240pp. 5⅜ x 8½. 29677-6 Pa. $6.95

THE HISTORY OF UNDERCLOTHES, C. Willett Cunnington and Phyllis Cunnington. Fascinating, well-documented survey covering six centuries of English undergarments, enhanced with over 100 illustrations: 12th-century laced-up bodice, footed long drawers (1795), 19th-century bustles, 19th-century corsets for men, Victorian "bust improvers," much more. 272pp. 5⅜ x 8¼. 27124-2 Pa. $9.95

ARTS AND CRAFTS FURNITURE: The Complete Brooks Catalog of 1912, Brooks Manufacturing Co. Photos and detailed descriptions of more than 150 now very collectible furniture designs from the Arts and Crafts movement depict davenports, settees, buffets, desks, tables, chairs, bedsteads, dressers and more, all built of solid, quarter-sawed oak. Invaluable for students and enthusiasts of antiques, Americana and the decorative arts. 80pp. 6½ x 9¼. 27471-3 Pa. $7.95

HOW WE INVENTED THE AIRPLANE: An Illustrated History, Orville Wright. Fascinating firsthand account covers early experiments, construction of planes and motors, first flights, much more. Introduction and commentary by Fred C. Kelly. 76 photographs. 96pp. 8¼ x 11. 25662-6 Pa. $8.95

THE ARTS OF THE SAILOR: Knotting, Splicing and Ropework, Hervey Garrett Smith. Indispensable shipboard reference covers tools, basic knots and useful hitches; handsewing and canvas work, more. Over 100 illustrations. Delightful reading for sea lovers. 256pp. 5⅜ x 8½. 26440-8 Pa. $7.95

FRANK LLOYD WRIGHT'S FALLINGWATER: The House and Its History, Second, Revised Edition, Donald Hoffmann. A total revision—both in text and illustrations—of the standard document on Fallingwater, the boldest, most personal architectural statement of Wright's mature years, updated with valuable new material from the recently opened Frank Lloyd Wright Archives. "Fascinating"–*The New York Times*. 116 illustrations. 128pp. 9¼ x 10¾. 27430-6 Pa. $11.95

AUTOBIOGRAPHY: The Story of My Experiments with Truth, Mohandas K. Gandhi. Boyhood, legal studies, purification, the growth of the Satyagraha (nonviolent protest) movement. Critical, inspiring work of the man responsible for the freedom of India. 480pp. 5⅜ x 8½. (USO) 24593-4 Pa. $8.95

CELTIC MYTHS AND LEGENDS, T. W. Rolleston. Masterful retelling of Irish and Welsh stories and tales. Cuchulain, King Arthur, Deirdre, the Grail, many more. First paperback edition. 58 full-page illustrations. 512pp. 5⅜ x 8½. 26507-2 Pa. $9.95

THE PRINCIPLES OF PSYCHOLOGY, William James. Famous long course complete, unabridged. Stream of thought, time perception, memory, experimental methods; great work decades ahead of its time. 94 figures. 1,391pp. 5⅜ x 8½. 2-vol. set.
Vol. I: 20381-6 Pa. $12.95
Vol. II: 20382-4 Pa. $12.95

THE WORLD AS WILL AND REPRESENTATION, Arthur Schopenhauer. Definitive English translation of Schopenhauer's life work, correcting more than 1,000 errors, omissions in earlier translations. Translated by E. F. J. Payne. Total of 1,269pp. 5⅜ x 8½. 2-vol. set. Vol. 1: 21761-2 Pa. $11.95
Vol. 2: 21762-0 Pa. $11.95

MAGIC AND MYSTERY IN TIBET, Madame Alexandra David-Neel. Experiences among lamas, magicians, sages, sorcerers, Bonpa wizards. A true psychic discovery. 32 illustrations. 321pp. 5⅜ x 8½. (USO) 22682-4 Pa. $8.95

THE EGYPTIAN BOOK OF THE DEAD, E. A. Wallis Budge. Complete reproduction of Ani's papyrus, finest ever found. Full hieroglyphic text, interlinear transliteration, word-for-word translation, smooth translation. 533pp. 6½ x 9¼. 21866-X Pa. $10.95

MATHEMATICS FOR THE NONMATHEMATICIAN, Morris Kline. Detailed, college-level treatment of mathematics in cultural and historical context, with numerous exercises. Recommended Reading Lists. Tables. Numerous figures. 641pp. 5⅜ x 8½. 24823-2 Pa. $11.95

THEORY OF WING SECTIONS: Including a Summary of Airfoil Data, Ira H. Abbott and A. E. von Doenhoff. Concise compilation of subsonic aerodynamic characteristics of NACA wing sections, plus description of theory. 350pp. of tables. 693pp. 5⅜ x 8½. 60586-8 Pa. $14.95

THE RIME OF THE ANCIENT MARINER, Gustave Doré, S. T. Coleridge. Doré's finest work; 34 plates capture moods, subtleties of poem. Flawless full-size reproductions printed on facing pages with authoritative text of poem. "Beautiful. Simply beautiful."—*Publisher's Weekly.* 77pp. 9¼ x 12. 22305-1 Pa. $6.95

NORTH AMERICAN INDIAN DESIGNS FOR ARTISTS AND CRAFTSPEOPLE, Eva Wilson. Over 360 authentic copyright-free designs adapted from Navajo blankets, Hopi pottery, Sioux buffalo hides, more. Geometrics, symbolic figures, plant and animal motifs, etc. 128pp. 8⅜ x 11. (EUK) 25341-4 Pa. $8.95

SCULPTURE: Principles and Practice, Louis Slobodkin. Step-by-step approach to clay, plaster, metals, stone; classical and modern. 253 drawings, photos. 255pp. 8¼ x 11. 22960-2 Pa. $10.95

CATALOG OF DOVER BOOKS

THE INFLUENCE OF SEA POWER UPON HISTORY, 1660–1783, A. T. Mahan. Influential classic of naval history and tactics still used as text in war colleges. First paperback edition. 4 maps. 24 battle plans. 640pp. 5⅜ x 8½. 25509-3 Pa. $12.95

THE STORY OF THE TITANIC AS TOLD BY ITS SURVIVORS, Jack Winocour (ed.). What it was really like. Panic, despair, shocking inefficiency, and a little heroism. More thrilling than any fictional account. 26 illustrations. 320pp. 5⅜ x 8½. 20610-6 Pa. $8.95

FAIRY AND FOLK TALES OF THE IRISH PEASANTRY, William Butler Yeats (ed.). Treasury of 64 tales from the twilight world of Celtic myth and legend: "The Soul Cages," "The Kildare Pooka," "King O'Toole and his Goose," many more. Introduction and Notes by W. B. Yeats. 352pp. 5⅜ x 8½. 26941-8 Pa. $8.95

BUDDHIST MAHAYANA TEXTS, E. B. Cowell and Others (eds.). Superb, accurate translations of basic documents in Mahayana Buddhism, highly important in history of religions. The Buddha-karita of Asvaghosha, Larger Sukhavativyuha, more. 448pp. 5⅜ x 8½. 25552-2 Pa. $9.95

ONE TWO THREE . . . INFINITY: Facts and Speculations of Science, George Gamow. Great physicist's fascinating, readable overview of contemporary science: number theory, relativity, fourth dimension, entropy, genes, atomic structure, much more. 128 illustrations. Index. 352pp. 5⅜ x 8½. 25664-2 Pa. $8.95

ENGINEERING IN HISTORY, Richard Shelton Kirby, et al. Broad, nontechnical survey of history's major technological advances: birth of Greek science, industrial revolution, electricity and applied science, 20th-century automation, much more. 181 illustrations. ". . . excellent . . ."–Isis. Bibliography. vii + 530pp. 5⅜ x 8¼. 26412-2 Pa. $14.95

DALÍ ON MODERN ART: The Cuckolds of Antiquated Modern Art, Salvador Dalí. Influential painter skewers modern art and its practitioners. Outrageous evaluations of Picasso, Cézanne, Turner, more. 15 renderings of paintings discussed. 44 calligraphic decorations by Dalí. 96pp. 5⅜ x 8½. (USO) 29220-7 Pa. $4.95

ANTIQUE PLAYING CARDS: A Pictorial History, Henry René D'Allemagne. Over 900 elaborate, decorative images from rare playing cards (14th–20th centuries): Bacchus, death, dancing dogs, hunting scenes, royal coats of arms, players cheating, much more. 96pp. 9¼ x 12¼. 29265-7 Pa. $11.95

MAKING FURNITURE MASTERPIECES: 30 Projects with Measured Drawings, Franklin H. Gottshall. Step-by-step instructions, illustrations for constructing handsome, useful pieces, among them a Sheraton desk, Chippendale chair, Spanish desk, Queen Anne table and a William and Mary dressing mirror. 224pp. 8⅛ x 11¼. 29338-6 Pa. $13.95

THE FOSSIL BOOK: A Record of Prehistoric Life, Patricia V. Rich et al. Profusely illustrated definitive guide covers everything from single-celled organisms and dinosaurs to birds and mammals and the interplay between climate and man. Over 1,500 illustrations. 760pp. 7½ x 10¼. 29371-8 Pa. $29.95

Prices subject to change without notice.

Available at your book dealer or write for free catalog to Dept. GI, Dover Publications, Inc., 31 East 2nd St., Mineola, N.Y. 11501. Dover publishes more than 500 books each year on science, elementary and advanced mathematics, biology, music, art, literary history, social sciences and other areas.

PHOTOGRAPHIC SKETCHBOOK OF THE CIVIL WAR, Alexander Gardner. · 100 photos taken on field during the Civil War. Famous shots of Manassas Harper's Ferry, Lincoln, Richmond, slave pens, etc. 244pp. 10⅞ x 8¼. 22731-6 Pa. $9.95

FIVE ACRES AND INDEPENDENCE, Maurice G. Kains. Great back-to-the-land classic explains basics of self-sufficient farming. The one book to get. 95 illustrations. 397pp. 5⅜ x 8½. 20974-1 Pa. $7.95

SONGS OF EASTERN BIRDS, Dr. Donald J. Borror. Songs and calls of 60 species most common to eastern U.S.: warblers, woodpeckers, flycatchers, thrushes, larks, many more in high-quality recording. Cassette and manual 99912-2 $8.95

A MODERN HERBAL, Margaret Grieve. Much the fullest, most exact, most useful compilation of herbal material. Gigantic alphabetical encyclopedia, from aconite to zedoary, gives botanical information, medical properties, folklore, economic uses, much else. Indispensable to serious reader. 161 illustrations. 888pp. 6½ x 9¼. 2-vol. set. (USO) Vol. I: 22798-7 Pa. $9.95
Vol. II: 22799-5 Pa. $9.95

HIDDEN TREASURE MAZE BOOK, Dave Phillips. Solve 34 challenging mazes accompanied by heroic tales of adventure. Evil dragons, people-eating plants, blood-thirsty giants, many more dangerous adversaries lurk at every twist and turn. 34 mazes, stories, solutions. 48pp. 8¼ x 11. 24566-7 Pa. $2.95

LETTERS OF W. A. MOZART, Wolfgang A. Mozart. Remarkable letters show bawdy wit, humor, imagination, musical insights, contemporary musical world; includes some letters from Leopold Mozart. 276pp. 5⅜ x 8½. 22859-2 Pa. $7.95

BASIC PRINCIPLES OF CLASSICAL BALLET, Agrippina Vaganova. Great Russian theoretician, teacher explains methods for teaching classical ballet. 118 illustrations. 175pp. 5⅜ x 8½. 22036-2 Pa. $5.95

THE JUMPING FROG, Mark Twain. Revenge edition. The original story of The Celebrated Jumping Frog of Calaveras County, a hapless French translation, and Twain's hilarious "retranslation" from the French. 12 illustrations. 66pp. 5⅜ x 8½. 22686-7 Pa. $3.95

BEST REMEMBERED POEMS, Martin Gardner (ed.). The 126 poems in this superb collection of 19th- and 20th-century British and American verse range from Shelley's "To a Skylark" to the impassioned "Renascence" of Edna St. Vincent Millay and to Edward Lear's whimsical "The Owl and the Pussycat." 224pp. 5⅜ x 8½. 27165-X Pa. $4.95

COMPLETE SONNETS, William Shakespeare. Over 150 exquisite poems deal with love, friendship, the tyranny of time, beauty's evanescence, death and other themes in language of remarkable power, precision and beauty. Glossary of archaic terms. 80pp. 5³⁄₁₆ x 8¼. 26686-9 Pa. $1.00

BODIES IN A BOOKSHOP, R. T. Campbell. Challenging mystery of blackmail and murder with ingenious plot and superbly drawn characters. In the best tradition of British suspense fiction. 192pp. 5⅜ x 8½. 24720-1 Pa. $6.95

THE WIT AND HUMOR OF OSCAR WILDE, Alvin Redman (ed.). More than 1,000 ripostes, paradoxes, wisecracks: Work is the curse of the drinking classes; I can resist everything except temptation; etc. 258pp. 5⅜ x 8½. 20602-5 Pa. $5.95

SHAKESPEARE LEXICON AND QUOTATION DICTIONARY, Alexander Schmidt. Full definitions, locations, shades of meaning in every word in plays and poems. More than 50,000 exact quotations. 1,485pp. 6½ x 9¼. 2-vol. set.
Vol. 1: 22726-X Pa. $16.95
Vol. 2: 22727-8 Pa. $16.95

SELECTED POEMS, Emily Dickinson. Over 100 best-known, best-loved poems by one of America's foremost poets, reprinted from authoritative early editions. No comparable edition at this price. Index of first lines. 64pp. 5¹⁵⁄₁₆ x 8¼. 26466-1 Pa. $1.00

CELEBRATED CASES OF JUDGE DEE (DEE GOONG AN), translated by Robert van Gulik. Authentic 18th-century Chinese detective novel; Dee and associates solve three interlocked cases. Led to van Gulik's own stories with same characters. Extensive introduction. 9 illustrations. 237pp. 5⅜ x 8½. 23337-5 Pa. $6.95

THE MALLEUS MALEFICARUM OF KRAMER AND SPRENGER, translated by Montague Summers. Full text of most important witchhunter's "bible," used by both Catholics and Protestants. 278pp. 6⅛ x 10. 22802-9 Pa. $12.95

SPANISH STORIES/CUENTOS ESPAÑOLES: A Dual-Language Book, Angel Flores (ed.). Unique format offers 13 great stories in Spanish by Cervantes, Borges, others. Faithful English translations on facing pages. 352pp. 5⅜ x 8½. 25399-6 Pa. $8.95

THE CHICAGO WORLD'S FAIR OF 1893: A Photographic Record, Stanley Appelbaum (ed.). 128 rare photos show 200 buildings, Beaux-Arts architecture, Midway, original Ferris Wheel, Edison's kinetoscope, more. Architectural emphasis; full text. 116pp. 8¼ x 11. 23990-X Pa. $9.95

OLD QUEENS, N.Y., IN EARLY PHOTOGRAPHS, Vincent F. Seyfried and William Asadorian. Over 160 rare photographs of Maspeth, Jamaica, Jackson Heights, and other areas. Vintage views of DeWitt Clinton mansion, 1939 World's Fair and more. Captions. 192pp. 8⅞ x 11. 26358-4 Pa. $12.95

CAPTURED BY THE INDIANS: 15 Firsthand Accounts, 1750-1870, Frederick Drimmer. Astounding true historical accounts of grisly torture, bloody conflicts, relentless pursuits, miraculous escapes and more, by people who lived to tell the tale. 384pp. 5⅜ x 8½. 24901-8 Pa. $8.95

THE WORLD'S GREAT SPEECHES, Lewis Copeland and Lawrence W. Lamm (eds.). Vast collection of 278 speeches of Greeks to 1970. Powerful and effective models; unique look at history. 842pp. 5⅜ x 8½. 20468-5 Pa. $14.95

THE BOOK OF THE SWORD, Sir Richard F. Burton. Great Victorian scholar/adventurer's eloquent, erudite history of the "queen of weapons"–from prehistory to early Roman Empire. Evolution and development of early swords, variations (sabre, broadsword, cutlass, scimitar, etc.), much more. 336pp. 6⅛ x 9¼. 25434-8 Pa. $9.95